TRACE-ORGANIC SAMPLE HANDLING

Vol. 10
in the Series now titled
METHODOLOGICAL SURVEYS — *sub-series* (A): Analysis

Series Editor: Eric Reid
Wolfson Bioanalytical Unit

Robens Institute of Industrial & Environmental Health & Safety
University of Surrey

Other titles, with retrospective assignment to the two sub-series

Below: **Sub-series (B): Biochemistry**
1971 SEPARATIONS WITH ZONAL ROTORS
[ranks as Vol. 1]
published by (and available from)
The Unit at Guildford

METHODOLOGICAL DEVELOPMENTS IN BIOCHEMISTRY
(Series title, now superseded by METHODOLOGICAL SURVEYS)
published by Longman Group Limited (Vols. 2–4; available only from the Unit)
or North-Holland Publishing Company (Vol. 5)

Sub-series (A): Trace-Organic Analysis

1973 SEPARATION TECHNIQUES (Vol. 2)
1973 ADVANCES WITH ZONAL ROTORS
(Vol. 3)
1974 SUBCELLULAR STUDIES (Vol. 4)

1976 ASSAY OF DRUGS AND OTHER
TRACE COMPOUNDS IN
BIOLOGICAL FLUIDS (Vol. 5)

METHODOLOGICAL SURVEYS [IN BIOCHEMISTRY — *optional phrase, now superseded]*

1977 MEMBRANOUS ELEMENTS
AND MOVEMENT OF MOLECULES
(Vol. 6; including muscle)
1979 CELL POPULATIONS (Vol. 8)
1979 PLANT ORGANELLES (Vol. 9)
1981 (intended) CANCER-CELL
ORGANELLES *(Vol. 11)*

1978 BLOOD DRUGS AND OTHER
ANALYTICAL CHALLENGES
(Vol. 7)
1980 TRACE-ORGANIC SAMPLE
HANDLING
(Vol. 10; *this volume)*

TRACE-ORGANIC SAMPLE HANDLING

Edited by
ERIC REID, Ph.D., D.Sc.
Head of Wolfson Bioanalytical Unit
Institute of Industrial and Environmental Health and Safety
University of Surrey

ELLIS HORWOOD LIMITED
Publishers · Chichester

Halsted Press: a division of
JOHN WILEY & SONS
New York · Chichester · Brisbane · Toronto

First published in 1981 by

ELLIS HORWOOD LIMITED
Market Cross House, Cooper Street, Chichester, West Sussex, PO19 1EB, England

The publisher's colophon is reproduced from James Gillison's drawing of the ancient Market Cross, Chichester.

Distributors:

Australia, New Zealand, South-east Asia:
Jacaranda-Wiley Ltd., Jacaranda Press,
JOHN WILEY & SONS INC.,
G.P.O. Box 859, Brisbane, Queensland 40001, Australia

Canada:
JOHN WILEY & SONS CANADA LIMITED
22 Worcester Road, Rexdale, Ontario, Canada.

Europe, Africa:
JOHN WILEY & SONS LIMITED
Baffins Lane, Chichester, West Sussex, England.

North and South America and the rest of the world:
Halsted Press: a division of
JOHN WILEY & SONS
605 Third Avenue, New York, N.Y. 10016, U.S.A.

British Library Cataloguing in Publication Data
Trace-organic sample handling. —
(Methodological surveys: sub-series (A), analysis; vol. 10).
    1. Chemistry, organic
    2. Chemistry, Analytic
    I. Reid, Eric   II. Series
    547'.34   QD271.4   80–49715

ISBN 0–85312–187–7 (Ellis Horwood Limited, Publishers)
ISBN 0–470–27071–3 (Halsted Press)

Typeset in Press Roman by Ellis Horwood Limited.
Printed in Great Britain by Butler and Tanner Ltd., Frome, Somerset.

Table of Contents

The NOTES & COMMENTS at the end of each section (headed 'NC') include some comments made at the Forum on which this book is based, together with some supplementary points besides those in the opening 'guide' (article #0).

Editor's Preface. 9

List of Authors 11

#0   **ANALYSES ENTAILING SAMPLE PROCESSING: AN OVERVIEW**
        E. REID 15

#A   **WORKPLACE AIR SAMPLES**
#A-1   The evaluation of occupational exposure — E. KING 32
#A-2   Setting up standards and field analysis for organic vapours —
        STEPHEN CRISP 39
#A-3   Passive sampling and dosimetry —
        A. BAILEY & P. A. HOLLINGDALE-SMITH 43
#A-4   Air sampling for trace-organics: an overview — J. G. FIRTH . . . 51

#NC(A)   **NOTES and COMMENTS related to Workplace Air Samples**
        including Notes on
#NC(A)-1 Approaches to trapping trace-organics and generation of
        standards — D. J. FREED 59
#NC(A)-2 The use of plastic bags for air sampling —
        P. A. HOLLINGDALE-SMITH. 60
#NC(A)-3 Vapour-sample storage in a rigid container —
        E. REID & A. D. R. HARRISON 61

#B   **WATER AND EFFLUENTS**
#B-1   Concentration and clean-up of organic solutes in water —
        JAMES S. FRITZ 68
#B-2   Sample handling as practised in the Water Research Centre —
        I. WILSON. 76
#B-3   Trace enrichment in the HPLC analysis of aqueous samples —
        R. W. FREI & U. A. TH. BRINKMAN. 86

#NC(B)   **NOTES and COMMENTS related to water and effluents**
            including Notes on
#NC(B)-1 Sparging, trapping and concentrating trace-organics in water
            samples — C. E. ROLAND JONES 96
#NC(B)-2 Methodology for pollutants, as in plant outfalls — JAMES S.
            SMITH 98

#C   **SAMPLES FOR RESIDUE AND OTHER NON-PHARMA-
            COLOGICAL STUDIES**
#C-1   Enzymic liberation of chemically labile or protein-bound drugs
            from tissues — M. D. OSSELTON .101
#C-2   Extraction and clean-up procedures for foods —
            N. T. CROSBY 111
#C-3   Detection of adventitious nitrogenous organics in food samples
            — C. L. WALTERS 119
#C-4   Work-up of crops and soils for residues of pesticides and their
            metabolites — A. P. WOODBRIDGE & E. H. McKERRELL. . . .128

#NC(C)   **NOTES and COMMENTS related to the foregoing topics**
            including Notes on
#NC(C)-1 Stable isotopes as MS tracers in a study of the metabolites and
            residues from a pesticide — P. HENDLEY & J. LAM 148
#NC(C)-2 Illustrative procedures for contaminants in foodstuffs — M. J.
            SAXBY 150
#NC(C)-3 Prevention of contamination in the residue laboratory — A. P.
            WOODBRIDGE 156
#NC(C)-4 *Analytical case history:* Measurement of a plasticizer (DEHA) in
            tissue — S. FAYZ & R. HERBERT 158
#NC(C)-5 *Analytical case history:* Assay of blood for traces of anionic and
            non-ionic surfactants — P. THACKERAY & D. HOAR 161

#D   **APPROACHES FOR BIOLOGICAL FLUIDS, ESPECIALLY
            DRUG-RELATED**
#D-1   Enzymatic approaches to trace-organic analysis —
            G. GUILBAULT 168
#D-2   Drugs in saliva — ROKUS A. DE ZEEUW 176
#D-3   Solvent-partitioning strategy in drug assays — KENNETH H.
            DUDLEY. 188
#D-4   Sample preparation for trace drug analysis —
            J. A. F. DE SILVA. 192

#NC(D)   NOTES and COMMENTS related to the foregoing topics
         including Notes on:
#NC(D)-1 Mass spectrometry and trace-organic analysis in biological fluids
         – L. E. MARTIN & R. J. N. TANNER 205
#NC(D)-2 Drug binding to plasma proteins – C. J. COULSON & V. J.
         SMITH 210
#NC(D)-3 Minimization of pre-HPLC plasma processing through column-
         switching and selective detection – JAN E. PAANAKKER,
         JIMMY M. S. L. THIO & Henk J. M. VAN HAL 214
#NC(D)-4 *Analytical case history:* Determination of ranitidine and meta-
         bolites in plasma and urine – P. F. CAREY & L. E. MARTIN 217
#NC(D)-5 HPLC determination of hydrolysis rates of corticosteroid esters
         in vitro and *in vivo* –
         M. E. TENNESON & T. COWEN 218

#E   AUTOMATED SAMPLE HANDLING
#E-1   Automation of small column chromatography –
         G. B. BARLOW. 221
#E-2   Automated sample pre-treatment – P. B. STOCKWELL 227
#E-3   Automated analysis in pharmacokinetic studies – W. PACHA &
         H. ECKERT. 240
#E-4   Automated analysis of chlorthalidone based on the inhibition
         of carbonic anhydrase – M. M. JOHNSTON, M. ROSENBERG,
         T. E. DORSEY & R. F. DOYLE. 253

#NC(E)   NOTES and COMMENTS related to the foregoing topics
         including Notes on:
#NC(E)-1 An automated fluorescence assay for indapamide – P. E.
         GREBOW, J. TREITMAN & A. YEUNG 259
#NC(E)-2 A multi-sample TLC applicator – J. P. LEPPARD & E. REID . .260

#F   ENSURING RELIABLE RESULTS, ESPECIALLY FOR
         DRUGS IN BLOOD
#F-1   Influence of blood-sampling procedures on drug levels in plasma
         – OLOF BORGÅ, INGA PETTERS & RUNE DAHLQVIST . . .265
#F-2   Some matrix-effects on the extraction of hydrophobic amines
         and a quaternary ammonium compound from serum and plasma
         – D. WESTERLUND & L. B. NILSSON 270
#F-3   Fatty acids and plasticizers as potential interferences in the
         bioanalysis of drugs – ROKUS A. DE ZEEUW & JAN E.
         GREVING. 274

8 Table of Contents

#F-4 Modifications in the work-up procedure for drugs due to the biological matrix — JÖRGEN VESSMAN 284

#F-5 Stability of drugs in stored biological samples — H. K. ADAM . . .291

#F-6 GLP in bioanalytical method validation — J. A. F. DE SILVA . .298

#F-7 Limits of detection — J. McAINSH, R. A. FERGUSON & B. F. HOLMES 311

#F-8 Assessment of assay precision from linear calibration graphs — J. P. LEPPARD. 320

#NC(F) NOTES and COMMENTS related to the foregoing topics
        including Notes on:

#NC(F)-1 Policies for internal standards — KENNETH H. DUDLEY 336

#NC(F)-2 The usefulness of internal standards in quantitative bioanalytical methods — JÖRGEN VESSMAN 341

#NC(F)-3 Critical factors in drug analysis — STEPHEN H. CURRY & ROBIN WHELPTON 347

#NC(F)-4 Problems in getting valid results: an overview — B. SCALES . . .353

#NC(F)-5 Problems in concentration of solvent extracts for drug analysis — L. E. MARTIN. 363

#NC(F)-6 Volume reduction under partial reflux of trace-organic extracts — W. DÜNGES, K. KIESEL & K.-E. MÜLLER 366

#NC(F)-7 A simple multi-tube solvent evaporator — E. REID, A. D. R. HARRISON & J. P. LEPPARD .366

The 'COMMENTS' in NOTES & COMMENTS largely represent remarks made at the Forum on which this book based. Insofar as they refer to particular contributions or groups of contributions, the following page guide will assist retrieval:

    #A-1 to A-4 [and #NC(A)-1], pp. 66-67;
    B-1 & B-3 [& NC(B)-1], p. 100;
    C-1 & C-2 [& NC(C)-1, -5], pp. 166-167;
    D-1 to D-3, pp. 219-220;
    E-2 to E-4 [& NC(E)-1], pp. 263-264.
    F-1 to F-8 [& NC(F)-1 & -2], pp. 347-363 [NC(F)-3 & -4]
    and 370-371.

CUMULATIVE INDEX OF COMPOUND TYPES (covering Vols. 5 & 7 also) . 372

GENERAL INDEX 378

Corrections to Vol. 7 383

Editor's Preface

This volume, in a series category now titled *Methodological Surveys (A): Analysis,* differs from its predecessors (Vols. 5 and 7; *see opposite title page*) in a notable respect: in common with the Institute which the Editor's laboratory has joined, there is coverage of environmental trace-organics as well as drugs in biological fluids. Amongst the analytical fraternity there is specialization in respect of type of analyte (a useful term for the compound to be analyzed!) and of substrate within a range which spans factory air, drinking water, food-stuffs and biological fluids. Even the specialist may find some illumination from articles that fall outside his area. Examples of a common thread are the use of solvent extraction or 'XAD-2' resin to isolate the analyte, as indicated in the opening 'perspectives' article [#0]. Moreover, the specialist faced on occasion with an unfamiliar analyte (not macromolecular) or substrate will, it is hoped, find this book a uniquely useful guide to strange territory.

Much editorial effort has been devoted to achieving integration, whilst leaving some overlap, of contributions which mostly represent material presented at the 3rd International Bioanalytical Forum held at the University of Surrey in September 1979. This editorial aim has entailed some sacrifice of the publication speed that would be desirable if the book were merely a symposium record, but there has been the benefit that important contributions received as late as May 1980 have been assimilated. Some such contributions, two being in a 'Notes & Comments' section, were in fact written largely in retrospect at the Editor's behest for the sake of perspective; the authors who shouldered this task are Dr. J. G. Firth, Mrs. I. Wilson, Dr. A. P. Woodbridge and — heroically distilling lore from Forum discussions — Prof. S. H. Curry and Dr. B. Scales (who gave valuable help in planning the Forum). Other authors, too, have earned the Editor's gratitude.

As is amplified in article #0, the main literature gap which this book aims to fill concerns sample 'handling' rather than actual measurement. Automatic sample-processing equipment (section #E) is belatedly becoming available, to help with ever-increasing workloads that were hardly envisaged when, in 1918,

H. D. Dakin (*Biochem. J.*, **12**, 290–317) referred to a 36-h extraction as being "surprisingly rapid"! However, measurement and interpretation also feature strongly, especially in the section [#F] on achieving reliable results. A plea is now made for abandonment of units such as 'ppm', in favour of, say, 'mg/m^3 (of air)', although *not* for adoption of SI policies.

Donations towards Forum expenses came from Beechams, Geigy (U.K.), Glaxo Holdings, ICI Pharmaceuticals, Sandoz (Basle), Shell Biosciences and Squibb International. Gratitude is expressed to the publisher's staff, especially for coping with difficult material in implementing the new policy that *Methodological Surveys* will be typeset in traditional format. Acknowledgement is made, where applicable, to sources from which illustrative or other material has been reproduced: the journals concerned include *J. Chromatog.* (Elsevier), *J. High Res. Chromatog. Chromatog. Commun.* (Springer-Verlag) and *J. Lab. Clin. Med.* (C. V. Mosby Co.).

Abbreviations
Varieties of chromatography have the conventional abbreviations TLC (prefix HP = high-performance), HPLC, and GC — a preferable term to GLC, as is pointed out in art. #0 [*see* (vii)(3)] which may be consulted also for the definition of the term TLV. Other abbreviations include ECD = electron-capture detection, MS = mass spectrometry, and UV = ultraviolet.

Guildford, 19 May 1980

List of Authors

'Notes' (headed 'NC . . .' in the text) are not distinguished from main contributions.

H. K. Adam – pp. 291-297
ICI Pharmaceuticals Ltd., U.K.

A. Bailey – pp. 43-50
Chemical Defence Establishment, U.K.

G. B. Barlow – pp. 221-226
Inst. of Child Health, London, U.K.

Olof Borgå – pp. 265-269
Huddinge Hospital, Sweden.

U. A. Th. Brinkman – pp. 86-95
as for R. W. Frei.

P. F. Carey – pp. 217-218
as for L. E. Martin

C. J. Coulson – pp. 210-213
May & Baker Ltd., U.K.

T. Cowen – pp. 218-219
Squibb International Development
Lab., U.K.

Stephen Crisp – pp. 39-42
Lab. of the Government Chemist., U.K.

N. T. Crosby – pp. 111-118
Lab. of the Government Chemist, U.K.

Stephen H. Curry – pp. 347-352
London Hosp. Med. College, U.K.

Rune Dahlqvist – pp. 265-269
as for Olof Borgå

J. A. F. de Silva – pp. 192-204 and
    298-310
Hoffmann-La Roche Inc., U.S.A.

Rokus A. de Zeeuw – pp. 176-187 and
    274-283
State Univ., Groningen, The Netherlands.

T. E. Dorsey – pp. 253-258
as for M. M. Johnston

R. F. Doyle – pp. 253-258
as for M. M. Johnston

Kenneth H. Dudley – pp. 188-191 and
    336-340
Univ. of N. Carolina, U.S.A.

W. Dünges – p. 366
J. Gutenberg-Univ., Mainz, W. Germany.

H. Eckert – pp. 240-252
Sandoz Ltd., Basle, Switzerland.

S. Fayz — pp. 158-160
as for R. Herbert

R. A. Ferguson — pp. 311-319
as for J. McAinsh

J. G. Firth — pp. 51-58
Health & Safety Executive Labs., U.K.

D. J. Freed — p. 59
Bell Labs., U.S.A.

R. W. Frei — pp. 86-95
Free Univ. of Amsterdam,
The Netherlands

James S. Fritz — pp. 68-75
Ames Lab., Iowa State Univ., U.S.A.

P. E. Grebow — p. 259
USV Labs., Tuckahoe, U.S.A.

Jan E. Greving — pp. 274-283
as for Rokus A. de Zeeuw

G. Guilbault — pp. 168-175
Univ. of New Orleans, U.S.A.

A. D. R. Harrison — pp. 61-66 and
366-368
as for E. Reid

P. Hendley — pp. 148-149
ICI Plant Protection Div., Jealott's Hill,
U.K.

R. Herbert — pp. 158-160
Sunderland Polytechnic, U.K.

D. Hoar — pp. 161-166
Unilever Research Ltd., U.K.

P. A. Hollingdale-Smith, pp. 43-50 and
60-61
as for A. Bailey

B. F. Holmes — pp. 311-319
as for J. McAinsh

M. M. Johnston — pp. 253-258
USV Labs., Tuckahoe, U.S.A.

C. E. Roland Jones — pp. 96-98
Redhill, U.K.

K. Kiesel — p. 366
as for W. Dünges

E. King — pp. 32-38
Natl. Occupational Hygiene Service Ltd.,
Manchester, U.K.

J. Lam — pp. 148-149
as for P. Hendley

J. P. Leppard — pp. 260-263, 320-335
and 366-368
Univ. of Surrey, U.K.

J. McAinsh — pp. 311-319
ICI Pharmaceuticals Ltd., U.K.

E. H. McKerrell — pp. 128-147
as for A. P. Woodbridge

L. E. Martin — pp. 205-209, 217-218
and 363-365
Glaxo Group Research Ltd., U.K.

K.-E. Müller — p. 366
as for W. Dünges

L. B. Nilsson — pp. 270-273
as for D. Westerlund

M. D. Osselton — pp. 101-110
Home Office Central Res. Estab., U.K.

Jan E. Paanakker — pp. 214-217
Organon International B. V.,
The Netherlands.

W. Pacha — pp. 240-252
as for H. Eckert

Inga Petters — pp. 265-269
as for Olof Borgå

E. Reid — pp. 15-31, 61-66, 260-263
    and 366-368
Univ. of Surrey, U.K.

M. Rosenberg — pp. 253-258
as for M. M. Johnston

M. J. Saxby — pp. 150-155
Leatherhead Food Res. Assocn., U.K.

B. Scales — pp. 353-363
ICI Pharmaceuticals Ltd., U.K.

James S. Smith — pp. 98-99
Allied Chemical Corpn., Morristown,
U.S.A.

V. J. Smith —   pp. 210-213
as for C. J. Coulson

P. B. Stockwell — pp. 227-239
Lab. of the Government Chemist, U.K.

R. J. N. Tanner — pp. 205-209
as for L. E. Martin

M. E. Tenneson — pp. 218-219
as for T. Cowen

P. Thackeray — pp. 161-166
as for D. Hoar

Jimmy M. S. L. Thio — pp. 214-217
as for Jan E. Paanakker

J. Treitman — p. 259
as for P. E. Grebow

Henk M. M. van Hal — pp. 214-217
as for Jan E. Paanakker

Jörgen Vessman — pp. 284-290 and
    341-347
AB Hässle, Sweden.

C. L. Walters — pp. 119-127
Leatherhead Food Res. Assocn., U.K.

D. Westerlund — pp. 270-273
Astra Läkemedel AB, Sweden

Robin Whelpton — pp. 347-352
as for Stephen H. Curry

I. Wilson — pp. 76-85
Water Res. Centre, Medmenham, U.K.

A. P. Woodbridge — pp. 128-147 and
    156-157
Shell Biosciences Lab., U.K.

A. Yeung — p. 259
as for P. E. Grebow

CORRECTIONS to Vol. 10, TRACE-ORGANIC SAMPLE HANDLING

p. 28, line 17:   chronic   *should read*   chromic

p. 106, line 5: 2^2-hydroxy   *should read*   2-hydroxy

p. 116, legend: stopped working   *should read*   working

p. 161, line 4:   Unilver   *should read*   Unilever

p. 206, line 9:   minimal   *should read*   nominal

p. 208, lines 1 & 3 of 3rd para.: after 10^{-6} and 10^{-5} *add*   s

p. 218, line 10: after   ranitidine   *add*   to its

p. 323, line 2: (s_y) *should read* (s_Y)

line 6: between   is   and   not   *insert*   often

legend to Fig. 2(b): S_{yx} *should read* s_{yx}

p. 326, line 6:   90%   *should read*   99%

p. 329, line 4: y in numerator   *should read* Y

p. 333, lines 8, 7 & 6 [each   for Y'...   entry] and 4 from foot, $s_{x'}$ *should read* $s_{X'}$

p. 373, line 7 of left Index column,   (DEHP): p. 167 lacks the citation — Lewis *et al.* (1978) *Clin. Chem.* **24**, 741–748.

p. 377, under SUMMARY: Non-amide   *should read*   Non-imide

p. 381, entry on   Mixing: *should read*   230–, 360

p. 382, entry on   Silanization,   & p. 383, Vessels: 372   *should read*   371

Book list facing title p.: entry for this book (Vol. 10) to be 1981.

#O Analyses Entailing Sample Processing: An Overview

E. REID, Wolfson Bioanalytical Unit,
Institute of Industrial & Environmental Health & Safety,
University of Surrey, Guildford GU2 5XH, U.K.

To help put the contents of this book and its predecessors into perspective, annotated guidance is now given with especial attention to bridge-building between the sections that follow. Such bridges include 'Good Laboratory Practice' (GLP), 'precision' (in its proper sense of 'reproducibility'; sometimes mis-termed 'accuracy'), use of organic solvents and, increasingly, of solid agents or merely protein-removing agents as aids to the work-up of the raw sample (solid, liquid or gaseous), alertness to obviating drudgery, and steering clear of pitfalls. The term 'sample handling' as now used covers sample preparation in the sense of work-up or processing, and also pre-processing measures such as storage.

Trace-organic analysis is a territory or, regrettably, a group of non-communicating territories that is still rather lacking in maps, in spite of a history that goes back at least to the mid-1940s when B. B. Brodie devised enterprising methods for assaying drugs in blood. A recent book with the colourful sub-title *A New Frontier in Organic Chemistry* [1] contains a few perspective articles, but none range over the whole area. The present selective sketch includes some citations which may repay study by investigators outside the author's field of interest; thus, much drug-related literature is of wider relevance. Surveys of sample-preparation for the non-specialist have been compiled by B. L. Karger *et al.* [2], D. A. Skoog & D. M. West [3], J. W. Dolan [in 4] and, with emphasis on biological fluids, by E. Reid [5,6].

Trace-organic literature, including that from manufacturers, is dominated by instrumentalists who, all too often, seem to be in search of applications and yet show runs with 'pseudo' samples rather than worked-up 'real' samples. Since, however, the choice of work-up procedures is indeed governed partly by the intended measurement technique, aspects of chromatography in its end-step role as considered by contributors to a previous volume [7] are touched on below. Firstly a question is put: will one's values stand up to scrutiny?

(i)   APPRAISING THE RELIABILITY OF ANALYTICAL RESULTS

The advent of so-called GLP in the drug-safety area and, imminently if not already, in the environmental area implies a nightmare reorganization in laboratories concerned, but warrants two cheers from analysts if not three. Disregarding those aspects that savour of defence against an over-fussy auditor, GLP is commendable as a spur to meticulousness in respects such as instrument custodianship, record-keeping and analytical quantitation [8].

In considering the bearing of GLP on quantitation, it has to be kept in mind that few work-up procedures give 100% recovery and that, accordingly, the curve against which test values are read has to be derived from 'determinate' or 'processed' standards (this author's preferred nomenclature [1,7]) spiked initially into the actual matrix. The guidelines adopted by one U.S. company as given below [#F-6] by J. A. F. de Silva perhaps err on the side of over-thoroughness and, moreover, over-reliance on 'processed internal standards' which this author [7] and S. H. Curry & R. Whelpton [7, and #NC(F)-3, *this vol.*] feel to be sometimes treacherous notwithstanding popularity that may often be justifiable [#NC(F)-1 & -2, *this vol.*]. This particular misgiving does not apply if isotopically labelled compound is used. In general, there is a risk that a particular laboratory setting up in-house guidelines will lean towards those of the most earnest laboratories, and go beyond the common-sense which now seems acceptable to the FDA in the analytical field.

An example of common-sense is tolerance of some day-to-day variability in the slope of the calibration line, as encountered and defended in the case of chlorpromazine assay by GC-AFID [9; *see also* #NC(F)-4]. Another is the permissibility of judging linearity and drawing the straight line by eye, as accepted by E. Fingl [in 10] and R. W. Burnett [11] who do, however, show the usefulness of statistical approaches if judiciously chosen. Fingl's article, and those by S. H. Curry & R. Whelpton, J. Chamberlain and J. J. de Ridder [7], are unusual in their relevance to multi-sample assay for trace-organics. Hoped-for confidence limits and, more generally, the assay design differ from those in traditional chemical assays that are the focus for most statistical expositions. Some of these [e.g. 12] nevertheless give useful guidance on aspects such as analytical quality control (not to be confused with product quality control!) and on rejection of 'outliers' ('rogues'). The latter aspect, as also dealt with in a NIOSH publication [13], is considered later [#NC(F)-3] in relation to drug assay.

In discussing how calibration data can furnish confidence limits, J. P. Leppard [#F-8] warns against an obsessive quest for high precision, as does E. King

[#A-1] in connection with analyses on workplace air. There is a nomenclature pitfall: the chemist or the layman is prone to use the term 'accuracy' where the intended meaning is 'precision' as distinct from the correctness of the estimate. An estimate may be systematically wrong (inaccurate) even if precise, for reasons such as a consistent manipulative error [12] or non-specificity resulting, say, from non-distinction between the compound of interest and a metabolite.

Accuracy and precision should evidently be ascertained in all assays, as various authors including J. A. F. de Silva [in 7] and E. Fingl [in 9] have emphasized. Also important is the sensitivity or – a less criticized term – the limit of detection, as probed statistically in a critique [14] which is, however, more concerned with 'instrument noise' than with 'sample noise'. J. McAinsh *et al*. [#F-7] give criteria which are arguably rather strict [#F-8, #NC(F)-3]. At least, however, there is no excuse for reporting a level as nil or non-detectable (and 'N. D.' could mean 'not determined!') rather than as < ... , i.e. below a limit value which should be stated.

Reliability as appraised in the GLP context obviously hinges on well documented methodology for the compound in question, and on good house-keeping and manipulations in the laboratory. Here, and especially for the latter, the analytical fraternity have become afflicted with the jargon expression 'Standard Operating Procedure' (SOP), much more concerned with laboratory practice than is the misleadingly named GLP. The SOP concept, amplified below [#F-6], seems admirable insofar as it calls for clear procedural compilations, readily accessible to laboratory staff, concerning solvent purification and numerous other requisites mentioned later in this article and elsewhere in this book.

(ii) SOME TRENDS IN METHODOLOGY AS INFLUENCED BY THE MEASUREMENT STEP

If a wide-ranging guide such as readers might hope for were to concern merely retrieval from pertinent literature in general or even from this and the previous volumes only, as distinct from expounding the actual state-of-the-art, the task of compiling it would still be daunting. An article by J. A. F. de Silva [7] does go some way towards putting in perspective the methodological scene. Only a few aspects are now touched on, firstly focussing on the concluding stage(s) of the assay.

In the trace-organic assay field there will be a continued key role for GC, suitably with packed rather than capillary columns except for complex environmental samples [cf. #B-2]. The waste of most of the worked-up sample has to be tolerated as a constraint inherent in the small injection volume. In HPLC (of which a new account has appeared [14]) this constraint has ceased to apply, now that the boldness of European chromatographers (e.g. R. W. Frei [7]) in injecting large volumes (say 100 μl) has found worldwide acceptance. The constraint that still besets HPLC is the difficulty of detecting trace compounds that lack UV absorption. Electrochemical responsiveness (oxidative or reductive) sometimes offers a detection approach, if the investigators have the requisite patience as well as equipment.

Those contemplating derivatization in connection with HPLC [cf. #B-3], GC or TLC, with possible conferment of detectability, now have the benefit of a handbook that deserves high praise [15] and, for GC of drugs, a review [16]. In HPLC, post-chromatographic derivatization as is commonly performed in TLC will gain momentum when, as a safe prediction, manufacturers offer post-column reactors. These ideally, despite the cost, should offer an extraction step as an option. Already there has been dramatic benefit to HPLC from the swing to microparticulate packing materials and to the reverse-phase (RP) type. The latter type is particularly suitable for polar solutes, such as drug metabolites, that are not readily analyzed by GC. There has also been a swing towards curtailed sample preparation, which at least should comprise deprotein-ization in the case of blood plasma; replaceable glass beads at the top of the precious column offer a convenient alternative to a pre-column as a means of protecting against insults. Pre-HPLC sample preparation and also a drug biblio-graphy are included in a recent review [17] which has a GC companion [18] that is likewise admirable for tracking down blood-assay methods.

For·trace amounts of chlorocarbons in water, by GC-ECD (with a coated packing), G. Brozowski et al. [in 1] prefer to inject the raw sample direct. With plasma, however, direct injection is even more deleterious to a GC column, than to an HPLC column, although tempting for a volatile compound present in relatively high concentration. One example of direct GC, of relevant to monitoring operating-theatre staff, is an assay for halothane, which can other-wise be readily extracted into heptane. Use of GC-ECD rather than GC-FID may facilitate the measurement (although ECD responsiveness is abysmal for some halo compounds, e.g. dichloro-methane or -ethane). Citations for halothane are to be found in a review [19 (in a series which gives assay methods for diverse drugs)] that now serves to indicate a methodological repertoire which has changed little, in respect of halothane, during the 8 years that has elapsed. Thus GC analysis of halothane in plasma or tissues may be performed after distillation into heptane or co-distillation with toluene [cf. 'Co-distillation' entries in the Index, this vol.], or by a 'headspace' approach, a reviving technique of which other examples are to be found elsewhere [cf. Index entries]. Turbi-dimetric methods [19] can be regarded as obsolete, and X-ray fluorescence as an 'over-kill'; but the IR approach applied to the vapour [19] is still a reasonable option. Insofar as an early method [19] hinged on spectrophotometry (A_{367}), HPLC could be contemplated also.

The feasibility of using a 'near-raw' sample, as with halothane, is a strong expectation where a trace-organic is to be measured by an 'affinity' method, as

in the immuno-assay of a range of drugs and hormones (V. P. Butler, in [6];
J. A. F. de Silva, in [7]. Such approaches, although debated at the 1979 Forum,
are thus given little mention in the present sample-handling book, apart from a
contribution by G. Guilbault [#D-1] which up-dates some enzymic approaches
surveyed in his classical book. Enzyme-inhibition approaches [as in #E-4]
have long been available [20], e.g. for pesticides which, possibly after chemical
transformation [21], inhibit particular cholinesterases with some specificity [20],
and for monoamine oxidase inhibitors such as iproniazid, isocarboxazide and
paraglyine [22].

Most trace-organic assays do, however, entail pre-concentration and/or
clean-up to remove interfering material. This is implicit in the following sketch,
which deals firstly with starting-material that is 'solid' rather than in liquid or
vapour form as usually desired by the analyst.

(iii) SOLID SAMPLES (INCLUDING PLANT AND ANIMAL TISSUES)

With food materials there is an initial task which falls outside the scope of
this book and which may entail mandatory procedures—the procurement of a
representative sample [1]. Then there is a need for an extraction step, the
nature of which depends on the nature of the analyte (hydrophilic or lipophilic)
and on its state in the matrix. Where there is tight binding that cannot be overcome by a stratagem such as subtilisin treatment [#C-2; cf.#C-1], a possible
remedy is pyrolytic estimation, as in the case of some pesticides [23].

Amongst the food-oriented contributions in section #C of this vol.,
several testify to the usefulness, for sample clean-up, of 'Florisil' (a magnesium
silicate) or alumina, of a TLC step, or of a 'C-18' silica possibly used in the
HPLC mode. The preferred approach depends partly on the starting-material,
ranging from soils and horticultural produce as in the penetrating survey by
A. P. Woodbridge & E. H. McKerrell [#C-4] to processed foods as in N. T. Crosby's
contribution [#C-2]. Also important is the polarity of the analyte, ranging from
pesticides and, as must also be considered, their 'metabolites' (transformation
products of possibly diverse origin) [#C-4], to basic or non-basic nitrosamines
[#C-3]. For aflatoxins [#C-2, #NC(C)-2] the novice can consult an outline by
S. Nesheim which takes account of FDA stipulations and which has a lengthy
bibliography [pp. 355–372 in 1].

Other books which deal, at least in part, with trace organics in solid foods
include a compendium of methods [24], and two books centred on measurement—
one with a good chapter on sample preparation [25], and another [26] with
interesting mention (by D. J. Manning) of cleaner GC patterns at above 150°
if, in headspace analysis, Tenax GC is used rather than Porapak Q for trapping
volatiles. For pollutants in solid wastes, F. C. McElroy et al. [in 1] give a useful
survey.

Where the starting material is an animal tissue [e.g. #NC(C)-4], reputable conditions for collection and storage as discussed by B. Scales [pp. 43–54 in 7] are of course important, but are difficult to ensure for human specimens, especially with forensic samples as considered by M. F. Osselton [#C-1] from the drug-measurement viewpoint. Whether the homogenizing medium should be aqueous or organic (or aqueous and then aqueous/organic; cf. #F-4) depends on whether the analyte is lipophilic. If the compound is known to associate with a particular subcellular fraction, the wise analyst will, after consulting a biochemist if necessary, use this fraction rather than a total homogenate for the isolation.

In general, it is necessary to get protein out of the way. In the special case of endogenous trace compounds such as nucleoside triphosphates, the propensity of tissue enzymes to cause degradation immediately after death can be thwarted by quickly freezing the tissue in liquid nitrogen, as also used when eventually the tissue comes to be ground to a powder in a mortar: this powder is dispersed, whilst still frozen, in 0.5 M perchloric acid in a Potter–Elvehjem homogenizer with a loose-fitting pestle, and the extract can be freed from perchloric acid by neutralization with KOH [27]. This approach might have facilitated the determination of a plasticizer susceptible to enzymic hydrolysis [#NC(C)-4]. Too long a precipitation time for the $KClO_4$ may result in some loss of analyte [e.g. 27a].

(iv) ASPECTS OF SOLVENT EXTRACTION

Aqueous acetonitrile is gaining favour both as an extractant for solid samples —possibly with a partitioning step which takes care of endogenous lipids [#C-2, #C-3 & #NC(C)-2] – and as a protein precipitant for plasma, especially where the ensuing step is reverse-phase HPLC. With plasma, besides lipid interferences [#F-3] there can be troubles due to incomplete protein precipitation, obviated by using 1.5 rather than 1.0 vol. of acetonitrile [28]. As shown in the author's laboratory with acetone as precipitant (1 vol.), acidification of the supernatant brings down a further small amount of protein; but trial of pre-acidified plasma indicated that protein which comes down at neutral pH is not acetone-precipitable at acid pH. Other observations with one-phase systems have been made by J. Vessman [#F-4].

Two-phase systems as widely used for lipophilic compounds occupy a prominent place in this book [notably #D-3] and other literature [e.g. 5–7]. J. W. Dolan [in 4] has added to the scattered lore [5; see also index entry in 6 & 7] on overcoming emulsion troubles, minimization of which may be an incidental benefit of adding isoamyl alcohol as an adsorption-suppressor. In practice it is 'amyl alcohol' that U. K. analysts should buy for this purpose as catalogue consultation (BDH Chemicals Ltd.) makes evident: it consists largely of 3-methylbutan-1-ol with some 2-methylbutan-1-ol. This explains an apparent discrepancy between two procedure descriptions for chlorpromazine and metabolites [9, and S. H. Curry & R. Whelpton in 7].

Basic compounds are traditionally solvent-extracted at an alkaline pH; yet weakly acidic compounds may also be extractable [#D-3], and at acid pH a protonated base such as flavoxate may show unexpected solvent-extractability. For acidic compounds the choice of an acid pH implies that plasma here is preferable to whole blood, in view of the massive precipitation when erythrocytes are present [#D-4]. Selectivity within a family of compounds may be achievable by judicious choice not only of solvent but also of pH, guided by pK_a values which may be determined by titration even where a basic form is water-insoluble at alkaline pH [29]. Temperature variation deserves trial as a means of achieving selectivity.

Achievement of reliable phase mixing has been considered elsewhere [*this vol.*, e.g. B. Scales, #NC(F)-4; also refs. 6 (observations indexed under 'Solvent extraction') & 5]. The reciprocating shaker used in our laboratory is somewhat quieter than those offered by manufacturers, who also neglect flammability risks that are reduced if the motor is of induction type with gear-down (e.g. as made by Parvalux Ltd., Bournemouth). Where, as with ether, the 'sealed' tubes are prone to creepage loss, a wrist-action shaker is advantageous, with the tubes no more than half-full.

Cartridge-type aids to solvent extraction (e.g. 'Extrelut', 'Extubes', 'Jetubes', 'Separatubes'), where the solvent is passed by gravity (or possibly centrifugally) over the sample dispersed in an inert support, may be a mixed blessing. Even if no interferences arise from their use, there is the complication of a progressively diminishing concentration of the analyte in the aqueous phase. Use of a large solvent volume, allowing ample time for equilibrium, could ensure exhaustive extraction but would furnish a rather dilute extract. Non-exhaustive extraction might vary in extent with analyte concentration, whereas with conventional batch extraction the proportion should be constant.

Partition behaviour

In the following outline, the analyte symbols are unorthodox but hopefully clear. For a compound 'C' distributed batchwise between two phases (each of which should be pre-saturated with the other), at a pH which must be stated unless the compound is uncharged, the *distribution constant* (partition isotherm) $K_{D(S)}$ for a particular 'species' S of the molecule is of less interest to the analyst than the *partition ratio* (distribution ratio or—a less happy term—extraction coefficient) D_C, viz. the ratio of total concentrations as analyzed. (Note that the numerator refers to the organic phase, notwithstanding contrary usage amongst physical chemists, and that activity rather than concentration values should really be used for theoretical calculations.) In determining D_C, it is a good precaution to work at 'realistically low' concentrations that may call for fluorimetric or radioisotopic rather than UV measurements. For a *weak base* Q (salt form = $Q^+ X^-$ or QX) having the dissociation constant K_a for the protonated form Q^+, D_C may be calculated from $K_{D(Q)}$ if only the species Q enters the organic phase:

$$\frac{1}{D_C} = \frac{1}{K_{D(Q)}} (1 + \frac{[H^+]}{K_a})$$

although a high D_C value would mitigate against establishing $K_{D(Q)}$ from the intercept of a pH curve or from K_a. (Complications could arise if the anion X^- were present in the buffer also, e.g. by use of HCl in studies with an amine hydrochloride). For a *weak acid* HX whose anion X^- does not enter the organic phase:

$$\frac{1}{D_C} = \frac{1}{K_{D(HX)}} (1 + \frac{K_a}{[H^+]}) .$$

As the following example illustrates in the simple case of a non-ionizable compound W, admixed with an unwanted interfering compound U and itself not overwhelmingly favouring the organic phase, it pays to carefully consider certain options.

(1) Use a relatively large solvent volume at the outset?

(2a) Subject the organic phase to a washing step? (Sometimes this is mis-termed 'back-extraction', a term appropriate for the transfer of an extracted base into aqueous acid). An alternative may be considered.

(2b) Using lower extraction volumes than in (1), perform a second extraction? Let the sample be 1 ml of blood containing 36 μg of W and 18 μg of U (W:U ratio = 2.0). Suppose that the partition ratio is 2.0 for W and 0.5 for U (hence *separation factor* = 4).

Then:

	W, μg in org. phase (& % recovery)	U, μg in org. phase	W:U in org. phase
(1) One extraction, 4 ml solvent	32 (89%)	12	2.7
(2a) After washing the 4 ml with 2.7 ml water	24 (67%)	5	4.7
(2b) Extract the blood a second time, only 1 ml solvent each time, and combine the extracts (2 ml); no wash	24 (67%) + 8 = 32 (89% total)	6 + 4 = 10	3.2 (cf. 1st extr., 4.0)

Evidently it is slightly advantageous to recovery, and disadvantageous for removing U, to use more than 1 vol. of solvent for extraction; 2 X 1 vol. is as good as 1 X 4 vol., but the inconvenience of extracting twice has to be weighed against the benefit of a less voluminous extract. A wash step is warranted if it is more important to remove U than to maximize the yield of W. The nature of interfering substances such as the hypothetical U is of course seldom known.

If U were non-ionizable as in the above example, and W were basic, back-extraction into a small vol. of acid after alkaline extraction as above could evidently furnish W in high yield and virtually free of U.

Ion-pair extraction can occur inadvertently with endogenous counter-ions [e.g. #F-2], but can be a valuable tool for hydrophobic analytes, especially if amphoteric, as documented in G. Schill's laboratory [e.g. in 6]. Empirical trial at different values of pH and counter-ion concentration need not await theorizing that is complicated by possibilities such as dimerization in the organic phase. Possible adduct formation can be due to excess counter-ion, or can be a deliberate consequence of adding, say, a hydrophobic organic acid where the compound of interest is basic. In practice, what can be ascertained as the equivalent of a distribution constant is a *conditional extraction constant* derived from observed distribution ratios. It is salutary to consider a plasma catecholamine assay [30] where 3-*O*-methyl biogenic amines were isolated by ether extraction in the presence of 'tetraphenylborate' (Ph_4BNa, the Na salt of tetraphenylboron): inspection of the data indicates that the improved partition ratios with complexing at near-neutral pH values scarcely attained 0.1. Even this modest success has not been repeatable in our laboratory (A. J. Thomson) or in another U. K. laboratory. Yet this unorthodox anionic agent, as previously used for extracting choline esters [31], may warrant rescue from apparent neglect.

Drying down solvent extracts
In few laboratories is a rotary evaporator the preferred means of drying down extracts in multi-sample assays. Some approaches are described at the end of this volume [#NC(F) −5, −6 & −7; comments in #NC(F)-4]. Many drug investigators make use of a nitrogen stream at atmospheric pressure, usually with warming. There is a volatility problem with certain drugs, exemplified by oxprenolol [E. Reid, in 7] and lidocaine [32], although losses (not observed with disopyramide or quinidine [32]) may be circumventable by minimizing the time and the heat applied. Another risk, where the drug has been extracted into the solvent from aqueous sodium hydroxide, is that traces of the latter may turn up in the dried-down residue [32]. A methanol rinse-down step may minimize wall losses [E. Reid, in 7].

Literature on environmental contaminants such as pesticide residues abounds in allusions to the 'Kuderna–Danish evaporator', although few laboratories seem to have it in daily use. This device, illustrated along with others in M. J. Saxby's contribution [#NC(C)-2], serves to concentrate extracts to, say, 5 ml or, if used with a 'micro-Snyder column' [33], to well below 1 ml, with minimal risk of losing solutes such as pesticides. Since evaporation is quick (and has to be watched), multiple samples can be handled serially with little inconvenience.

(v) SOLID AGENTS FOR ISOLATING SOLUTES
Amongst agents that have been in use for some years [5, also A. A. A. Aziz

et al., in 6], especially for analytes that are not readily solvent-extractable, the choice sometimes falls on an ion-exchange resin or XAD-2, although seldom is either applied directly to raw plasma [but see #F-3]. XAD-2 (or a more crosslinked polystyrene resin, XAD-4) is now an established tool for trace-organics in water [#B-1, #B-2]; prior clean-up of the resin to obviate GC interferences [#B-1] might be omitted now that a purified grade has become available (from Applied Science Europe B. V.). Aliphatic versions (XAD-7, XAD-8), and also activated carbon (of coal origin), have been favoured for some trace-organics in water [34]. XAD resins are again mentioned below in connection with air sampling. The sample need not be a liquid.

In general, any solid agent should be checked with the analytes of interest in case recovery is incomplete, especially if a batch change has to be made. 'Florisil' as used in pesticide studies is included in this warning. This and other problems with solid agents are considered by F. C. McElroy *et al.* [in 1] in a useful survey of methods for environmental pollutants.

An agent of growing popularity is C_{18}-silica (ODS-silica), whose merits for residue analyses are described by A. P. Woodbridge (#C-4). With the aid of a syringe, the sample may be forced through a bed of the particles (possibly coarser than used for HPLC) which may be bought in cartridge form (Waters Associates; Applied Science), or which is introduced directly into the syringe as was done in pioneer studies [35]. These were concerned with hormonal peptides in plasma, which was used direct or after deproteinization with 30% trifluoroacetic acid (which, unlike trichloroacetic acid, is volatilizable); a suitable eluting agent was methanol/5% trifluoroacetic acid (4:1 by vol.). In the author's laboratory, a drug of dicarboxylic amino acid type resisted elution, but satisfactory recoveries were observed (by Linda Kelly) for atenolol, practolol, and—at least in some experiments—metoclopramide. The possible usefulness of the agent for the pre-GC clean-up of basic drugs in plasma deserves deeper exploration. A possible alternative [cited in 36] is a cellulose column, as investigated for tricyclic antidepressants.

(vi) SOME METHODOLOGICAL PITFALLS

Analysts in the drug field may find this and the previous books [6,7] to be helpful [*see* Check-list at the end of this article, which anticipates #NC(F)-4] in strengthening the 'sixth sense' that is vital to minimize the risk of disasters or irreproducibility in multi-step analyses on body fluids. Attention is now drawn to relevant articles [e.g. #D-4] which have now been signposted and reinforced by S. H. Curry & R. Whelpton [#NC(F)-3] and by B. Scales [#NC(F)-4], and to book chapters which summarize lore from the laboratories of C. E. Pippenger [10] and A. H. Beckett [37], e.g. on sample collection and on preserving labile drugs and metabolites. Thus, hydroxylamines are prone to oxidize in alkaline solution or in certain solvents (not diethyl ether, the purification of which is discussed [37] with especial attention to aldehyde impurities); moreover, basic compounds such as methadone may become alkylated by chlorohydrocarbon solvents [37]. Dichloromethane needs scrutiny for traces of cyanogen chloride that may cause artifacts [38]. There is scattered lore on solvent purification [indexed in 6, 7]. Lability troubles may, of course, arise in a final chromatographic step [7, 37] as well as in sample preparation.

Poor drug recoveries from blood or plasma after deep-freeze storage [#F-5 & #NC(F)-3] may be due not to decomposition but to increasing tenacity of *in vitro* protein binding, as suspected in the author's laboratory for the acid 'MFCA' (formed *in vivo* from flavoxate). Another effect has been found with methadone spiked into blood: fresh blood bound it more strongly than aged blood [37]. Storage of blood samples at −15° rather than 2° impaired the extractability of metoclopramide [J. P. Leppard & E. Reid, in 7].

(vii) AIR AND PERSONAL SAMPLES IN RELATION TO HEALTH RISKS

Where an airborne trace-organic could entail a risk to health, ideally an analytical study should entail measurement not only of the air but also of appropriate samples taken from each individual, notwithstanding the present dearth of epidemiological data correlated with antecedent body levels of the substance. Whether or not this ideal is reasonable [E. King, #A-1], analysts in the occupational hygiene area will increasingly be faced with samples of possibly unfamiliar type, e.g. blood. In principle the drug-assay methodology that dominates this book should give them guidance and courage, with the reservation that the very volatility of the substances concerned can add to the difficulty of getting reliable results.

For a range of offending substances, excellent guidance is to be found in a monograph [39] which deals with blood, urine and breath samples—although not with saliva, useful though this may become if correlations between salivary and blood levels as sometimes found for drugs [#D-2] prove to have counterparts amongst airborne contaminants. Where benzene is the contaminant, exposure may be assessed by assaying urine for phenol as a main metabolite. Commonly this is done by ether extraction followed by GC which is complicated by endogenous interfering substances found even in non-exposed individuals [39]. This problem was circumvented in the author's laboratory by use of a supposedly archaic method [cited in 39], entailing steam distillation of the phenol and subsequent colorimetric determination. The moral is that GC may on occasion be a fetish.

GC is, however, rightfully the dominant measurement method for organics in air samples. The novice faced with deciding how to collect these has hitherto been handicapped by the paucity of critical, up-to-date surveys of approaches. A recent book [40] gives excellent guidance, whilst not being a full basis for making judgements on sampling methods and equipment. The following remarks are intended merely to complement the perspective given in a later article [#A-4], to which the reader could usefully turn at this stage, and in an excellent review of sorbent methods [41]. The remarks are prompted by personal experience of building up a range of equipment and methods whilst barely being able to see the wood for the trees.

Questions that arise, some inapplicable to 'passive' sampling [#A-3, #A-4], may include the following.

(1) What type of wearable sampling pump is appropriate for trace-organics?— The rate is appropriately lower than for airborne particles, not only to minimize the risk of incomplete pick-up if an adsorbent is used but also because the quantity needed for analysis is relatively small (particularly if, with an adsorbent, GC is done not on a CS_2 eluate but on a vapour concentrate obtained by 2-stage thermal desorption) and is suitably spread over hours rather than minutes unless, as is useful on occasion, a 'grab' sample is taken. Amongst pumps that have the advantages of adjustable average rate (constancy is not vital) and a lapsed-volume meter, two may be mentioned: a 'Sipin' model, and the ingenious orifice-regulated 'Accuhaler'. A splitter available for the latter enables replicates to be collected, so that a poor GC run after thermal desorption is not a calamity.

(2) How should the sample be collected and (if not analyzed at the collection point) transported?—Some organics may be kept in the vapour state, in a plastic bag [#NC(A)-2] or an appropriate rigid vessel [#NC(A)-3]; but there is a general preference for adsorption [#A-4], if not agreement on the choice amongst a range of agents [41, 42]. Porous polymers, which are rightly gaining favour over charcoal and which include XAD resins [41], are particularly amenable to thermal desorption and re-use; Tenax is less prone than Porapak Q to give spurious GC peaks with strong heating [citations in 26]. Silica gel may be advantageous [43].

(3) What type of GC column suits best?—Porous polymers seem to be ousting coated packings for some organics (hence 'GC' is a better general term than 'GLC').

(4) How should calibration be done?—Whilst it does seem to be valid to spike an organic liquid directly onto an adsorbent, the capability to make a standard atmosphere is desirable or, if the sample is in vapour form throughout, essential. A 'dynamic' standards generator [40] is an asset. Only where the compound is being assayed routinely does it seem worth buying a 'custom-prepared' cylinder or incurring the purchase cost or setting-up trouble of a permeation-tube device or, if the compound is rather involatile, a diffusion-tube device. Often it suffices to have a static atmosphere, as can be prepared merely by injecting the liquid compound (say 1 µl) into a Winchester bottle *via* a septum which is normally covered with a film of mercury [*see* J. G. Firth, #A-4]; overnight equlibration is,

in our experience, essential, but thereafter the atmosphere may stay constant for weeks.

Contaminants such as vinyl chloride monomer (VCM) [42] present special problems because of their chemical reactivity. Problems also arise where the persons at risk are not employees but the public at large, insofar as the levels to be determined are far below those allowed in places of employment. This analytical challenge was successfully met, in respect of street air and of expired breath, in an investigation which dealt with contaminants such as benzene [43]. Amongst a collection of method reviews [1] which deal also with flue emissions, that of F. C. McElroy *et al.* is notable.

It is, however, workplace atmospheres that are emphasized in the following section [#A] and in present-day regulations centred on 'Threshold Limit Values'. In brief, for the benefit of readers not acquainted with this field, the TLV for a particular substance is the allowable prevailing concentration as averaged over a working week (Time-Weighted Average, TWA). For most offending agents the TLV may be modestly exceeded on occasion ('permitted excursion'). For some substances where exposure lasts for minutes rather than hours, a 'Short-term Exposure Limit' (STEL) may be specified.

(viii) DARE ANALYSTS LOWER THEIR SIGHTS?

As remarked by book contributors in the case of workplace pollutants [#A-1] and medicinal drugs [#F-8], there may be justifiable misgivings about the bearing on human health of laboriously obtained analytical values. Certainly there is the possibility, in the GLP climate [(i), *above*] that analysts may feel obliged to produce values which, especially in respect of precision, emulate those of over-earnest members of the analytical fraternity. At least analysts are hardened to unawareness by their customers of the time and skill needed for back-room beavering. May morale survive!

(ix) POSTSCRIPT

Retrieval of lore will be helped by the General Index (as in [6] &]7]) and by the retrospective surveys on pp. 347 & 353. Concerning a point mentioned in both, viz. the risk in trace-organic analysis of adsorptive losses onto glassware, both this problem and a notorious counterpart — 'creep-out' from the tube wall in a later run — have now been encountered with a chlorpromazine metabolite in the author's laboratory. The trouble lies in the drying-down tubes (cf. p. 23) and is curable by pre-treating the tubes with chromic cleaning mixture. This remedy turned out not to be novel (cf. J. P. Desager, p. 326 in [7]). To the above lore on acetonitrile deproteinization (p. 20) may be added the useful finding [44] that salt addition gives two phases.

**'SOP' CHECK-LIST FOR PROCESSING BIOLOGICAL FLUIDS:
SOME CHOICES***

Project ref., date, & initials of supervisor:

#THE SAMPLE AND THE TEST COMPOUND, viz.:

   any storage lability: other lability (e.g. light):

Whether storage at $\sim2°$ rather than deep-frozen: *always/o.k. if brief*
                                                        (e.g. overnight)

Initial exposure to plastic such as polypropylene [presumably PTFE O.K.] :

   little risk/not worth the risk

Medium for dissolving pure compound,   which *is not/is* a salt, viz.:

   — calculation factor $\dfrac{\text{free cpd.}}{\text{salt form}}$ =

   water/aq. methanol (1:1 by vol. ?)/OTHER:
   —containing/adjusted with [SPECIFY] :

#VESSELS (shun lubricants!) — Make it clear which stages(s) concerned

Surface: *glass/silanized glass/polycarbonate*/OTHER:
      — whether need to pre-wash with appropriate solvent(s): *no/yes*
      — whether exceptional precautions (tongs !) to avoid lipids: *no/yes*
      — whether chronic pre-clean needed: *no/yes*
Shape etc: whether advantageous to have conical bottom: *no/yes*
      whether pressure or escape risk: *no (no cap or stopper needed)/*
                                    yes, mild/yes, needs reaction vial

                                                        [*continued*

---

* This form is proving helpful for drug assay benchwork in the Editor's laboratory.

#SOLVENTS (Intentions, and action date + initials)

SOLVENT (list any mixture, besides pure solvents) & GRADE (list supplier & batch no. — *the bench worker's responsibility*)	DESIRABLE TESTING (e.g. GC-ECD after drying down) & OUTCOME/ REF. to record of test	TREATMENT(S), e.g. (1) distil (b.p.?); (2) alumina; (3) shake with water or... [SPECIFY] ; (4) IF MIXTURE, re-check saturation	STORAGE LIMIT & ANY PRECAU-TIONS (e.g. exclude light)

#DRYING-DOWN STEP(S) [SPECIFY] where losses could occur

 Use N_2 stream/other mode [SPECIFY; finally over desiccant ?] :
  — *vital/desirable* to keep below stated temp.limit & to stop as soon as 'dry'
  — *no need/need* to rinse down with small vol. of methanol and re-dry

#PERMITTED PAUSES (say overnight)—Allowable to LEAVE BLANK until question arises
 Bench/cupboard/refrigerator, at stage [SPECIFY; desiccant or other precaution ?] :

#OTHER POINTS NEEDING ATTENTION, e.g. concerning buffers or deproteinizing agents
 —

References

[1] Hertz, H. S. & Chesler, S. N. (eds.) (1979) *Trace Organic Analysis: A New Frontier in Organic Chemistry*, National Bureau of Standards Special Publication 519, Washington DC, 817 pp.

[2] Karger, B. L., Synder, L. R. & Horvath, C. (1973) *An Introduction to Separation Science*, Wiley, New York.

[3] Skoog, D. A. & West, D. M. (1980) *Principles of Instrumental Analysis*, 2nd edn., Holt-Rinehart-Winston, London.

[4] Snyder, L. R. & Kirkland, J. J. (1979) *Introduction to Modern Liquid Chromatography*, 2nd edn., Wiley, New York.

[5] Reid, E. (1976) *Analyst*, 101, 1–18.

[6] Reid, E. (ed.) (1976) *Assay of Drugs and other Trace Compounds in Biological Fluids* (Vol. 5, Methodological Developments in Biochemistry*), North-Holland, Amsterdam, 254 pp.

[7] Reid, E. (ed.) (1978) *Blood Drugs and Other Analytical Challenges* (Vol. 7, Methodological Surveys in Biochemistry*), Horwood, Chichester, 355 pp.

[8] Smith, R. V. (1976) *Am. Lab.*, 8, 47–53.

[9] Bailey, D. N. & Guba, J. J. (1979) *Clin. Chem.*, 25, 1211–1215.

[10] Pippenger, C. E., Perry, J. K. & Kutt, H. (eds.) (1978) *Antiepileptic Drugs: Quantitative Analysis and Interpretation*, Raven Press, New York, 367 pp.

[11] Burnett, R. W. (1980) *Clin. Chem.*, 26, 644–646.

[12] Calus, I. M. (1980) *Anal. Proc.*, 17, 120–123.

[13] Taylor, D. G., Kupel, R. E. & Bryant, J. M. (1977) *Documentation of the NIOSH Validation Tests*, DHEW (NIOSH) Publication 77–185, Cincinnati.

[14] Barney II, J. E. (1967) *Talanta*, 14, 1363–1366.

[15] Knapp, D. R. (1979) *Handbook of Analytical Derivatization*, Wiley-Interscience, New York, 741 pp.

[16] Nicholson, J. D. (1978) *Analyst*, 103, 1–28 & 193–222.

[17] Meffin, P. J. & Miners, J. O. (1980) *Progr. Drug Metabolism*, 4, 261–307.

[18] Kaye, C. M., (1980) *Progr. Drug Metabolism*, 4, 165–259.

[19] Daley, R. D. (1972) in *Analytical Profiles of Drug Substances*, Vol. 1 (Florey, K., ed.), Academic Press, New York, pp. 119–147.

[20] Townshend, A. (1973) *Process Biochem.*, March issue, pp. 22 & 24.

[21] Giang, P. A. & Schechter, H. S. (1960) *J. Agric. Food Chem.*, 8, 51–54.

[22] Curry, A. C. & Mercier, M. (1970) *Nature*, 228, 281–282.

[23] Balba, H. M., Still, G. G. & Mansager, E. R. (1979) *J. Assoc. Off. Anal. Chem.*, 62, 237–240.

[24] Gallay, W., Egan, H., Monkman, J. I. Truhaut, R., West, P. W. & Widmark, G. *Analytical Methods for the Determination of Selected Pollutants*, Butterworths, London, 296 pp.

* Retrospectively termed (as for this vol.) Methodological Surveys (A): Analysis

[25] MacLeod, A. J. (1973) *Instrumental Methods of Food Analysis,* Elek Science, London, 802 pp.

[26] King, R. D. (ed.) (1978) *Developments in Food Analysis Techniques,* Applied Science Publishers, London.

[27] Hurlbert, R. B., Schmitz, H., Brumm, A. F. & Potter, V. R. (1954) *J. Biol. Chem.,* **209**, 23-39.

[27a] Kim, B. K. & Koda, R. T. (1977) *J. Pharm. Sci.,* **66**, 1632-1634.

[28] Chu, S. Y., Oliveras, L. & Deyasi, S. (1980) *Clin. Chem.,* **26**, 521-522.

[29] Po, A. Li Wan & Irwin, W. J. (1980) *Lab. Pract.,* **29**, 21-25.

[30] Da Prada, M. & Zürcher, G. (1976) *Life Sciences,* **19**, 1161-1174.

[31] Fonnum, F. (1969) *Biochem. J.,* **113**, 291-298.

[32] Flood, J. G., Bowers, G. N. & McComb, R. B. (1980) *Clin. Chem.,* **26**, 197-200.

[33] Burke, J. A., Mills, P. A. & Bostwick, D. C. (1966) *J. Assoc. Offic. Anal. Chemists,* **49**, 999-1003.

[34] Van Rossum, P. & Webb, R. G. (1978) *J. Chromatog.,* **150**, 381-392.

[35] Bennett, H. P. J., Hudson, A. M., McMartin, C. & Purdon, G. E. (1977) *Biochem. J.,* **168**, 9-13.

[36] Scoggins, B. A., Maguire, K. P., Norman, T. R. & Burrows, G. D. (1980) *Clin. Chem.,* **26**, 5-17.

[37] Beckett, A. H. & Cowan, D. A. (1978) in *Drug Metabolism in Man* (Gorrod, J. W. & Beckett, A. H., eds.), Taylor & Francis, London, pp. 237-257.

[38] Franklin, R. A., Heatherington, K., K., Morrison, B. J., Sherren, P. & Ward, T. J. (1978) *Analyst,* **103**, 660-662.

[39] Piotrowski, J. K. (1977) *Exposure Tests for Organic Compounds in Industrial Toxicology,* DHEW (NIOSH), Cincinnati, 138 pp.

[40] Thain, W. (1980) *Monitoring Toxic Gases in the Atmosphere for Hygiene and Pollution Control,* Pergamon, Oxford, 159 pp.

[41] Crisp, S. (1980) *Ann. Occup. Hyg.,* **23**, 47-76.

[42] Hill, R. H., McCammon, C. S., Saalwaechter, A. T., Teass, A. W. & Woodin, W. J. (1976) *Anal. Chem.,* 1395-1397.

[43] Gage, J. C., Lagesson, V. & Tunek, A. (1977) *Ann. Occup. Hyg.,* **20**, 127-134.

[44] Mathies, J. C. & Austin, M. A. (1980) *Clin. Chem.,* **26**, 1760.

#A Workplace Air Samples

#A-1   THE EVALUATION OF OCCUPATIONAL EXPOSURE

E. KING, The National Occupational Hygiene Service Ltd.,
12, Brook Road, Fallowfield, Manchester, M14 6UH, U.K.

The field function of occupational hygiene is to evaluate work exposure to adverse environmental factors with a view to ensuring that the risk to health or comfort is, or can be made, 'acceptable'. For organic vapours, this evaluation may be by blood or urine samples, late afternoon or early morning exhaled air samples, 'personal' air samples or static air samples. The uncertainty in 'personal' air sampling is governed by the relevance of the sample as taken to other times and tasks, by the vapour collection, the sample transport and the vapour preparation for analysis, and finally—and usually the least important—by the sensitivity and accuracy of the analytical method used. The task of 'evaluation' must therefore be considered as a whole, and not fragmented.

One of the nicer things about television programmes is their ability to entertain despite the attempts of the producers to make them serious. One such episode in my field was the occasion when a Professor of Chemistry was seen with a GC and a bottle of air purporting to have come from a factory, putting the two together and then pontificating on the risk to health in that factory. This was the 'have instrument, will measure' syndrome which, if allied to the 'have pump, will sample' syndrome, causes so much chaos, and often risk to health, on the shopfloor.

In this country, and unlike America, the monitoring of the factory atmosphere started as an expertise coming out of the analytical laboratory. This approach is typified by the 1937 to 1940 DSIR pamphlets, and by the ICI methods developed in the war years and published in 1956 under the general direction of the Analytical Chemists' Committee of ICI Ltd. The emphasis was on the analytical procedures, understandably since without them no progress could be made, but it retarded the development of the occupational hygiene proper. The old axiom of 'to measure is to know' is valid only if the right parameters are being measured. The analysis, no matter how accurate, is only one step in the process of the assessment of an occupational environment, and should not dominate the whole, but rather serve it.

For the present purpose I am taking a limited view of occupational hygiene, just one aspect of its field function, viz. the evaluation of occupational exposure to adverse environmental factors. In this day and age, with Health and Safety Executive Inspectors, health-and-safety representatives, their union experts, not to mention the press and the television, all prepared to make instant judgements from inadequate data, this has become the sharp end of the profession, and we must take care that such data as we produce are real, and properly qualified. It is to be noted that I am excluding the other types of measurement made in my field, those for legal and engineering purposes for instance, and limiting myself to 'evaluation'.

THE FRAMEWORK

Our problem is quite simple—essentially that the 'factory environment' and 'environmental monitoring' are catch phrases that rarely have reality. A factory, or any shop in that factory, or any process in any shop, consists of a number of micro-environments. Very often it is a case of each worker having his own, very specific, micro-environment even if averaged over a long time, it being a function of his own movements within a general environment, which itself varies with position and time, and his own actions on the job which may produce a highly personal environment. Because of these variables we can monitor individuals in nominally uniform working groups, by breath assessment or biochemical means, and find remarkable consistency within individuals, sometimes over years in the case of lead workers, for instance, allied to a factor of five or more between men doing the same jobs. The 'dirty worker' is no capitalist myth—I myself am of that ilk compared with my wife, in that when we paint a room I finish with paint over hands and clothes, and she is spotless—and he is not to difficult to observe if one cares to use the excuse of 'air sampling' for spending an hour or two on the shop floor simply watching men. Traditionally, the night shift differs from the day shift, and so on. The hygienist's lot is sometimes a sleepless one.

Let us consider the types of 'toxic materials' that exist in these micro-environments. First the 'dusts', which show wide variations in size–frequency distribution with distance from the site where they become airborne, depending upon the air movements at the time. For fibrogenic dusts the problem is minor, since we are only concerned with the particles of Stokes diameter 7 μm and below, hence the 'fall-out' of the larger ones is immaterial. We used to argue this way with the toxic dusts—lead, cadmium etc.—but recent evidence suggests that we are dealing with a high percentage absorption of the usually small percentage of the fine dust which reaches the alveoli, allied to a low percentage absorption of the far greater mass of the larger particles which deposit in the bronchi. Hence the very poor correlation of 'dust in air' with biochemical evidence of absorption. Also, of course, there is no agreed definition of 'total' airborne dust—this is a theoretical time/space concept which cannot be measured. Eventually this will be defined by reference to sampling head geometry, but this is still under debate.

This can be neatly illustrated by the current long-running Eurofarce about lead in air. It started by taking the U.S.A. 'standard' of 0.1 mgPb/m^3, as corresponding to a blood limit of 60 μg Pb/100 ml. So the first draft had the simple value 0.1 mg Pb/m^3. This is, of course, neither necessary nor achievable in the major lead industries. In the next drafts we had: 0.1 mg Pb/m^3 'respirable' (a factor of, say, 5); 0.1 mg Pb/m^3 'respirable', or below 10μm aerodynamic diameter sampled at less than 4 m/sec (by sampling at 0.5 m/s with a downward facing filter, a factor of, say, 50); then, by the ninth draft, 0.1 mg Pb/m^3 sampling up to 15 μm aerodynamic diameter. After all, what is a factor of 'x' provided the 0.1 mg Pb/m^3 is preserved? So, in addition to the space/activity micro-environments, 'dust' adds another dimension of uncertainty.

The situation is rather simpler with 'fume'—in our jargon, freshly oxidized metallic vapour—provided that we sample on a man near the source in a well-ventilated shop. At this point, we have sub-micrometre agglomerates of even smaller particles. However, once the 'fume' has become airborne, it starts flocculating, and the much larger particles have neither the lung penetration nor the pharmacological properties of the initial material. This represents another complication for air sampling and interpretation.

Fortunately, my main concern here is with organic materials, and I shall further limit this to organic gases and vapours and their liquid phases.

SAMPLES FROM THE EMPLOYEE

My medical colleagues sometimes use a phrase to the effect 'when in doubt, examine the patient', and this is not such a bad principle. Admittedly, as in medicine, first-hand knowledge of the problem may well confuse what were previously firm opinions; but an hour or two spent watching the process concerned, especially if fore-armed with process layout, floor diagrams and job breakdowns etc., may well save several hours of sampling and analysis. There is little point in a complicated evaluation of a grossly low exposure or a grossly high one, although some 'samples' may have to be taken for the record. There is no point in taking multiple air samples when the real potential absorption route is the skin. There is no point in sampling at all if the conditions are not 'typical or worse', and so on. But out of all this comes, or should come, a programme of some kind. At this point the statisticians' eyes light up, and we can start talking of distribution and confidence limits and so on, testing for various compliances. This is all good fun, and we all do it occasionally; but in this day and age we are concerned not with 5% or even 1%, but with protecting the individual workers in individual factories. And while we could solve the unemployment problem in the scientific world by allocating one environmental scientist and one analyst to each worker, so that he is continuously monitored, I feel that this would exceed all the cash limits known to man or woman.

We must, therefore, make the most of the various tools we possess, projecting our findings backwards and forwards with time and our understanding of the plant and process, to give an informed opinion of the long-term risk to the workers' health without attempting to enforce, for instance, the U.S.S.R. concepts of permissible air levels, most of which are quite unachievable in Russia or elsewhere in run-of-the-mill industry.

What personal measurements are open to us in this context of organic gases and vapours? If we exclude 'fringe' specimens that have been experimented with for other materials—tears, sweat, saliva etc.—we are left with the following possibilities.

(1) Levels of the materials or their metabolites in blood or urine samples, these of course relating directly to absorption by the individual workers, and reflecting skin as well as lung absorption . This approach is very attractive, and is widely used in those cases where from long experience we know what we are doing. We monitor benzene exposure by urinary phenol, trichloroethylene exposure by urinary trichloroacetic acid etc., and only rarely use air sampling except for detection purposes. A complicating factor if looking at urinary metabolites as our lodestone is that some materials—notably styrene in general industry— have two or more metabolic routes, giving different metabolites and different intermediaries, one of which could conceivably present a greater risk than the others. And the ratios vary.

(2) Levels of the material in exhaled air relate to current blood level, and superficially this is an attractive approach. However—and this applies to blood and, to a lesser extent, urine—this reflects both current intake and the equilibrium between the body storage—usually fatty tissues in our case—and the blood. So we are all busy arguing whether we should be taking end-of-Thursday-shift samples of blood, urine or exhaled air, to give us the 'highest' body burden, or early-morning level to give the residual body burden. This is all tied in with our limited knowledge and agreement on the dynamics of the toxicity of the solvents, some of them having both narcotic and systemic toxicity, and we therefore fall back on what is essentially a much cruder form of worker monitoring, but one which offers less opportunity for interdisciplinary confusion, namely air sampling.

(3) Air sampling is a time-honoured means of assessing environments. In my time in the profession, I have seen it change from an almost experimental status in the 1940s, through a medical/scientist confrontation position in the 1950s, increasing development in the 1960s, and universal and uncritical popularity in the 1970s—uncriticized, that is, except by the professionals in the field. Most of them have been horrified by the misuse of our sampling procedures and TLV concepts by people who have no real understanding of

either, and we are afraid lest they thereby fall into disrepute and we lose useful tools.

TOOLS FOR AIR SAMPLING

Let us first eliminate some versions of air sampling. Disadvantages of one approach, the indicator tube, include poor calibration and specificity, and snap sampling usually atypical of real worker-exposure. However, indicator tubes often be used to demonstrate very good or very bad conditions to the sharp end of the community—shop stewards and works managers—on site. They can be held in reserve by the foreman to check leakages, or levels in confined spaces before entry, and so on. In brief, their educated use has possibly saved more lives than more 'scientific' methods. The only qualification is that batches must be checked for 'activity' before use, and that not too much must be read into the intermediate readings.

Next, we have the other end of the financial range, the multi-point infra-red recorder. This is the true 'monitor' of spatial levels, in that it can register changes and thereby process faults or losses; but it can rarely give an accurate measure of the personal exposure of the individual workmen, for the reasons previously discussed. A massive leak of vapour to a worker's face when dipping a tank or opening a blind, does not register upwind of him, and only registers in a diluted form downwind.

More recently, the 'gas badge', has become available, a carbon absorber into which vapours can diffuse. A number of us have experimented with these, and we seem to all agree that, to date, it is an experimental approach having to be validated by more orthodox methods in any given situation. We have indeed recently carried out such an exercise, on 6 men exposed to 4 solvents, monitored by serial personal air samples, serial badges, urine and exhaled air samples, etc., etc. We are still trying to make some sense of the data. I suspect we never shall.

At the end is the true personal air sample, taken by a pump worn by the worker with the sampling area around the lapel. This is the norm to measure 'that to which the workman is exposed' for comparison with the various air limits which are thus defined. It all seems straightforward. There are snags. The first is that the sampling time should be related to the time of biological action of the material concerned. There is no point in proving that the 8-hour exposure to toluene was acceptable if the worker was unconscious at the end of the shift, having received the whole of his daily dose in the last 10 min. While one can evade any personal decision on this by referring to 'C' values, or 'STELs' etc., etc., the more responsible way is to read up the derivation of the standard, and the toxicology of the material, and be able to assure not only the worker but also yourself that you are doing the right thing. You may decide that the 'right' thing is serial 20-min samples, or 21 per man day, or 210 per 10 man days, or some 100 h of GC time. This is to measure one day's exposure for ten men. All to within ±1%, because this is the accuracy of the GC; or is it?

AIR SAMPLING AND INTERPRETATION

We have an unofficial check list of about 30 items for site investigations, a list which enables us to make a reasonable guess as to whether we have typical conditions on the day, and possible variations from them. This list includes, for instance, the wind, weather and general ventilation (doors and windows) and the effect total variations in these are likely to have on the general workshop conditions, and the specific processes and forced-air extraction systems. We note the housekeeping, which sometimes seems too good to be typical. We note the jobs not being carried out, and their possible contribution to the general. We want comments by the workman, as well as by the managers, on how typical the conditions are and how they can vary. We get suspicious when we find the man who normally does the job is doing something else, and a shop steward or managerial nominee is there in his place.

With solvents, we must ascertain what are the ambient and working temperatures at the time, to be later related to the temperature/vapour pressure curves, and so on; these factors can cause our end results, to differ by an order of magnitude, not merely by a percentage.

Making the usually false assumption that we can cover all these variables —we can otherwise shut both eyes and ignore them—let us look at the methodology, firstly the hardware. Is the sampling pump calibrated for air flow with the sampling head to be used in the field, and will it maintain that air flow over time? We assume the gauge fitted to the pump to be of little value except to show ON/OFF flow. Concerning absorbent—silica gel or carbon—we must ask whether it has been prepared properly and, in the case of silica gel, will last out the sampling time planned under the factory conditions. This entails serial sampling heads being tested under laboratory conditions, but at the factory flow rate, using increase in weight as the measure of absorption. Next, can you desorb it, not merely immediately after collection, but 48 h after it has been left stoppered up? In other words, can you really be sure that the material going into the GC— and this is still the general instrument in use—relates reasonably to that in the factory air as sampled?

Is the carbon disulphide that you will almost certainly have to use to desorb some vapours from charcoal going to present a bigger risk to the health of the laboratory staff, or those unfortunates downwind of the usually unscrubbed fume-cupboard exhaust, than exists in the factory you are assessing? Or would it not be better to have a lower collection/desorption efficiency with a less retentive medium and a less toxic desorber? Or can you get the claimed efficiency out of thermal desorption that most of us have failed to achieve? Can you push the GC to allow a doubled throughput at the cost of, say, 10%

precision? And at the end of the day, can you look at the whole exercise, from your observations in the factory to the output of the instrument, and give an informed opinion as to the possible risk to health of individual workers in that factory from when you left it up to the time of your next visit? And if you cannot say the risk is acceptable, what is the rough cost in jobs or money of making it so?

All this is a very far cry from the bottle of factory air and the GC; yet really this is the nub of the matter. Occupational disease is caused, and can be prevented, on the shop floor. The evaluation of the exposure of workmen to risk materials in their environment is far too serious a matter for it to be treated as an exercise in analytical chemistry, or even air-sampling technology, but must be looked at as a whole, and the various individual parts subjugated to that end.

#A-2   **SETTING UP STANDARDS AND FIELD ANALYSIS FOR ORGANIC VAPOURS**

STEPHEN CRISP, Laboratory of the Government Chemist,
Department of Industry, Cornwall House, Stamford Street,
London SE1 9NQ, U.K.

The principal methods of generating standards atmospheres of organic vapours are discussed. Techniques considered are atomization/injection, air flow over or through the test substance, diffusion cells and permeation devices. Field tests yielding rapid results are reviewed, including the use of detector tubes, chemical field tests, and instruments.

GENERATING STANDARDS

A standard atmosphere is a source of air containing a gas or vapour of known concentration. The many techniques available to generate such atmospheres can be classified into two main categories, static and dynamic methods [1,2]. In static methods the gas mixture is stored in a closed vessel. With these methods the main problems are sorption of the compound on the container walls or other surfaces, the effect of pressure change on withdrawing an aliquot, and that only a limited volume can be prepared. Some of these problems are less serious at higher concentrations. For work where large volumes of gas mixtures are required, dynamic methods of atmosphere generation are advocated, as the concentration of the mixture can readily be checked and re-set to different levels, thus obviating the need for separate standards. In the case of organic vapours the most commonly used techniques are atomization/injection and vapour saturation. In the atomization/injection technique the organic compound (or a suitable solution) is fed to a nebulizer from a low-speed motor-driven syringe into a metered gas stream. The required concentration is achieved by dilution in up to three dilution stages by mixing with metered streams of air. This technique has been used to produce atmospheres of ethanol and acetone [3].

The vapour-saturation approach

Vapour saturation can take a variety of forms. A stream of air may be passed over a volume of liquid or through it; with the latter procedure aerosol formation must be avoided. For solid organic compounds of suitable volatility, air can be passed through a mixture of the compound and a carrier, such as sand. A liquid, placed in a container connected to the main gas stream by a capillary tube, diffuses vapour at a constant rate, provided the apparatus is thermostatically controlled [4]. This diffusion procedure has been used for toluene-2, 4-diisocyanate (TDI) and benzene. A further vapour-saturation technique involves suspending a plastic bag containing the organic liquid in an air stream. After an initial stabilizing period a constant rate of permeation is attained. The technique has been used for many ketones, but could not be applied for acetone [5]. Another device, more usual for inorganic gases, is the permeation tube [6]. In this the liquid is trapped in a tube by tightly fitting plastic spheres through which vapour permeates. Designs of these devices suitable for organic liquids are also available [7,8].

Ascertaining the concentration

The generator requires standardizing by the analysis of an air sample. Although it is possible to calculate the concentration of organic vapours generated by some techniques, e.g. diffusion tubes, this is not recommended as a standard-ization; however, the calculation may be useful in aiding generator design. The outlet from the generator should lead to a sampling point at atmospheric pressure. The atmosphere can be sampled with an appropriate collecting device and the concentration determined after analysis by any suitable technique. In the case of permeation tubes there is a trend to regard these as absolute standards and to base concentrations on the weight loss from the tube over long periods, typically a week. However, this procedure entails assumptions, e.g. that the organic liquid is 100% pure, that the permeation rate remains constant, that there are no leaks between the tube and sampling point, that no chemical changes occur to the compound in the generator, and that the flow-measuring devices are accurate. It is better to establish the concentration analytically, and only if validated should the weighing procedure be used.

FIELD ANALYSIS

If a fairly immediate analysis of the concentration of a substance in the workplace air is required then there are three main options: detector tubes, simple field tests, and instrumental measurement. The actual selection depends on availability of suitable tests or instruments and on economics. Suitable instruments for organic compounds include portable gas chromatographs and IR spectrometers.

Detector tubes are of several types, viz. stain length, colour matching, and appearance of colour. They are simple to use and tubes are available for many organic compounds. The tubes are not highly accurate and in some instances have proved unreliable. However, several specifications require an accuracy [really 'precision'–*Ed.*] of ± 35% at ½ TLV and ± 25% over the range 1 to 5 TLV [9]. Other constructional and stain-diffusion requirements are dealt with in a British Standard [10].

A number of field tests are available in the so-called Red Booklet series published by HMSO on behalf of the Health and Safety Executive [11]. These tests are designed to be sufficiently simple to be carried out by unskilled operators, to be more reliable than indicator tubes, and to give rapid results. Currently methods are available for the following organic compounds: benzene, toluene, xylene, styrene, carbon disulphide, phosgene, aniline, halogenated hydrocarbons, trichlorethylene, acrylonitrile, aromatic isocyanates, acetone, isophorone, cyclohexanone and methylcyclohexanone. The publication of methods for formaldehyde and 1-naphthylamine is in preparation. A few of these tests for the more complex substances at very low concentrations may not be considered simple to perform.

Field tests generally comprise of four phases, viz. air sampling, sample collection, reaction development, and quantitative estimation. Where the volume of air required is 500 ml or less, a hand aspirator can be used for air sampling; but for larger volumes a calibrated, battery-operated pump is used. The organic substance is collected from the sampled air either on a chemically impregnated paper held in a suitable holder or in a suitable solvent contained in a bubbler or midget impinger. At flow rates up to 200 ml/min a simple bubbler is sufficient, but for higher flown up to 5 l/min a sintered bubbler or a midget impinger is required. The organic compound is detected by a suitable colour reaction which may occur *in situ* on the paper or in solution. However, in some cases further treatment with chemical reagents is required for colour development.

The intensity of the colour produced gives a measure of the airborne concentration of the compound under investigation. The concentration is evaluated visually, by comparison with colour standards which are printed papers provided with the method, with a set of coloured solutions made from a mixture of inorganic salts or with a set of coloured glass discs viewed through a colour comparator. Normally the colour standards are provided for 0, ½, 1 and 2 TLV.

REFERENCES

[1]   Nelson, G. O. (1971) *Controlled Test Atmospheres: Principles and Techniques,* Ann Arbor Science Publishers, Ann Arbor, Michigan.
[2]   Smith, A. F. (1981) Chapter 6 in *The Detection and Measurement of Hazardous Gases* (Firth, J. G. and Cullis, C. F., eds.). To be published.
[3]   Smith, A. F. and Wood, R. (1970) *Analyst,* **95,** 683–690.
[4]   As for 1, p. 126f.
[5]   Andrew, P. and Wood, R. (1968) *Chem. Ind.,* 1836–1838.
[6]   O'Keeffe, A. E. and Ortman, G. C. (1966) *Anal. Chem.,* **38,** 760–763.

[7]   Teckentrup, A. and Klockow, D. (1978) *Anal. Chem.*, **50**, 1728.

[8]   Godin, J. and Boudene, C. (1978) *Anal. Chim. Acta*, **96**, 221–223.

[9]   NIOSH Certified Equipment List (July 1, 1978) U.S. Department of Health, Education and Welfare, Washington D.C.

[10] British Standards Institution (1976) *Gas detector tubes*, BS 5343.

[11] Health and Safety Executive HM Factory Inspectorate, *Methods for the Detection of Toxic Substances in Air*, Booklets nos. 1–26, HMSO, London.

#A–3 PASSIVE SAMPLING AND DOSIMETRY*

A. BAILEY and P. A. HOLLINGDALE-SMITH,
Chemical Defence Establishment, Porton Down,
Salisbury, SP4 0JQ, U.K.

The essential components of a diffusion dosimeter are a diffusional resistance of known characteristics and a gas or vapour sink. Given the diffusional transfer constant of the resistance, the dosimeter exposure time and a suitable means of analysis of the contents of the sink, the dosage (concentration × time) or the mean concentration can be calculated. Samples may be of badge type employing a permeable or porous membrane usually with an air gap between the membrane and the sink, or of open-ended tube type. Sinks can be chemical layers specifically reactive towards particular gases or vapours, or non-specific adsorbents such as charcoal, molecular sieves, silica gel or polymers such as Porapak. A knowledge of sink capacity is needed to avoid the risk of saturation. Reproducible recovery of adsorbed samples, particularly after storage, is important. Effects of varying temperature and humidity should be considered.

The introduction of the Health & Safety-at-Work Act has increased the need for means of monitoring workplace atmospheres, particularly at the individual level, to ensure compliance with relevant threshold limit values (TLVs). Individually worn devices for collecting atmospheric samples are comparatively bulky and inconvenient to wear for some types of job. They are also expensive, and require their own power supply, admittedly self-contained, which adds to their weight; the pump is usually strapped to the waist, with a tube leading to the sampling point close to the worker's breathing zone. A sampling device which obviates the need for pumps, batteries and tubes would offer obvious advantages from the worker's point of view and should also be cheaper, and therefore able to be deployed in greater numbers at reasonable cost. Work on such passive sampling devices has been proceeding for several years in various establishments, and some versions are being marketed.

A passive sampling device for personal atmospheric monitoring should be capable of sampling for at least a working day at any concentration likely to be encountered, without being overloaded; its sampling rate should be constant with time and directly proportional to concentration, and the effects of temperature, humidity and air movements on the sampling rate should be within acceptable limits. What these limits should be is arguable, but ±20% would probably be acceptable for most purposes. Finally, the results obtained from passive samplers should be comparable with those from dynamic samples taken in parallel.

The basis of the sampling method is the transfer of a gas or vapour from the air being sampled to a sink, at a rate which is defined by a diffusional resistance of some type. Membranes, diffusion tubes and perforated discs have all been used as the diffusional resistance [1-5]. Given the transfer constant of the resistance, the exposure time and a suitable means of determining the amount of material collected, then the dosage (concentration X time) or the time-weighted average concentration over the exposure period may be readily calculated.

SYSTEMS

In the work carried out the Chemical Defence Establishment (CDE) [1] silicone rubber membranes were originally used; but owing to supply problems porous polypropylene film is now used instead and has also been found satisfactory. The CDE diffusion sampler (Fig. 1) employs charcoal cloth as the adsorptive sink. This material has similar adsorptive capacity to charcoal granules, and is much more convenient to handle and incorporate in a sampler of this type.

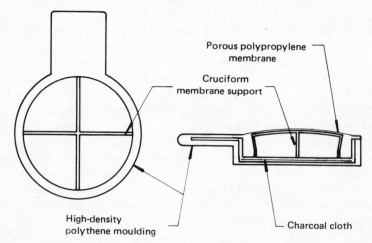

Fig. 1 – Porton diffusion sampler (flush membrane type). The external diameter is 3.8 cm.

The West badge (Louisiana button) [2] incorporates granular charcoal as the adsorbent, and a silicone rubber membrane (Fig.2). The Abcor (Walden) gas badge (not illustrated) has an air gap similar in thickness to that in the CDE sampler, with a cellulose-type membrane protected by a wire mesh. The 3M passive sampler [4] is similar in design and dimensions to the CDE version, with a porous polypropylene film membrane, air gap and charcoal paper adsorbent.

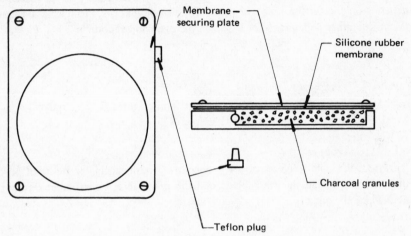

Fig. 2 – West badge. The width dimension is 4.2 cm.

Palmes & Gunnison [3] have used a tube-type sampler with the adsorbent at one end and the other end left open, so that the tube itself constitutes the diffusional resistance (Fig.3). Workers at ICI Ltd., BP Ltd., and HSE have used tube-type [5] and badge-type samplers, with a silicone rubber membrane, an air gap which can be varied in depth, and either porous polymer granules or Tenax as the adsorbent.

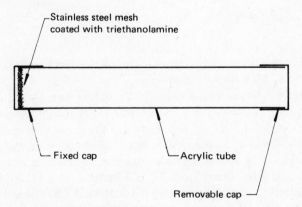

Fig. 3 – Palmes–Gunnison tube sampler [3]. Its o.d. is 11 mm.

VARIATIONS IN SAMPLING RATE WITH CHANGES IN AMBIENT CONDITIONS

Temperature

a. *Porous polypropylene membranes*

Diffusion of gases through porous polypropylene membranes is analogous to diffusion along capillaries, and is governed by Fick's law, so that the rate depends on the diffusion coefficient D for a particular gas. From kinetic theory the temperature dependence of D for simple gases and vapours is

$$D = D_0 \left(\frac{T}{T_0}\right)^{\frac{3}{2}}$$

so that the ratio of diffusion fluxes J_1, J_2 at temperatures T_1, T_2 is $J_1/J_2 = (T_1/T_2)^{\frac{1}{2}}$ which gives a variation of less than 0.2% per °C.

b. *Silicone rubber*

Diffusion through silicone rubber proceeds by permeation, or by solution of the diffusing species in the rubber, and the permeability coefficient P may be defined by the equation

$$-P = -SD = \frac{S.JL}{p_1 - p_2}$$

where $p_1 = C_1/S$ and $p_2 = C_2/S$; S is Henry's Law Constant, L is the membrane thickness, p_1 and p_2 are the vapour partial pressures at the outer and inner membrane surfaces and C_1 and C_2 are the corresponding concentrations.

The parameters D and S satisfy the Arrhenius relationship so that

$$P = P_0 \exp(-E_p/RT), \qquad \text{and} \quad E_p = \Delta H + E_D$$

where E_D is the activation energy of diffusion, E_p is the activation energy of permeation and ΔH is the heat of solution.

If a non-condensable gas is the diffusing species then E_D and ΔH are both positive, and E_p is positive. If a condensable gas is the diffusing species then ΔH may be negative owing to contributions from the heat of condensation so that negative values of E_p may result. Hence membrane permeabilities increase with temperature for non-condensable gases, but may decrease with temperature for condensable gases. Barrer *et al.* [6] and Barrer & Chio [7] have shown that for many substances in silicone rubber E_D is about 4 kcal/mol and ΔH is in the range 5-7 kcal/mol. The variation in P with temperature is therefore small.

Humidity

Measurements of uptake rate with the CDE passive sampler using halothane showed no measurable variation over a wide range of relative humidities, from about 10% up to >90%. Measurements made at BP Ltd. using tube-type samplers with Porapak adsorbent and silicone rubber membranes showed no change in the uptake rate of benzene when the humidity was varied from 'dry' to 90%.

Sampling rates measured at various relative humidities ranging from 15% to 80% made at ICI Ltd. with trichloroethylene and several types of passive sampler were generally unaffected by humidity changes, except for the Abcor badge where the uptake rate was found to decrease with increasing humidity.

High humidities may well have an effect when passive samplers are used with compounds which are not strongly held on the adsorbent, e.g. vinyl chloride. The effect is more likely to manifest itself as losses from the sample prior to analysis than as a change in the sampling rate.

Air velocity

In samplers employing a membrane there will be a boundary layer at its surface where a concentration gradient will exist due to sampling taking place. This boundary layer decreases in thickness with increasing air flow over the face. It is important, therefore, that sampler design be such as to minimize the effect of changes in thickness of the boundary layer, and in the case of the CDE sampler an important aspect (Fig. 4) is the mounting of the membrane flush

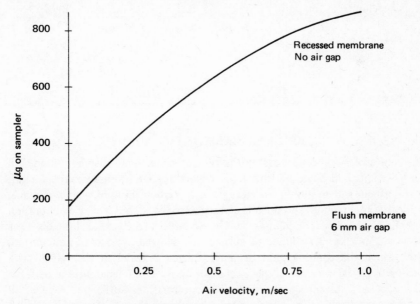

Fig. 4 – Sampling-rate variation with air-velocity. The CDE sampler was exposed to methyl *iso*butyl ketone (150 μg/l) for 1 h periods.

with the surface of the sampler, giving an air gap of about 6 mm between membrane and adsorbent. For tube samplers without membranes, Palmes & Gunnison [3] found that, provided the ratio of length to diameter exceeded about 7:1, variations in air velocity over the sampler had little effect on sampling rate.

STORAGE STABILITY

Some delay between sampling and analysis is inevitable in most surveys of the type for which passive sampling devices are intended, and reproducible recovery of adsorbed samples after varying periods is obviously important. Table 1 shows pertinent results.

Table 1. Retention of adsorbed compounds during storage. The results were obtained using charcoal cloth. The carbon tetrachloride was recovered by elution with n-heptane. Carbon disulphide was used as eluent in the case of acrylonitrile, and the Refrigerant 12 was thermally desorbed from the cloth.

Compound and exposure concentration	Storage period, days	Sample recovery, %
Carbon tetrachloride, 40 μg/l	0	91
	5	94
Acrylonitrile, 160 μg/l	0	100
	1	107
Refrigerant 12 (CCl_2F_2),	0	100
1200 μg/l	1	97
	4	91
	6	94

SAMPLE RECOVERY FROM ADSORBENTS FOR ANALYSIS

The choice of method depends on the type of sampler and adsorbent used, and on the compound being measured. Most work at CDE has used charcoal cloth as adsorbent and a solvent to elute, usually carbon disulphide or n-heptane. Difficulty in obtaining reproducible elution led to the adoption of an equilibration technique for some compounds such as styrene. The charcoal-cloth disc was immersed in a known volume of solvent and allowed to come to equilibrium, after which the concentration of the extracted material in the solvent was measured. It is, of course, necessary to measure the equilibrium between adsorbent and solvent over a range of concentrations sufficient to cover all sample levels likely to be encountered; the system then follows a Freundlich isotherm.

Sample recovery by thermal desorption has attractions in that the whole sample can be used for analysis, e.g. by direct transfer to a GC column, thus improving the sensitivity of the method without need to resort to evaporation of the solvent as would be necessary with the elution approach. It is also the method most likely to be adaptable to automation, particularly in the case of tube-type samplers (other than the Palmes–Gunnison type). Since, however, not all compounds can be thermally desorbed with high efficiency, the method needs to be checked with individual compounds to ascertain its suitability.

EXAMPLES OF THE USE OF PASSIVE SAMPLERS

The Palmes–Gunnison tube with triethanolamine as adsorbent has been used to monitor NO_2 levels in houses [7]. The CDE sampler has been used to monitor several compounds in different situations, including furfuraldehyde and diethyl sulphate in a graphitization plant, halothane in operating theatres, methyl ethyl ketone as used in a cleaning procedure for plastic strips, and styrene in workshops during laying up of glass fibre hulls.

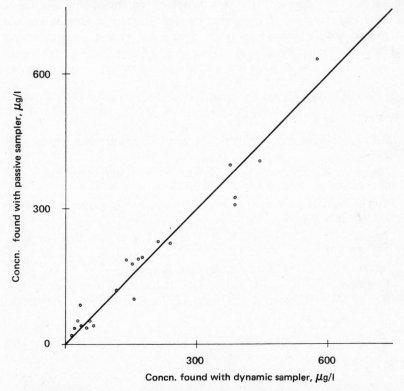

Fig. 5 – Comparison of passive and dynamic sampling: paired results for methyl ethyl ketone.

For the methyl ethyl ketone survey, CDE passive samplers were paired with conventional personal samplers operating at a measured flow rate of 2 l/min and likewise having charcoal cloth as the adsorbent material. The passive sampler was fastened to one lapel of the worker's overall and the sampling head of the conventional sampler to the other. Solvent elution by carbon tetrachloride was followed by IR determination of the methyl ethyl ketone in solution. The 21 paired values (Fig. 5) show some scatter, but are close to the theoretical line over the range 30–630 μg (slope of regression line 0.91; $R = 0.94$).

For styrene, recovered from the charcoal cloth by equilibration in *n*-heptane and measured by UV (247 nm) absorption, paired results (not illustrated) again showed some scatter in the results and, moreover, a slight bias towards higher concentrations from the passive samplers, as was likewise observed in a vinyl chloride survey carried out using the West passive monitor paired with OSHA charcoal sampling tubes [8].

CONCLUDING COMMENTS

Provided that the design of the passive sampler is optimized to minimize the effect of changes in ambient conditions on the sampling rate, the results obtained from passive and dynamic sampling are comparable. Passive sampling devices are cheaper than conventional personal samplers and more convenient to wear. They may therefore help make occupational hygiene surveys more informative.

References

[1] Bailey, A. & Hollingdale-Smith, P. A. (1977) *Ann. Occup. Hyg.*, **20**, 345–356.
[2] Nelms, L. H., Reiszner, K. D. & West, P. W. (1977) *Anal. Chem.*, **49**, 994–998.
[3] Palmes, E. D. & Gunnison, A. F. (1973) *Am. Ind. Hyg. Assoc. J.*, **34**, 78.
[4] US Patent (1975) 3924219.
[5] Tomkins, F. C. & Goldsmith, R. C. (1977) *Am. Ind. Hyg. Assoc. J.*, **38**, 371–377.
[6] Barrer, R. M. (1939) *Trans. Faraday Soc.*, **35**, 595.
[7] Barrer, R. M. & Chio, H. T. (1965) *J. Polymer Sci. Part C. (10)*, 111.
[8] West, P. W. & Reiszner, K. D. (1978) *Am. Ind. Hyg. Assoc. J.*, **39**(8), 645.

#A-4   AIR SAMPLING FOR TRACE-ORGANICS:
        AN OVERVIEW*

J. G. FIRTH, Health and Safety Executive,
Occupational Medicine and Hygiene Laboratories,
403 Edgware Road, London, NW2 6LN, U.K.

The various types of sample collected in inhalation-exposure investigations are typically analyzed at a distance, often by GC. Ways to collect and hold the samples are outlined, with emphasis on solid (ad)sorbents. Passive samplers will play an increasing role.

The last few years have seen an increasing rate of development and introduction of instruments for measuring airborne concentrations of harmful gases and vapours at concentration levels of interest to industrial hygienists. However, no single inexpensive measuring technique has been developed which is capable of measuring a very wide range of harmful gases over concentration ranges from fractions of a ppm, up to a few hundred ppm. Hence, the majority of measurements of hazardous gas concentrations involve obtaining a sample of the gas at the particular location of interest and returning this sample to a laboratory for subsequent analysis. The development of sampling techniques is progressing rapidly because in this area it is possible to develop standard techniques which are applicable to a wide range of gases and vapours.

In general, three types of measurement of atmospheric concentrations of gases and vapours are made, viz. short-term measurements of the concentration at a point in space and time [cf. S. Crisp's article, #A-2, *this vol.* – Ed.] ; measurements of average concentrations at a given point, and continuous measurements of the variations of concentrations with time at a given point. The first type of measurement is generally used for rough checks of concentration levels in an atmosphere and, with the correct statistical distribution of measurements in space and time, for assessing whether or not a particular hygiene

---

* *Editor's note.*–Some background is given at the end of article #0. Remarks made at the Forum have now been expanded into a survey, applicable to gases as well as organic vapours, which represents the 'core' of section #A. The author's term 'accuracy' has not been altered to 'precision' [but see #0]. For readers unused to 'ppm', styrene and dichlorvos serve as examples: 4.4 and 10 mg/m³ respectively correspond to 1 ppm.

standard is being achieved. The second type of measurement is widely used for measuring exposure of personnel by mounting portable samplers on the worker in the breathing zone. The third type of measurement is used to measure changes in concentrations from the normal accepted levels due to the occurrence of leaks etc., and can of course be used in conjunction with alarm systems. This type of measurement relies almost completely on instrumentation but the first and second types, although utilizing instrumentation wherever convenient systems exist, generally involves sampling followed by analysis.

MODES OF SAMPLING

The simplest method is to draw the sample into a suitable container. These can be pre-evacuated bottles or bags, or the atmosphere can be pumped into a bottle or bag, which is then sealed. Use may be made of an ordinary glass bottle with a screw cap, having a rubber septum through which a sample of the atmosphere in the bottle can be removed for analysis* and containing a small amount of mercury which, when the bottle is stored upside-down, effectively seals the bottle neck/cap system. Plastic bags have also been used. In both cases

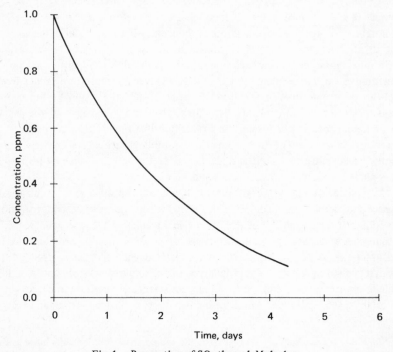

Fig. 1 — Permeation of SO_2 through Mylar bag.

* Or an aliquot of liquid compound injected if a standard atmosphere is wanted, as indicated in sect. (vii) of #0.—*Ed.*

the maintenance of the integrity of the sample prior to analysis is a major problem. In general, the more reactive gases or vapours need to be measured at lower concentration levels, and hence adsorption of the gas onto the walls of the bottle, or into the plastic of the bag, is a major problem. This is illustrated in Fig. 1. Vapours of polar molecules, e.g. alcohols and aldehydes, will adsorb onto the surface of glass, as will reactive small molecules such as NO_2. It is sometimes not possible to desorb this adsorbed gas prior to analysis. Organic vapours will permeate through the plastic of the bag, and attempts to minimize this problem have involved incorporating aluminium layers in the plastic. [*see later*, # NC(A)-2.–Ed.] . Even so, reactions can still occur in the gas phase between the vapour of interest and other contaminants present in the atmosphere if the sample is left for extended periods before analysis (Fig. 2).

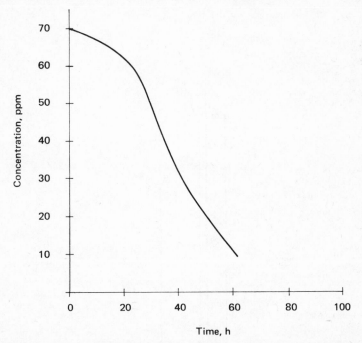

Fig. 2 – Change of NO_2 concentrations in high humidity atmosphere in aluminized plastic bag.

However, the main problem with the above type of sample is that very little of the vapour is available for analysis. Accordingly, sampling techniques based on a concentration method have been developed. These are essentially gas filtering systems in which a large amount of atmosphere is sampled and the vapours of interest are captured in a suitable medium and are then available for analysis. The level that prevailed in the atmosphere may readily be calculated from the amount of atmosphere sampled and the amount in the filter medium.

The oldest form of capturing system involved passing the atmosphere through a suitable liquid absorbent. The liquid is one in which the vapour is very soluble, i.e. one in which the equilibrium of the vapour between the liquid and gas phases lies heavily towards the liquid side. This can be achieved by simple solubility of the vapour in the absorbent medium, e.g. non-polar organic molecules in toluene, or by chemical reaction of the vapour with the absorbent, or with a reagent dissolved in the absorbent. Examples of such chemical systems are the absorption of HCl in water, the absorption of NO_2 into an azo-dye forming solution, and the absorption of formaldehyde into an aqueous solution of 3-methyl-benzothiazol-2-one hydrazone hydrochloride.

Liquid absorption agents are generally used in simple bubbler systems (Fig. 3). The atmosphere is drawn through the reagent using a small sampling

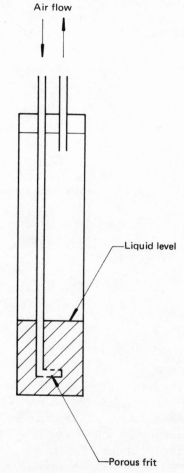

Air flow

Liquid level

Porous frit

Fig. 3 – Bubbler sampler.

pump. To ensure efficient transfer of the vapour from the gas to the liquid phase, it is important that the interfacial area be kept as high as possible, and maintained for as long as possible. This means that the bubble size through the liquid should be kept as small as possible and the flow rates kept as small as is compatible with obtaining sufficient sample for analysis. In general, bubblers using porous frits are used to reduce bubble size, and flow rates vary between 200 ml/min and 2 l/min. Although glass bubblers are still widely used, these are being replaced by bubblers made from plastic, such as polypropylene. These are obviously much less likely to break, which is important, particularly in personal sampling systems. The absorption efficiency, i.e. the ratio of the amount of material adsorbed to the amount of material passing through the adsorbent, is generally high in liquids containing a reagent because a large excess of reagent is used. Where the trapping mechanism is a solution of the vapour in the adsorbent, then at higher sampling speeds slippage can occur and it is often necessary to use a back-up bubbler.

Use of solid adsorbents along with a pump

The disadvantages of liquid absorption systems, in particular the mounting of the bubbler onto personnel and the handling of the reagent after sampling, have led to their widespread replacement by systems employing adsorption onto a suitable solid. As for liquid absorption systems, the solid adsorbent can involve physical adsorption of the vapour onto the surface and into the pores of the solid, chemisorption of the vapour onto functional groups on the surface of the solid or adsorption onto reagents with which the solid has been impregnated. The most widely used solid matrix for impregnation has been paper. Chemically impreganted papers can be used to give direct colorimetric estimation of the amount of material adsorbed, e.g. the interaction of H_2S on lead acetate-impregnated papers to produce black lead sulphide, or they can be used in systems involving subsequent analysis, e.g. the absorption of arsine on silver nitrate-impregnated paper, followed by the determination of arsenic by X-ray fluorescence spectrometry. Adsorption efficiencies for such systems are high (as in liquid systems) so that the short path-length of the gas through the paper and, hence, short contact time is still sufficient to ensure 100% adsorption of the vapour.

By far the most widely used solid adsorption systems are tubes containing a high surface area or porous solid. Various types of charcoal are used for sampling a very wide range of organic vapours. The most widely used absorption tube is the NIOSH [National Institute of Occupational Safety and Health-*Ed*.] charcoal tube (Fig. 4) [1]. The atmosphere is drawn through the tube at about 200 ml/min and the vapour adsorbed directly onto the charcoal. A second charcoal plug is used behind the first one so that if slippage occurs through the first charcoal plug, or it is overloaded by high concentrations, this can be verified by the presence of vapour on the second plug. Obviously, many variations on this basic design are possible, with larger or smaller amounts of charcoal, depending on the heat of adsorption of the gas onto the charcoal. It is also possible to employ rates as low as a few ml per min, so that sampling periods can be extended over many hours.

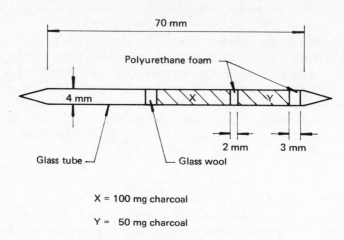

X = 100 mg charcoal

Y = 50 mg charcoal

Fig. 4 – Charcoal tube sampler with breakable tips.

In general, the sample is removed from the charcoal by washing with a suitable solvent such as CS_2. It is feasible to perform thermal desorption, achieved by heating the charcoal to a suitable temperature, where the equilibrium of the vapour between solid and vapour phase lies well over to the vapour side; but there should be wariness with charcoal since charcoal itself can have catalytically functional groups on its surface and also impurities such as transition metal ions which will act as catalysts for reaction of the adsorbed vapour. However, solvent desorption necessarily dilutes the sample, and hence systems where thermal desorption can be used have in fact been developed. These systems rely on a adsorption onto a porous polymer such as Tenax, Chromosorb, etc., of the types used in GC columns. The absorption systems take the form of small metal tubes similar in dimensions to the NIOSH charcoal tube, containing a few mg of porous polymer [2]. The vapour is desorbed by enclosing the tube in a small temperature-programmed furnace and passing a gas such as N_2 through it. The vapour desorbs and can be injected directly onto a chromatograph or, alternatively, can be trapped in a cold trap, which is then flash-heated so that the vapour arrives as one pulse into the chromatograph, thereby giving a 10-fold increase in sensitivity.

Obviously it is as important to know the amount of atmosphere which is being sampled in these filter systems as it is to know the amount of material adsorbed in the filter. The volume of atmosphere sampled is determined by the sampling pump and indeed, inaccuracies in the sampling pump are the main sources of inaccuracies in estimating the airborne concentrations of vapours. In vapour sampling systems it is not particularly important to have a constant rate of flow through the absorption system especially where solid adsorbents are used. Accuracy is mainly affected by the accuracies with which the overall volume passing in a certain time through the pump, is known. Hence, vane pumps, bellows pumps, etc., can be used for gas sampling. In most portable sampling pumps, battery problems, especially premature discharge of the batteries, are the main source of errors; but there is no reason why a well designed and maintained pump should not have an accuracy better than ±5%. It is also worth noting that in most applications, it is necessary to have a pump which is certified as safe for use in flammable atmospheres.

Passive samplers

The problems with pump accuracies and the costs of sampling pumps have led to the development of passive samplers. In these devices, the rate of sampling is controlled by the diffusion of the vapour across a membrane or a well defined air gap. This then eliminates the need for sampling pumps and makes the devices much more convenient to use. Two variations of passive samplers have been produced as illustrated in the preceding article: the badge sampler and the tube sampler [3, & #A-3, *this vol.*] .

The badge sampler is a flat device, similar in appearance to a radiation badge. The rate of sampling of the vapour is controlled by its passage through a membrane having a high diameter-to-thickness ratio. Where a solid membrane is used, e.g. silicone, then the rate of sampling is controlled by permeation, which is a complex mechanism involving dissolution and diffusion of the vapour in the membrane. Permeation is markedly affected by temperature, and also the permeability of membranes can be affected by other gases such as water vapour. Hence, many of the badge systems now employ porous membrane filters where the sampling is by diffusion through the gas phase in the pores of the membrane. Gas-phase diffusion is much less temperature-sensitive and is generally not affected by the precence of other gases at low concentrations. After passing through the membrane, the vapour is adsorbed onto a solid material. Charcoal cloth, porous polymers and impregnated paper can all be used. The effectiveness of the device depends on efficient adsorption onto the adsorbent so that the effective concentration in the gas phase immediately above the solid is zero. This ensures that the concentration gradient across the membrane, which is the factor governing the rate of diffusion through it, depends only on the concentration of gas in the ambient atmosphere.

Tube-type passive samplers utilize gas-phase diffusion down a tube with a fixed length (say 80 mm) and cross-sectional area where the diameter-to-thickness ratio of the air column is low. This ensures that the effects of turbulence at the open end of the tube caused by air flow past it have a minimal effect on the sampling rate; i.e. they do not reduce the effective diffusion path. Variations in ambient air flow have little effect on a well designed tube sampler. The adsorption material, which occupies the bottom third of the tube, can again be porous polymer, charcoal, or impregnated solid. Again, it is important that the effective gas-phase concentration immediately above the solid be zero.

As with pumped tube absorption systems, the vapour can be removed from the solid by solvent desorption, or by thermal desorption with the advantage, besides increased sensitivity, that the vapour can be removed without dismantling the sampler; it can thus be used repeatedly without re-calibration and be employed in automated desorption analytical systems [4].

The effective sampling rates of the badge-type sampler and the tube-type sampler are respectively about 50 ml/min and 2 ml/min. Therefore, only a small amount of vapour is actually adsorbed onto the adsorbent; but with the increasing sensitivity of modern analytical methods, it is very likely that passive samplers will replace all other types of sampling systems for most applications involving the estimation of atmospheric vapour concentrations.

References

[1]   NIOSH Manual of Sampling Data Sheets (1977), DHEW (NIOSH) Publication No. 77–159, Cincinatti.
[2]   Brown, R. H. & Purnell, C. J. (1979) *J. Chromatog.* **178**, 79–90.
[3]   Palmes, E. D., Gunnison, A. F., DiMatteo, J. & Tomczyk, C. (1976) *Am. Ind. Hyg. Ass. J.,* **37**, 570–577.
[4]   Brown, R. H. Charlton, J. & Saunders, K. J. (1980) *Am. Ind. Hyg. Ass. J.,* in press.

#NC(A) Notes and Comments related to Workplace Air Samples

#NC(A)–1 *A Note on*
APPROACHES TO TRAPPING TRACE-ORGANICS AND GENERATION OF STANDARDS*

D. J. FREED, Bell Laboratories, Murray Hill, NJ 07974, U.S.A.

In our version of a 'co-condensation' system [1], *n*-pentane vapour is introduced into the air stream during sample collection. The gases are mixed and then passed through a trap kept in liquid N_2, which freezes out the pentane. Trace-organics are trapped in the solid pentane and thereby concentrated, by up to 3 orders of magnitude depending on the flow rate of the air-sampling stream and the original concentration of the gaseous pollutant. Thus, an air stream containing 1 ppm by vol. of benzene was sampled at a flow rate of 100 ml/min together with a pentane stream of a flow rate of 10 ml/min together 10 min. Thereby approx. 100 mg (150 μl) of solid pentane was collected. It was allowed to warm to ambient temperature and subsequently analyzed by GC. The trapping efficiencies for various common atmospheric pollutants all exceed 90%. The technique is of particular value in the case of reactive materials or those which are poorly recovered from charcoal tubes [2].

Standards of acrolein, acrylonitrile, vinyl chloride and nitrosamines for direct GC injection can be directly generated from conversion reagents in pre-columns [3]. With 5 ng amounts, the precision is better than 5%, and there is good recovery over a wide range. The method is convenient for the preparation of precise standards, especially with high-risk compounds.

References

[1] Freed, D. J. (1979) in *Trace Organic Analysis: A New Frontier in Organic Chemistry* (Hertz, H. S. & Chesler, S. N., eds.), National Bureau of Standards Special Publication 519, Washington DC, pp. 95-99.

[2] Mindrup, R. F. (1977) *Determination of Organic Vapors in the Industrial Atmosphere,* Bulletin 769, Supelco, Inc., Bellafonte, PA, U.S.A.

[3] Freed, D. J. & Mujsce, A. M. (1977) *Anal. Chem.,* **49,** 139-141.

* Editor's precis based on a Forum Abstract.

#NC(A)-2 *A Note on*
THE USE OF PLASTIC BAGS FOR AIR SAMPLING

P. A. HOLLINGDALE-SMITH, Chemical Defence Establishment, Porton Down, Salisbury, SP4 0JQ, U.K.

It is often convenient to take whole-air samples (grab samples) for subsequent analysis, and when the analytical method to be used is vapour-phase infra-red which normally requires sample volumes of several litres, this type of sampling is essential. Collapsible plastic containers, which can be inflated by means of a battery-operated low-pressure pump, offer a means of taking this type of sample which has been successfully used in our laboratory for a number of years. As with any sampling method, it is important that the composition of the sample should not change during the period between sampling and analysis, or, if changes do occur then they should be quantifiable so that corrections to the analytical results may be made.

There are two obvious ways by which losses may occur from air samples stored in plastic bags, assuming of course that no holes exist: by permeation through the walls of the bag, and by adsorption losses onto the walls. In the early work carried out at CDE, bags made from a laminate of Mylar and polythene were used, and permeation losses were found to be appreciable for many gases. To overcome this, a 3-ply laminated material (LAMOFOIL M1275 available from Otto Neilson (UK) Ltd., Campfield Road, St. Albans) was adopted for the sampling bags. This comprises a layer of aluminium foil sandwiched between high-density polythene and terylene, the total thickness of the laminate being 50 μm. This material is readily heat-sealable to make up the sampling bags, and is sufficiently flexible for the bags to be re-used a number of times before the aluminium foil begins to break up. The presence of the aluminium foil interlayer eliminates all permeation losses from samples; but the possibility of adsorption losses still exists, and the only way to deal with this satisfactorily is to measure the losses occurring from known concentrations of compounds of interest over the longest period likely to elapse between sampling and analysis (Table 1).

More recently, gas sampling bags fabricated from materials such as polyvinylidine fluoride have become available. These are claimed to have a very low permeability and adsorption losses, but the writer has so far had no experience of their use.

A better method of filling the sample bags than direct inflation by means of a pump is to use a sealable box in which the deflated bag is placed with its inlet protruding, and to use the pump to exhaust the air from the box, thus inflating the bag. This method avoids the dual possibilities of losses by adsorption in the pumping system and contamination of the sample by materials in the pumping system, e.g. lubricants.

Table 1 Some typical losses over a 24 h period from Lamofoil bags of 64 l
volume

Compound	Concentrations (ppm)	
	Initial	*After 24 h*
Carbon tetrachloride	0.7	0.55
Chloroform	0.8	0.4
Ethanol	4.5	1.4
Freon 12	1.6	1.6
Freon 113	0.8	0.65
Halothane	0.2	0.15
Nitrous oxide	1.3	1.3
Toluene	0.6	0.5

In summary, gas sampling bags offer a convenient means for taking whole-air
samples; but checks should be made for storage losses with the compounds for
which they are to be used, particularly from low concentrations.

#NC(A)-3 *A Note on*
VAPOUR-SAMPLE STORAGE IN A RIGID CONTAINER

E. REID and A. D. R. HARRISON*, Wolfson Bioanalytical Unit,
Institute of Industrial and Environmental Health and Safety,
University of Surrey, Guildford, GU2 5XH, U.K.

The traditional practice of keeping atmospheric trace-organics in vapour form
right up to the GC measurement step has been recommended for chlorinated
solvents [1], and does not deserve to be entirely eclipsed by the adsorbent
approach, if only because "each new contaminant to be [thus] sampled at TLV
levels represents a sizable project" [2]. However, there is little documentation
to help in the choice of portable vessel. Inflatable bags [P. Hollingdale-Smith,
opposite] may, at least if made of 'Tedlar' (PVF), allow fair preservation of
some compounds [3]; polythene was ineffective [3]. Rigid vessels may take
the form of a tube, cylinder or flask [3-5], possibly pre-evacuated and sometimes
provided with a closure at each end—obviously not a conventional greased
stopcock; sample withdrawal may entail, as a minor penalty, slight ingress of
laboratory air unless the device is of syringe type. Dimethylphosphonate vapour
showed losses onto surfaces with a Teflon-barrelled glass syringe and even with a

* Present address: Albury Labs, Ltd., Albury, Surrey.

stainless steel needle [6] –which may not surprise gas chromatographers. Several organic compounds were, however, little diminished after long-term storage in stainless steel bottles [7], and there is a reasonable hope of minimal loss with a smooth glass surface, at least if silylated [6].

There was satisfactory retention of methane (and of H_2 and CO_2) by glass syringes with specially made butyl rubber plungers, although not with polypropylene or glass syringes having a 'Teflon' O-ring [8]. Vinyl chloride, endflurane and halothane were well retained by an aluminium syringe-type device (Boehringer Labs., model 5100) in which, as a seemingly error-prone step, a cap and lead washer are substituted for the plunger after collection [9]. The following description concerns an effort in our laboratory, partly successful, to adapt a commercially available all-glass syringe for sample collection and storage, if not for direct introduction of an aliquot into a GC column (an idea which was not pursued).

USE OF AN ADAPTED SYRINGE

Our device had assumed the form shown in Fig. 1 at the time when the design exercise was suspended. The 'Record'-type tip of an all-glass syringe ('Elvan' or 'Beacon'; e.g. from A. Horwell Ltd., London NW6) is capped, once the air sample has been taken, with a silicone-rubber disc (cut by cork-borer from a Hewlett-Packard GC septum) pressed up to the glass within a sleeve of silicone-rubber tubing (2.5-mm wall, 2.5-mm i.d.); in this *septum-terminated type,* withdrawal into a 1 ml syringe for GC injection entails needle insertion through the septum. An earlier *tubing-terminated type* entailed needle insertion through the wall of the tubing, which had been closed off by a glass rod inserted so as to stop *ca.* 5 mm short of the tip of the large syringe. At the other end of the syringe, leakage of any vapour that may have diffused between the smooth wall of the barrel and the ground-glass surface of the plunger is minimized during storage, as illustrated, by a 'medium nitrile' (hardness grade '80') O-ring pressed with the aid of an adapted 'bulldog' clip into the groove between the barrel and the top of the plunger. For sample collection or withdrawal, this essential O-ring has to be prised up, somewhat inconveniently, after removing the clip and so releasing the loose gland rings (Fig.1, *top right*).

VAPOUR STORAGE TESTS

With the tubing replaced by a needle, 20-ml vapour-containing air samples were taken from a septum-closed 2.5 l bottle. GC measurement was done on 1 ml sub-samples withdrawn as indicated above, the plunger of the large syringe descending of its own accord as the plunger of the injection syringe was slowly withdrawn. GC-FID was performed at 70° with a 3 m X 2 mm column containing a 2.5% coating of OV-17 on 60–80 mesh Chromosorb W. Typically the vapour

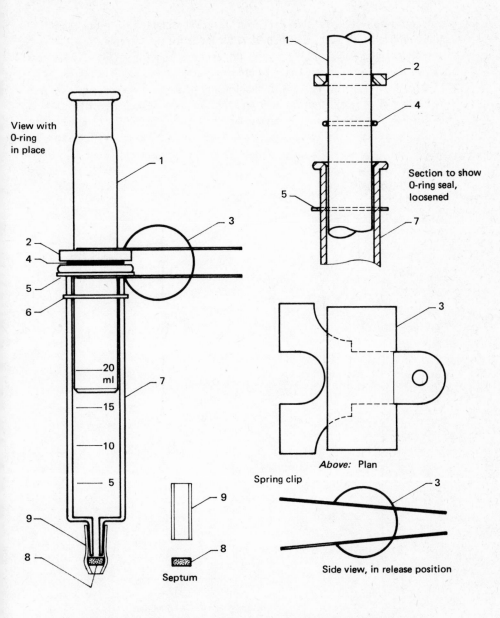

Fig. 1 — Septum-tipped version of the syringe device for sample collection and storage. 1 = Plunger; 2 and 5 = aluminium rings, loosely fitting; 3 = adapted office-type spring clip; 4 = O-ring; 6 = plastic ring to prevent the ring 5 from falling off when 3 is removed; 7 = syringe barrel; 8 = septum, held in a sleeve 9. For other explanations, see text.

concentration was such that the 1 ml injected contained no more than 1 nl liquid equivalent, little higher than the TLV levels for the solvents studied.

With toluene (b.p. 111°) in septum-terminated syringes (Fig. 2), the loss was 1 ±6.4% (S.E., N = 6) at $\frac{3}{4}$ h and 10 ±4.1% (N = 5) at 3 h, increasing to 15 ±6.1% (N = 4) at $4\frac{1}{2}$ h. With dichloromethane the loss was slight at $\frac{3}{4}$ h (8 ±4.4%; N = 5) but marked at 3 h (21 ±2.6%; N = 5) and especially at $4\frac{1}{2}$ h (34 ±3.3%; N = 4). Results for xylene (b.p. 139–144°), not illustrated, paralleled those for toluene: the respective values were +3 ±10.0%, −7 ±9.8% and −15 ±8.3%.

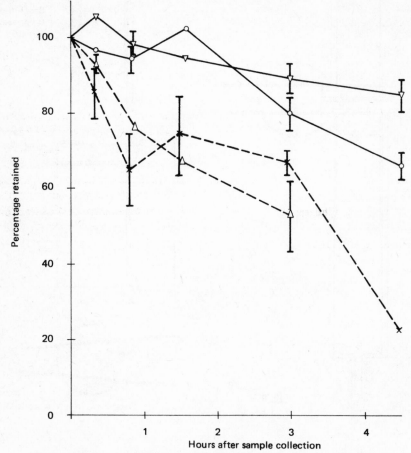

Fig. 2 – Vapour retention in septum-terminated syringes (*solid lines*) and in tubing-terminated syringes (*broken lines*). Toluene, ∇ and △; dichloromethane, o and x. Where there is no vertical bar to represent the S. E. as given in the text, the number of observations was only 1 or 2. Apparent fluctuations with time merely reflect the pooling of observations from a number of experiments, few of which had the full range of time points as plotted. The starting concentrations were of the TLV order, typically 0.4 µg of liquid solvent/l of air.

For vapours in tubing-terminated syringes the losses were significantly greater (Fig. 2). With toluene the 34% loss ($\pm$10.7%; $N = 8$) observed at $\frac{3}{4}$ h scarcely worsened up to 3 h (31$\pm$3.7%; $N = 6$), but with dichloromethane the 25% loss ($\pm$5%; $N = 3$) had worsened to 46% ($\pm$10.6%; $N = 3$) at 3 h. The retentions for xylene (not shown; e.g. 16% loss at 3 h, 38% at $4\frac{1}{2}$ h) were better than for toluene.

Although the lag period usually observed before losses became appreciable argued against the postulate of adsorption-prone surfaces, pre-saturation was tried by injecting 0.1 μl of ethylbenzene into the syringe 10 min before sample introduction. The results showed no influence on dichloromethane or toluene loss (and have been incorporated in the above mean values for the two versions of the syringe device). The losses are inferred to be diffusive, and are not attributable to the syringe tip in the case of the septum-terminated version: there seems to be vapour creepage past the plunger and escape past the O-ring, as judged by measurements after introducing mercury to obviate escape at one end or the other depending on the orientation of the syringe during storage. It was of no advantage to use a syringe having a particularly tight fit of the plunger in the barrel.

CONCLUSIONS

In principle there are advantages, already summarized, in using a syringe-type vessel for short-term collection and transport of air samples. The present device is inexpensive and, with a correction factor which may be as little as 10%, allows a 3-h interval between sample collection and GC analysis. Even a 5-mm length of narrow-bore plastic tubing, as in the earlier 'tubing-terminated' version, entails unacceptable losses of organic vapours, evidently by diffusion rather than by adsorption as known to occur, for example, onto silicone rubber in GC injection ports.

Rather than seek an inert plastic, e.g. neoprene as favoured for vinyl chloride [10], it seemed better to minimize exposure to plastic surfaces, as in the 'septum-terminated' version. Here the residual loss of organic vapours evidently occurs by diffusion past the plunger to the top of the syringe, where the pressed-down O-ring is an essential but rather inconvenient palliative. Mercury would obviously be an inappropriate sealant for field use, and silicone fluid has, not surprisingly, proved ineffective; but use of a low-melting alloy not containing mercury warrants exploration, as does low-temperature rather than ambient storage.

In the rather thin literature on organic vapour containment, there seems to be no pointer to a conclusion now reached, with dichloromethane as the most difficult model solvent—that storage difficulties are inversely related to boiling-point, at least where the losses are not adsorptive. Yet, insofar as persisting losses in the version of our device where there is minimal permeation through plastic are attributable to diffusion *via* a narrow gap, the extensive literature on diffusion devices for generating standard atmospheres [e.g. 11] turns out to be relevant to the correlation now found between losses and volatility.

Acknowledgement

Miss Barbara E. Brockway gave skilled and enthusiastic help.

References

[1] Bureau International Technique des Solvants Chlorés (1976) *Anal. Chim. Acta*, **82**, 1-17.
[2] Ballou, E. V. (1976) in *Second NIOSH Solid Sorbents Roundtable* (Ballou, E. V., ed.), DHEW, Washington, D. C., pp. 3-22 (*vide* p. 20).
[3] Company brochure (1978) *Catalog of Chromatography Supplies*, Supelco Inc., Bellafonte, PA, U.S.A.
[4] Katz, M. (1969) *Measurement of Air Pollutants*, World Health Organization, Geneva.
[5] Pickering, W. F. (1977) *Pollution Evaluation*, Dekker, New York.
[6] Lonnes, P. (1976) as for 2., pp. 177-208.
[7] Eaton, H. G., Williams, F. W., Wyatt, J. R., DeCorpo, J. J., Saalfeld, F. E., Smith, D. E. & King, T. L. (1979) in *Trace Organic Analysis: A New Frontier in Analytical Chemistry* (Hertz, H. S. & Chesler, S. N., eds.), National Bureau of Standards Special Publication 519, Washington, D. C., pp. 213-218.
[8] Lang, H. W. & Freedman, R. W. (1969) *Am. Indust. Hyg. Ass. J.*, **30**, 523-526.
[9] Lecky, J. H. & Andrews, R. (1977) *Internat. Lab.*, Jan.-Feb., 63-65.
[10] Miller, B., Kane, P. O. & Robinson, D. B. (1978) *Analyst*, **103**, 1165-1172.
[11] Nelson, G. O. (1971) *Controlled Test Atmospheres: Principles and Techniques*, Ann Arbor Science Publishers, Ann Arbor, Mich., U.S.A.

Comments on #A-1 *to* #A-4 — WORKPLACE AIR SAMPLES [*see also* (vii) in #0]

Analytical methodology sufficiently simple to be performed by staff with little training would be a boon; thus, thermal desorption could be preferable to elution by the toxic agent CS_2 (*remarks by* G. W. Collins, S. Richards). For 'TDI' (toluene-2,4-di*iso*cyanate), colorimetric methods that can cope with the proposed lowering of the TLV are becoming available, but they will not be simple or cheap (*answer to* M. J. Hiller). Gathering data has to be complemented by reassurance of the employees through a continuing public relations exercise (*remark by* T. R. Andrew). There is the possibility (*raised by* J. Ramsey) of 'intentional' exposure, as through solvent- or glue-sniffing.

Comments by R. H. Brown (Health & Safety Executive) concerning badge-type passive samplers (i.e. the original design with a recessed polypropylene membrane). — Work at Cricklewood bears out findings in the article from Porton [#A-3] : there is a significant face-velocity effect below 15 ft/min with styrene as test substance, and good correlations have been observed between the passive sampler and the conventional charcoal tube (for styrene, halothane and acrylonitrile); but like the charcoal in the latter, the charcoal cloth in the passive sampler may become saturated, particularly when sampling low-boiling organic compounds at high ambient humidity.

Editor's citation: Evidence is being gathered [1] concerning the effect of wind on diffusion-sampler performance.

[1] Palmes, E. D. (1979) *Ann. Occup. Hyg.*, 22, 85 (cf. 86).

Comments on #NC(A)-1, D. J. Freed — CO-CONDENSATION

D. J. Freed, *replying to* J. Leahy. — With pentane, being water-immiscible, there is no problem due to freezing out of water vapour in the trap, in which the water forms a separate globule. *Reply to* M. G. Bell. — It is a reasonable suggestion that, with a view to rendering the trapping assembly suitable for personal sampling, the co-condensation be with a liquid which condenses at room temperature and could be supplied as a heated vapour; but such a liquid might interfere in the GC determination. (Another suggestion was use of argon, which could be condensed by liquid nitrogen.) *Answering* J. S. Fritz, D. J. Freed said he had no experience concerning the reactivity of (e.g.) vinyl chloride or acrylonitrile with XAD polymeric adsorbents; but trial of Tenax had shown somewhat slow desorption.

#B   Water and Effluents

#B-1   CONCENTRATION AND CLEAN-UP OF ORGANIC SOLUTES IN WATER*

JAMES S. FRITZ, Department of Chemistry and Ames Laboratory, Iowa State University, Ames, Iowa 50011, U.S.A.

Some methods for concentrating organic solutes from aqueous samples are discussed. The advantages and limitations of solvent extraction and gas purging are covered briefly. The multi-stage procedure entailing XAD-resin sorption is effective for concentrating and determining a wide variety of organic solutes, although after elution (with ether) the evaporation step may entail losses of volatile compounds, and the GC injection has to be confined to a small fraction of the concentrate. Thermal desorption from XAD-resin avoids these difficulties. Effective procedures are described using either a packed GC column or a capillary column. A new ion-exchange procedure is described for selective concentration and GC separation of acidic substances in aqueous samples.

Organic contaminants in drinking water can impart a bad taste and odour. Their long-term health effects in water are largely unknown, although in at least one case a statistical link with the incidence of cancer was established for a polluted water [1]. Analysis of waste water effluents for organic compounds is necessary if industries are to meet environmental standards. Aqueous biological samples often must be analyzed for drugs and various other organic compounds.

The analytical task is often formidable. Water samples often contain less than one part in 10^9 of an organic pollutant and thus require efficient concentration procedures. More concentrated aqueous samples may contain very many organic compounds, and it becomes necessary to isolate and measure only the compounds of interest. This review will discuss in a selective manner some methods of concentrating and separating organic solutes that occur in aqueous samples†

---

* Under contract with the Department of Energy.

† See also the survey below by C. E. R. Jones, #NC(B)-1.–*Ed.*

SOLVENT EXTRACTION

Solvent extraction is essentially a very simple technique for isolating organic solutes from aqueous solution. Its effectiveness is enhanced by various techniques for continuous extraction, although these often require more manipulation.

The simplicity and practical effectiveness of solvent extraction is demonstrated by a method of determining halocarbons in drinking water [2]. A 10-ml water sample is extracted with 1 ml of purified pentane, an aliquot of which is then analyzed by GC-ECD. Extraction efficiency for the various trihalomethanes ranges from about 80% to 96%, but the percentage of each substance extracted is reproducible over the concentrations usually encountered in drinking water.

Unfortunately, for some classes of organic compounds only a low percentage can be solvent-extracted, making it difficult to use the technique for analysis of a broad spectrum of solutes. Further losses of solutes may occur during evaporation of the extracting solvent. In trace analysis, impurities in the solvent itself can be a limiting factor.

GAS PURGING

The distribution of volatile organic solutes between the aqueous phase and a gas phase can be very useful for analysis. McAuliffe [3] developed a method in which a water sample is shaken with nitrogen or helium in a glass syringe. Then a portion of the gas is injected into a gas chromatograph and the organic solutes are separated. Aliphatic hydrocarbons below about C_{10} are $> 98\%$ in the gas phase in a single equilibration. More polar compounds (even aromatic hydrocarbons) require several equilibrations for complete transfer to the gas phase.

Zlatkis and coworkers [4], Bellar and Lichtenberg [5], the Grob's [6], and others have made extensive use of a multi-stage water–gas equilibration method that is called gas stripping, gas purging, or head-space analysis. A gas is bubbled continuously through the aqueous sample for a given period of time and carries with it the volatile solutes. These are retained on a small column of Tenax, which permits the gas and water vapour to pass through. Finally, the volatiles are desorbed into a GC (usually with intermediate cold-trapping) for analysis.

Purging methods are certainly a valuable analytical tool. But, inevitably, there are borderline cases of solutes of limited volatility or with high water-solubility. In such cases, the method either fails or gives low recoveries.

RESIN SORPTION METHOD

This is an effective multi-stage concentration method in which an aqueous sample is passed through a small column containing XAD-2 or XAD-4 resin [7]. The resin is non-ionic and sorbs molecular organic compounds while allowing inorganic ions and some highly polar organic compounds to pass through.

The essential stages of this procedure for concentrating organic compounds from water are:

(1) *Sorption.* A water sample ranging from 1 l up to 100 l or more is passed rapidly through a column containing XAD-2 macroporous resin as a bed about 1 cm × 10 cm.

(2) *Elution.* The sorbed organics are eluted with 25–30 ml of diethyl ether.

(3) *Drying and evaporation.* The water in the ether effluent is frozen out and the ether is evaporated carefully to 1 ml in a special apparatus.

(4) *Separation.* A 2-μl aliquot of the ether is injected into a gas chromatograph and the individual organic solutes are separated.

(5) *Identification.* The solutes in an aliquot of the ether are identified by GC-MS.

This procedure has been thoroughly tested on nearly 100 model compounds added to water and has been found to give excellent recoveries of compounds that are amenable to chromatography.

For any satisfactory analysis at the trace level, careful attention must be paid to experimental detail. This procedure is no exception. The XAD-resin must be carefully purified to remove naphthalene, ethylbenzene, benzoic acid, and other impurities that are in the resin itself. This is best done by Soxhlet extraction with organic solvents, followed by heating the resin in a tube to approximately 190° while nitrogen or helium slowly passes through. Once pure, the resin should be kept in a sealed tube or under methanol to prevent sorption of organic impurities from the atmosphere.

Proper sample handling is also essential. Sample containers must be carefully cleaned and free from grease. The design of the glassware used in the evaporation step is also critical. A container with a cone-shaped tube at the bottom is used so that the walls will be continuously washed down with condensed ether during the evaporation and, thus, prevent loss of the solutes onto the walls of the container. Finally the container is sprayed with acetone to give a final washdown with condensed ether.

The XAD-resin sorption procedure is reliable and works for all classes of compounds surveyed. It does have the drawbacks that volatile solutes may be lost during the evaporation, and that the GC injection is limited to about 1/500th of the ether concentrate. The latter inefficiency greatly increases the size of the water sample needed.

Thermal desorption from XAD-resin

This is a procedure in which organic solutes in an aqueous sample are concentrated onto XAD-4 resin contained in a very small tube. Then the organic solutes are thermally desorbed into a GC, where the individual compounds are separated. This procedure is more effective than the earlier resin sorption method because organics from the entire sample are desorbed into the GC. This permits use of a much smaller aqueous sample. By sealing the ends of the XAD-4 sample tube with 'parafilm' after passing the water sample, field sampling is feasible, and the sealed tubes can be kept for several days before analysis in the laboratory.

In the thermal desorption method [8], the tube of XAD-4, still wet, is connected to a system housed in a box on front of the GC (Fig. 1). The solutes are desorbed for 10 min at 180°-200° using helium carrier gas. The vapours pass through a heated zone (200°-220°) to a small Tenax tube (~45°). The organic solutes are retained, but water vapour passes through. Then the Tenax tube is closed off by means of a 4-port valve (zero-volume fittings). It is heated to 275°-280°; then the valve is opened to inject the organic solutes into the GC. [*Editor's note:* another thermal desorption procedure is mentioned earlier in this book in connection with analyzing air for trace-organics.]

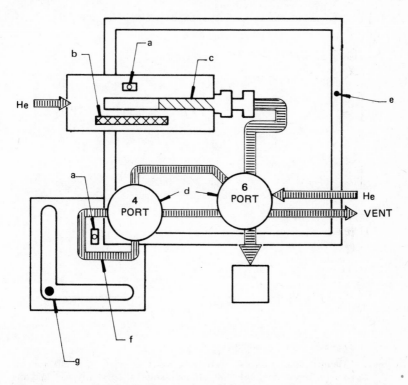

Fig. 1 – Thermal desorption apparatus. Components: a, thermocouple; b, cartridge heater; c, XAD-4 tube; d, valves; e, heating cord; f, Tenax pre-column; g, heating cord. *From ref. [8], by permission.*

This procedure uses a packed GC column of 6 ft., packed with Carbowax 20M on Chromosorb W, and programmed from 50° to 75° to 190°. A 200-ml sample of Ames (Iowa) tap-water analyzed by this procedure shows distinct peaks for several aromatic hydrocarbon impurities known to be in the water

(Fig. 2). Recoveries of model compounds added to water are generally quite good (Table 1). The results reported for methyl nonyl ketone are for a desorption temperature of $225°$: the results at $200°$-$210°$ were somewhat low. The absolute recoveries of several model compounds studied (ratio of peak height in the desorption procedure to the peak height obtained by direct injection) were also good.

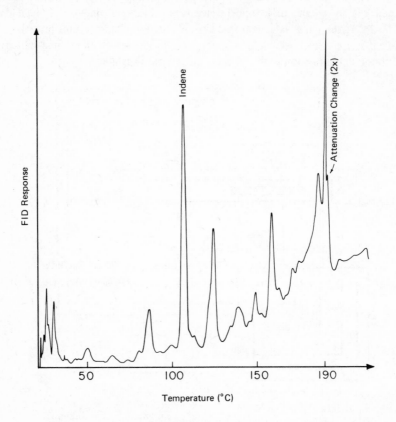

Fig. 2 – Thermal desorption GC analysis of Ames tap-water (200 ml) with a packed column. Desorption: 180–200°, 10 min. Column: 10% Carbowax 20M on Chromosorb W, 6ft. x 2 mm i.d. Temp. program: from 50° to 190° over 14 min.

The thermal desorption procedure has been adapted for use with a glass capillary GC column in order to obtain better resolution of the sample solutes [9]. The arrangement of the XAD-4 sample tube and the Tenax tube is similar to that previously used (Fig. 1). However, the approximately 10-fold difference in the diameters of the Tenax loop and the capillary chromotographic column necessitated

Table 1 The recovery of compounds from water by XAD-4. The compounds were determined in 20 ml samples containing 200 ng of each (10 ppb).

| Compound | *Desorption* | | *Recovery, %* |
	Time, min	Temp, °	
Toluene	15	210	88
Ethylbenzene	15	210	79
Indene	10	180–200	96
Napthalene	15	210	90
1-Methylnaphthalene	10	180–200	95
Hexane	8	175	88
Chloroform	4	175	93
Dibromomethane	4	175	88
Cyclohexanol	5	200	98
n-Heptyl alcohol	6	200	100
Benzyl alcohol	13	220	83
Methyl isobutyl ketone	8	220	100
Amyl isopropyl ketone	8	200	102
Methyl nonyl ketone	13	220	96
p-Methylacetophenone	13	220	99
Ethyl heptanoate	10	200	96
Octyl acetate	10	200	61
Bromobenzene	10	200	106
o-Dichlorobenzene	10	200	102

the use of an inlet splitter. The splitter is connected by a nut-and-ferrule combination to a glass injection-port sleeve (6 mm o.d. × 1.8 mm i.d.). The splitter consists of a tee (1/8 in. − 1/16 in. reducing union) connected to a split exit valve, restrictor and vent.

The capillary column procedure has been studied primarily for the concentration and determination of aromatic hydrocarbons, but it should also be applicable to the analysis of other organic impurities in aqueous samples. Fig. 3 shows much better resolution of the organic impurities in a tap-water sample compared with that shown in Fig. 2 for the packed-column procedure.

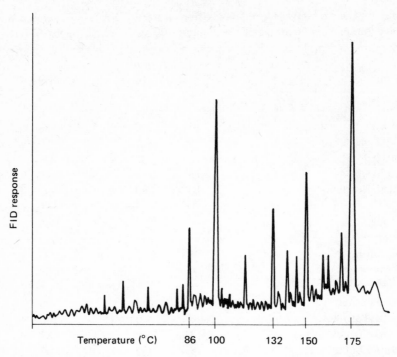

Fig. 3 – Thermal desorption GC analysis of Ames tap-water with a capillary column (cf. Fig. 2), of WCOT type; initially 50–75°.

AN ANION-EXCHANGE PROCEDURE FOR ACIDIC COMPOUNDS

Aqueous samples often contain so many organic compounds that selective concentration becomes necessary. Acids are commonly isolated by solvent extraction followed by back-extraction of the organic extract with aqueous base. However, this procedure misses many hydrophilic acids, and the initial organic extraction may result in difficult emulsions. It is better to use a small anion-exchange column [10]. A glass tube 0.8 cm × 12 cm is filled with 2 g (∿5 ml) of an anion-exchange resin in the hydroxyl form. (The resin is prepared from Rohm & Haas XAD-4 by chloromethylation and amination.)

Up to 1 l of aqueous sample is passed through the resin column: then the neutral compounds are removed by washing with methanol and with diethyl ether. Elution of the acids is accomplished by conversion into the moleundissociated form, which is eluted with an organic solvent. This is done by passing 20 ml of ether, saturated with HCl gas, through the column, pouring back and shaking to remove bubbles, and passing through again. Then 10 ml of fresh HCl-ether is passed through the column. The ether effluent is evaporated (steam bath, then N$_2$), 2–3 ml of diazomethane is added, the excess diazomethane is destroyed and the solution is diluted to volume. A portion of this solution, which contains methyl esters, is injected into a capillary-column GC.

For diverse model compounds (10 or 100 μg/l) there were excellent per cent recoveries:

Carboxylic acids: acetic, 92; propionic, 89; *iso*butyric, 91; butyric, 100; *iso*valeric, 95; valeric, 107; *iso*caproic, 113; caproic, 85; caprylic, 94; capric, 97; lauric, 103; myristic, 101; palmitic, 99; o-phthalic, 65; tetrachloroterephthalic, 83; 2, 4-dichlorophenoxyacetic, 93;

Sulphonic acids: benzenesulphonic, 100; toluenesulphonic, 95; *n*-dodecylbenzene-sulphonic, 89; *n*-dodecylsulphonic, 100; 3-sulphophenylbutyric, 93; 3-sulpho-phenylheptanoic, 90; 4-chlorobenzenesulphonic, 93; sulphoacetic, 75.

Phenols: 2-Chloro-, 87; 2, 4-Dichloro-, 100; 2, 4, 5-Trichloro-, 95; 2, 4, 6-Tri-chloro-, 96; Pentachloro-, 91; 2-Nitro-, 90; 4-Nitro-, 86; 2, 4-Dinitro-, 75.

The procedure has been used for the analysis of actual samples, and shows almost perfect selectivity for acids and acids salts. More than 50 individual organic acids have been isolated, separated and identified (by GC-MS) in water taken from a pesticide disposal pit [11]. Comparison of the GC traces produced by the ion-exchange procedure with those obtained using solvent extraction shows that many compounds are partially or completely lost in the other method.

References

[1] Dowty, B., Carlisle, D. & Laseter, J. L. (1974) *Science,* **187**, 75–77.
[2] Richard, J. J. & Junk, G. A. (1977) *J. Am. Waterworks Assoc.,* **69**, 62–64.
[3] McAuliffe, C. (1971) *Chem. Technol.,* **1**, 46.
[4] Zlatkis, A. & Tishbee, H. A. (1973) *Chromotographia,* **6**, 67–70.
[5] Bellar, T. A. & Lichtenberg, J. J. (1974) *J. Am. Waterworks Assoc.,* **66**, 739–742.
[6] Grob, K., Grob, K. Jr., & Grob, G. (1975) *J. Chromatog.,* **106**, 299–315.
[7] Junk, G. A., Richard, J. J., Grieser, M. D., Witiak, D., Witiak, J. L., Arguello, M. D., Vick, R., Svec, H. J., Fritz, J. S. and Calder, G. V. (1974) *J. Chromatog.,* **99**, 745–762. [*See also* Taleda, A. & Fritz, J. S. (1978) *J. Chromatog.,* **152**, 329–340.–*Ed.*]
[8] Ryan, J. P. & Fritz, J. S. (1978) *J. Chromatog. Sci.,* **16**, 488–492.
[9] Ryan, J. P. & Fritz, J. S. Unpublished work (1979).
[10] Richard, J. J. & Fritz, J. S. (1980) *J. Chromatog. Sci.,* **18**, 35–38.
[11] Richard, J. J. & Avery, M. Unpublished work (1980).

#B-2   SAMPLE HANDLING AS PRACTISED IN THE WATER
         RESEARCH CENTRE*

I. WILSON, Water Research Centre, Medmenham, Marlow, Bucks
S17 2HD, U.K.

*Amongst the techniques in use [cf. the surveys in #B-1 and #NC(B)-1–Ed.],
the commonest is solvent extraction. Head-space methods are preferred, with
resin methods as an alternative, where there is to be a high-resolution separation
by capillary GC or by HPLC, and final MS identification. A heat-desorption step
can be advantageous. Sample handling in connection with mutagen testing is
considered, e.g. freeze-concentration or freeze-drying.*

For the isolation and concentration of organic compounds in raw and treated
waters the Water Research Centre (WRC) has used a wide range of concentration
techniques including solvent extraction, resin adsorption, head-space concentration and freeze-drying. During the past few years the development of head-
space concentration and XAD-resin adsorption (see J. S. Fritz, #B-1, *this vol.*)
has had a major impact on the analysis of organic compounds in water. The
application of the individual techniques depends on the purpose of the analysis
and the sensitivity of the detection method used. As far as handling of samples
is concerned the work at WRC can be divided into three broad categories:
(1) determination of individual compounds or groups of compounds;
(2) identification of organic compounds in water; (3) detection of organic
mutagens in water.

DETERMINATION OF INDIVIDUAL COMPOUNDS OR GROUPS OF COMPOUNDS

A number of specific compounds (or groups of compounds) which are potentially
hazardous or have undesirable organoleptic properties have been found in water
as the outcome of accidental pollution, gradual build-up (accumulation) or
chemical reactions during water treatment. Methods for the determination of
such compounds usually involve selective isolation and concentration, often

* This survey is an expansion of remarks made at the Forum.–*Ed.*

followed by further purification, separation by GC, TLC or HPLC and quantitative estimation using a non-selective or (preferably) selective detector. Examples
of such methods which have been used in the WRC are shown in Table 1.

Table 1   Some methods used in the Water Research Centre for the determination
of individual compounds or compound groups using solvent extraction.

Determinand	Extractant	Vol. of sample	Method of detection	Detection limit
Haloforms	Petroleum ether	100 ml	GC-ECD	0.05 μg/l
Phenoxyacetic acid herbicides	Diethyl ether	500 ml	GC-ECD	1 μg/l
Organochlorine pesticides	Petroleum ether	500 ml	GC-ECD	0.3—5.0 ng/l
Acrylamide	Diethyl ether (extraction of a bromo- derivative)	100 ml	GC-ECD	0.1 μg/l
Synthetic steroids	Dichloro- methane	10 l	GC-FID	0.1 μg/l

Solvent extraction is still the most common technique for concentration
and isolation. It will most likely remain the most useful technique where relatively
high concentrations are determined, particularly when utilizing selective detection
as in the case of haloforms. Some methods where large volumes of samples are
required (as in the case of synthetic steroids) would, however, benefit if resin
adsorption techniques were employed under appropriate conditions of adsorption
and elution.

Various head-space methods [1, 2] have also been developed for the determination of haloforms in water; but methods using solvent extraction are
preferred, at least in the U.K., for their simplicity and speed.

IDENTIFICATION OF ORGANIC COMPOUNDS IN WATER

During the past few years the WRC has studied the organic content of drinking
waters and its significance to health. This involves concentration of a wide range

of organic compounds using the least selective techniques available, followed by high-resolution GC or HPLC separation and identification by MS. Head-space concentration and resin adsorption techniques are an essential part of this work.

Apparatus for purging (sparging) and heat-desorption, which is based on principles described in section #A [*see especially* #A4 – *Ed.*], has been constructed in our laboratories. A schematic diagram of the apparatus is shown in Fig. 1. Since the organic compounds desorbed from the Tenax trap are

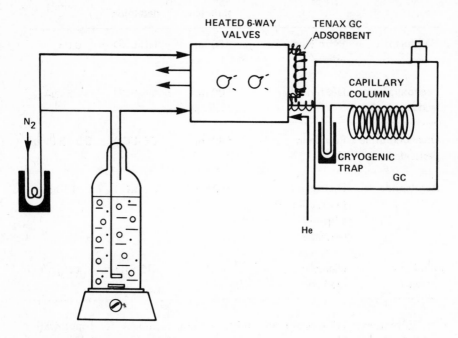

Fig. 1 – Apparatus for purging (sparging) and heat desorption in conjunction with capillary column GC.

separated on a capillary (WCOT) glass column, it is necessary to incorporate a second trapping stage before GC analysis to improve the introduction of the sample onto the column. The technique looks promising, as can be seen from Figs. 2 and 3 which illustrate the sensitivity and reproducibility of the method, respectively. However, the apparatus still requires further evaluation and possibly some technical improvements, and as yet it has not been used for GC-MS analysis.

For routine preparation of samples for GC-MS analysis the Grob-type purging method [3] is preferred to concentrate volatile compounds, mainly because this method does not require any modification of the GC instrument.

A schematic diagram of the apparatus is shown in Fig. 4. There are three basic differences compared to the heat-desorption method described above, viz. (a) the purging takes place in a closed circuit where a small volume of air is circulated by a special pump; (b) a small amount of activated carbon (2 mg) is used instead of Tenax GC as the adsorbent, and most important (c) the compounds are eluted with a small volume of a solvent (instead of by heat desorption). The eluate can be then analyzed by GC or GC-MS in the usual way without further concentration. For a 5 l sample of water the detection limits are in the region of 1–10 ng/l. This sensitivity is adequate for most purposes.

Comparing the two techniques for concentration of volatile compounds in water, it can be said that heat desorption is more sensitive for a given volume of sample, and more flexible (e.g. wider choice of adsorbents, and wider application such as the modification described by J. S. Fritz [#B-1, *this vol.*] where another heat-desorption step from XAD-resin replaces the purging of an aqueous sample); moreover this technique is more amenable to automation. Therefore it will most likely be the technique of the future especially if commercial instruments become more readily available. On the other hand, Grob's method is less expensive to set up since the apparatus itself is inexpensive and there is no need to modify the GC instrument. The method has been used routinely for several years in a number of laboratories including ours, it is well established, and is therefore likely to remain popular especially in smaller laboratories.

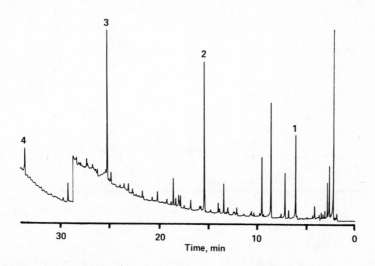

Fig. 2 – Sensitivity achievable by capillary GC following purging and heat desorption of a 500 ml sample. The numbered peaks correspond to compounds spiked into purified tap-water (50 ng/l): 1: 1-Cl-*n*-butane; 2: 1-Cl-*n*-hexane; 3: 1-Cl-*n*-octane; 4: 1-Cl-*n*-decane.

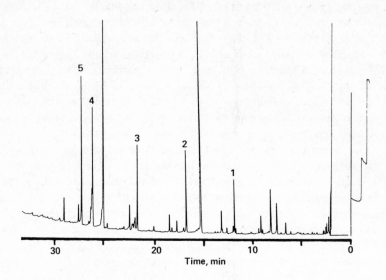

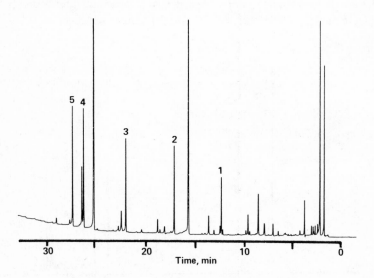

Fig. 3 – Reproducibility achievable by capillary GC following purging and heat desorption, exemplified by runs with two spiked samples. The numbered peaks, of height as given parenthetically for the upper and lower runs respectively, correspond to: 1: *n*-nonane (51 mm and 54 mm); 2: *n*-decane (77 mm and 82 mm); 3: *n*-undecane (80 mm and 88 mm); 4: *n*-dodecane (111 mm and 112 mm); 5: dibutylketone (137 mm and 113 mm).

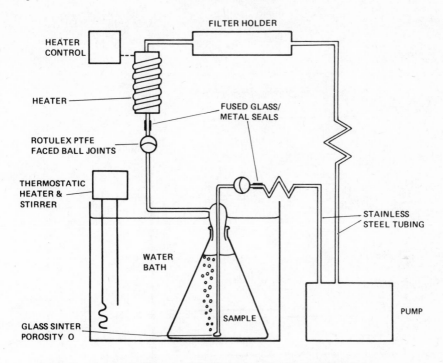

Fig. 4 — Schematic diagram of Grob's head-space concentration apparatus.

Although all of the head-space techniques are ideally suited to GC analysis, they suit only for a very limited range of organic compounds of relatively high volatility and low water-solubility. To extend the range other concentration techniques need also to be employed. For routine GC-MS identification of organic compounds in drinking water the WRC uses simultaneous concentration by Grob's method and resin adsorption. The resin adsorption method with XAD-2 is more or less identical to that described by Junk and others [4]. For semi-quantitative estimation of concentration, several deuterated compounds are added to the original sample as internal standards.

These compounds, which do not appear naturally, can be separated from the corresponding hydrogenated compounds on capillary column GC and easily detected by GC-MS. Fig. 5 shows a total ion chromatogram of an XAD-2 resin extract (in diethyl ether) from a treated water sample containing 100 ng/l of each of the deuterated internal standards.

Using the two concentration techniques and capillary columns, a hundred or more organic compounds can be identified by GC-MS in one sample. Even so this represents only 10% or less of the total organic matter present in water. Another approach to evaluate the significance of the organic content of drinking water to human health is mutagenic testing.

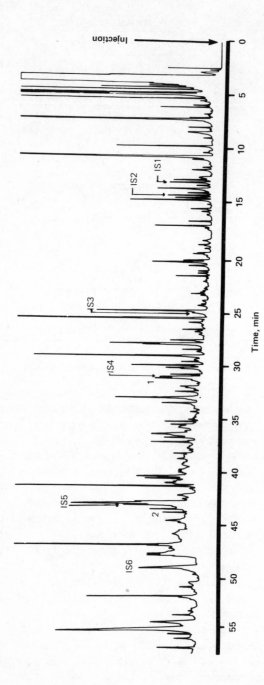

Fig. 5 – Total ion current chromatogram of an XAD-2 resin extract (in diethyl ether) from a spiked water sample. Spiking was done with 100 ng/l of each of the deuterated internal standards: IS1–chlorobenzene; IS2–p-xylene; IS3–phenol; IS4–napthalene; IS5–n-hexadecane;IS6–phenanthrene. 1–napthalene;2–n-hexa-decane.

DETECTION OF MUTAGENS IN WATER

Short-term bacteriological test for mutagenicity developed by Ames [5] are increasingly being applied in water laboratories. The WRC has been using mainly the fluctuation test (a modification of the Ames test) which was developed by M. H. L. Green and his colleagues at Sussex University [6], because this test is generally more sensitive and more suitable for testing aqueous samples.

The two main questions which need to be answered are: (1) are mutagenic compounds present in drinking water? (2) if so, what is their origin and identity?

To answer the first question it is desirable to test the whole organic content of the sample. Although direct testing of aqueous samples is feasible, as with physico-chemical detection the sensitivity of the test is often insufficient and the sample has to be concentrated. This inevitably results in a loss of some part of the organic content.

Processing of samples prior to mutagenic testing is similar to that for chemical analysis. Several concentration techniques have been under investigation in the WRC. These need to be evaluated in terms of recoveries of various types of organics, achievable concentration factor and suitability of the concentrates for mutagenic testing.

For concentrating to a moderate extent (up to 10 times), freeze-concentration has been proposed as a non-selective technique. There is, however, increasing evidence that recoveries of organic carbon after concentrating more than 5 times are rather poor and, more importantly, that the technique is selective. Thus the mean recoveries of TOC* were 54.1% and 24% with concentration factors of 6.0–11.4 X and 50.7–56.5 X respectively, as compared with 92.7% after partial freeze-drying to furnish a solution of the solids in 1/10th volume. In the latter technique, which is evidently promising and is being further evaluated, the aqueous sample is frozen in a thin layer, then water is sublimed under vacuum in a freeze-drying apparatus until the sample is reduced to, say, 10% of the original volume, whereupon it is thawed.

For the investigation of identities of any mutagens detected, relatively large samples are required to enable further fractionation and retesting of the fractions. Resin adsorption and freeze-drying, as compared in Table 2, appear to be suitable concentration techniques for this purpose. Resin adsorption is already used routinely in some laboratories for the preparation of concentrates for mutagenic testing, because it is a practical and convenient method and environmental mutagens have been detected in XAD extracts [7]. On the other hand, XAD-resin adsorption is applicable only to a small part of the total organic material present in waters; therefore one must investigate whether other mutagens are present which are not detected by XAD-resins. Solvent extracts of the solids remaining after freeze-drying of large volumes of water contain a number of compounds not detected in resin extracts. For example,

* Total organic carbon

Table 2   Comparison of resin adsorption and freeze-drying.

	Resin adsorption	Freeze-drying
Sample processing	Unlimited volume up to 150 l/24 hours	Maximum 15-60 l per week
Cost	Inexpensive	High initial cost
Accumulation of total organic carbon TOC	10-20% removed from water	>60% present in freeze-dried solids
Recovery of organic material	Selective elution with organic solvents	Selective extraction with organic solvents
Type of compounds lost	Very volatile compounds; at neutral pH: most acidic and basic compounds: hydrophilics	Most volatiles

Figs. 6(a) and 6(b) show the differences between HPLC profiles of methanol extracts of samples of water from the same source but concentrated by freeze-drying and resin adsorption, respectively. More detailed comparison of the two methods of concentration using chemical analysis and mutagenic testing is being made.

The concentration techniques described above are those employed in the Medmenham Laboratory of the WRC. Sample handling in the Stevenage Laboratory, which is mainly concerned with aspects of waste waters and sewage effluents, is generally similar.

References

[1] Bellar, T. A. & Lichtenberg, J. J. (1974) *J. Am. Waterworks Assoc.*, **66**, 739-744.
[2] Piet, G. J., Slingerland, P., de Grunt, F. E., van den Heuvel, M. P. M. & Zoetman, B. C. J. (1978) *Anal. Lett.*, **A11**, 437-448.
[3] Grob, K. & Zurcher, F. (1976) *J. Chromatog.*, **117**, 285-294.
[4] Junk, G. A., Richard, J. J., Grieser, M. D., Witiak, J. L., Witiak, M. D., Arguello, M. D., Vick, R., Svec, H. J., Fritz, J. S. & Calder, G. V. (1974) *J. Chromatog.*, **99**, 745-762.

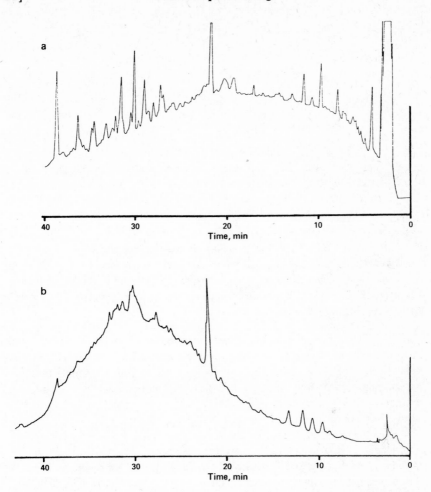

Fig. 6 – HPLC profiles of methanol extracts of treated water samples concentrated by (a) freeze drying, (b) resin adsorption. Column: Spherisorb-ODS: Mobile phase: linear gradient from 1% MeOH in 0.1% aq. acetic acid to 90% MeOH in 0.1% aq. acetic acid. Detection: UV (254 mm). (a) Freeze-dried extract (sample volume 15 I); 10 μl from 1 ml of MeOH extract; (b) XAD-2 extract (sample volume 30 I); 4 μl from 100 μl of MeOH extract.

[5]   Ames, B. N., McCann, J. & Yamasaky, E. (1975) *Mutat. Res.*, **31**, 347–364.
[6]   Green, M. H. L., Muriel, W. J. & Bridges, B. A. (1976) *Mutat. Res.*, **38**, 33–42.
[7]   Chriswell, C. D. (1977) *Chemical and Mutagenic Analysis of Water Samples*, NTIS, Springfield, VA, Rep. No. IS-M-126.

#B-3   TRACE ENRICHMENT IN THE HPLC ANALYSIS OF
        AQUEOUS SAMPLES

R. W. FREI and U. A. TH. BRINKMAN, Department of Analytical
Chemistry, Free University, de Boelelaan 1083,1081 HV Amsterdam,
The Netherlands

*Trace-enrichment techniques on solid surfaces (suitable HPLC packings)
combined with HPLC, as surveyed in an earlier volume [1], are further
discussed. On-column and pre-column technologies are compared, and examples
of applications (including water pollutants) are given for both approaches.
Their potential for selective sample clean-up, on-column derivatization, sample
storage and automation is indicated.*

As discussed and documented in a previous volume [1], the use of large injection
volumes, 100 μl or more gives improved detection limits in HPLC, e.g. for
polynuclear hydrocarbons [2]. This applies particularly where solubility problems
or compound instability preclude classical techniques such as freeze-drying,
extraction and/or evaporation, or steam distillation, which in any case have
serious limitations in terms of recovery and capacity.

Trace enrichment through adsorption onto solid surfaces has long been
exploited for GC assay. A well known example is the carbon adsorption method
(CAM), recently reviewed [3]. Grob [4] has developed a closed-circuit carbon
adsorption system, which has found wide use for the trace enrichment of organics
from aqueous samples. Unfortunately, CAM provides poor recoveries for certain
groups of compounds. Hydrophobic-layer materials have therefore been introduced
for relatively low-polarity organics. Junk *et al.* [5] have reported excellent
recoveries (90–100%) for compounds such as pesticides,alkylbenzenes,halogenated
compounds, polynuclear aromatics and phenols. They used a macroreticular
XAD-2 resin, packed into glass tubes, for sampling of litre quantities of water
samples. Other materials such as XAD-4 resin and Spherocarb have been discussed
by J. S. Fritz (*this volume*). In the HPLC context, both non-polar and polar
materials have been used, almost invariably based on a silica gel matrix.

ON-COLUMN TECHNIQUES

The use for trace-enrichment purposes of chromatographic-grade untreated
silica gel, and also alumina and diatomaceous earth, has long been exploited for

clean-up in classical column chromatography, as well as in TLC where intermediate drying steps can be of further help. The potential of trace-enrichment
techniques has been considerably expanded with the emergence of chemically
bonded stationary-phase materials. For example, hydrophobic surfaces such as
are present on the commercially available non-polar C_8 or C_{18} reversed-phase
(RP) materials lend themselves nicely to the collection of a wide range of organics.
In 1974, Kirkland [6] discussed the possibility of using pre-concentration
phenomena on the top of analytical HPLC columns for the purpose of trace
enrichment. Good separations have been achieved even with injection volumes as
large as 300 ml, e.g. with peptides or relatively non-polar compounds applied as
aqueous solutions to RP columns [1]. The explanation may be as follows.

When relatively non-polar organic species are injected from an aqueous
solution onto a hydrophobic surface they will become immobilized until the
elution strength of the solvent mixture is increased. This means that the
compounds are concentrated into a very small zone on top of an RP column
(trace-enrichment effect). The components can then be eluted with a suitable
eluent with very little band broadening. Similar phenomena have been observed
in adsorption systems for relatively polar compounds injected in non-polar
solvents [7].

Fig. 1 documents an example [1] based on four nonapeptides differing
in capacity factor (k'). When the injection volume was 340 instead of 34 μl
(containing the same amount of solutes, and dissolved in the mobile phase),
a significant reduction in peak height due to band broadening was observed
(Fig. 1 in [1]) for the nonapeptides of low k'. Obviously they start moving even
during the injection step, whereas for compounds with $k' > ca.$ 10 the mobility
in the mobile phase is low. If, alternatively, the compounds were injected from a
purely aqueous phase, no peak-height reduction, i.e., *no* additional band broadening, was observed with 340 μl (Fig. 1).

One benefit of this trace-enrichment phenomenon, of which a further
example has been given (Fig. 2 in [1]), is the potential sensitivity enhancement,
especially with several ml, which would permit quantitation of some minor
components barely visible with a small injection. Loop injection is precluded
with several ml; instead the sample is pumped onto the column in a first step
followed by isocratic or gradient elution with the same pumping system in the
next step(s). A good reproduction of repetitive large-volume injections can be
obtained using a timer-controlled step-gradient apparatus (Fig. 3 in [1]), which
can also be used to carry out a gradient separation of the pre-concentrated
compounds. That this approach is particularly useful for trace analysis in complex
matrices is shown in Fig. 2 for an ergot alkaloid isolated from urine without
pre-injection clean-up.

For an optimal trace-enrichment effect, injections from aqueous media are
best done on RP materials. Hence there can be direct injection of, e.g., aqueous
pharmaceutical preparations, biological fluids and particularly aqueous samples

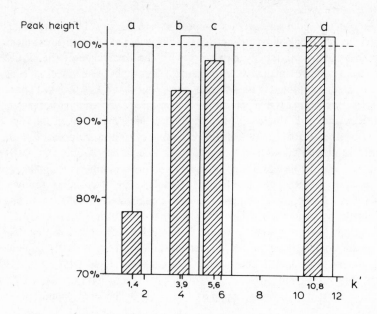

Fig. 1 – Effect of injection volume on peak height. ---, 34-µl injection taken as 100%. *Open blocks*, 340-µl injection from aqueous sample. *Shaded blocks*, 340-µl injection of sample dissolved in mobile phase. a, b, c and d: different peptides ranging in *k'* from 1.4 to 10.8. *From ref. [8], by permission.*

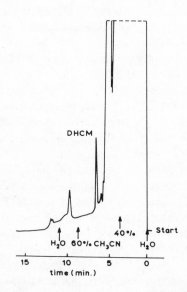

Fig. 2 – Chromatogram of dihydroergocristine mesilate (DHCM) in urine at the level of 2.74 µg/ml. *From a paper (J. Chromatog.) cited in [1].*

in connection with water-pollution studies. This has been nicely demonstrated by Creed [9], who analyzed a number of samples such as industrial effluent water, polluted river water, and well water. Otsuki [10] has achieved trace enrichment of phthalate esters—which are abundantly used as plasticizers—from aqueous samples on a μBondapak C_{18} column, with elution by a water–methanol gradient.

PRE-COLUMN TECHNIQUES

The direct injection of large amounts of sample and/or very dirty samples onto an analytical column can tax this column; after a few injections, e.g. of biological fluids or their extracts, the separation performance of the column may decrease drastically. The use of short pre-columns has therefore been recommended in order to protect the precious analytical column [e.g. 8, 11] and, simultaneously, to provide a partial clean-up of the sample. Besides, the pre-column can efficiently be used in the trace-enrichment step and can possibly be employed for field-sampling methods and even for on-column derivatization. Thus, Maitra and coworkers [12] performed off-line derivatization of gentamicin in human serum.

An interesting application of the pre-column concept has been proposed by Lankelma and Poppe [13] for the analysis of the cytostatic drug methotrexate in plasma. Pre-concentration is carried out on C_8 reversed-phase material and, using a back-flushing technique, the separation is next performed on a silica gel-based, chemically bonded anion exchanger. Huber and Becker [7] have recently advocated the use of so-called displacement HPLC for trace enrichment. Polar compounds are pre-concentrated on a silica gel pre-column, while RP materials are used with hydrophobic model compounds. Using a displacement solvent of suitable polarity, high enrichment factors are obtained, typically 1000–5000.

For the trace enrichment of phthalate esters from aqueous samples. Ishii et al. [14] use miniaturized HPLC equipment [15]; the dimensions of their pre-column are 0.35 mm i.d. and 18 mm length. This limits the sampling rate to about 100 μl/min. Sampling occurs off-line, and the pre-column is dried with air prior to its connection to the separation column. The long sampling times required with such a device for the assay of phthalates in polluted river water render this miniaturized version rather unsuitable for such problems: yet it may be valuable for bioanalytical applications where smaller sample volumes are used. This has been demonstrated by Ishii et al. [16] for corticosteroids in serum samples in the concentration range 20–130 ng/ml: thus, 200 μl of the serum are pre-concentrated as above on 20-μm Hitachi gel No. 3010 and separated, after air drying, on a silica gel column with dichloromethane-methanol (97:3 by vol.). The air-drying step has distinct advantages when using mobile phases of widely different polarity for the trace-enrichment and the separation steps.

The above procedures illustrate essentially off-line use of pre-columns. The real value of the pre-column approach is, however, in an on-line mode, which can be automated for routine analysis of large series of samples. Such an approach has, for example, been discussed by May *et al.* [17] and Eisenbeiss *et al.* [18]. The former group of workers reported on the analysis of hydrocarbons in marine sediments and seawater. Pellicular Bondapak C_{18} is used as solid surface in the pre-column, while μBondapak C_{18} is the stationary-phase material in the analytical column. Eisenbeiss and coworkers have developed a special adsorbent for the enrichment step, which allows them to elute and separate pre-concentrated polycyclic aromatic hydrocarbons under isocratic conditions, viz. with methanol–water (85:15).

In all the above cases relatively long pre-columns packed with small-diameter paticles have been used. Often this may well be unnecessary, since solutes displaying high capacity factors under trace-enrichment conditions will be pre-concentrated on a top layer of, say, a few mm of the pre-column. Long columns moreover have the disadvantage of a slow sample throughput due to high back-pressure. In a recent study [19] an attempt has therefore been made to work with very short pre-columns having a sorbent layer thickness of only 2 mm. Fig. 3 shows the design of the pre-column, which is hand-packed with a dense

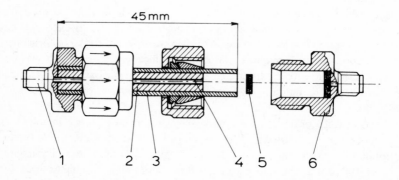

Fig. 3 – Pre-column design. 1. Swagelok, 1/4 in × 1/16 in. 2. varying-length PTFE rod, 4.6 mm o.d. 3. Stainless-steel tube, 45 × 4.6 mm i.d. 4. Stainless-steel capillary, 1/16 in o.d. 0.25 mm i.d. 5. PTFE or stainless-steel 20-μm frit. 6. Swagelok, 1/4 × 1/16 in., containing a 0.5 or 2-μm outlet frit. The space between 5 and the outlet frit of 6 is filled with packing material. *From ref. [19], by permission.*

slurry of the stationary-phase material in methanol using a micro spatula. Phthalate esters have again been used as model compounds, and samples of up to 1 l of e.g., tap-water, demineralized water and mineral water typically containing 0.1–2 ng/ml of phthalate esters are pre-concentrated on 5-μm LiChrosorb RP-18 with

95–100% solute recovery, using pumping speeds of 5–25 ml/min. The analytical column is also packed with LiChrosorb RP-18, and methanol–water mixtures are used for elution. Reproducibility from one pre-column to another is very satisfactory (Table 1). The largest column-to-column difference is about 10%, while the relative S.D. calculated from the data for all five columns is 4% ($n = 47$).

Table 1   Column-to-column reproducibility of on-line trace enrichment of di-n-butyl phthalate.

Conditions: trace enrichment from 16-ml sample volumes containing 194 ng DBP per ml at a pumping speed of 5 ml/min on a 2 × 4.6 mm pre-column. No. of observations denoted n.

Column	Signal		n
	peak height, mm/µg	rel. S. D., %	
1	49.5	3.9	10
2	47.9	5.3	7
3	45.2	5.4	10
4	49.9	4.7	10
4*	49.2	5.5	10

* after re-packing.

Rather surprisingly, this result hardly differs from that calculated for each single column (4–5.5%). Analogous results have been obtained with a flow-rate of 25 instead of 5 ml/min. A typical HPLC chromatogram recorded for a sample of mineral water (found to contain 1.2 ng/ml of di(ethylhexyl)phthalate and 0.08 ng/ml of di-n-butyl phthalate) is shown in Fig. 4a.

The principle outlined above has also successfully been used [19] for the trace enrichment of chloroanilines and polychlorinated biphenyls (PCBs). Fig. 4b,c nicely illustrates the negligible contribution of the short pre-column to band broadening, using a complicated PCB mixture as model compound.

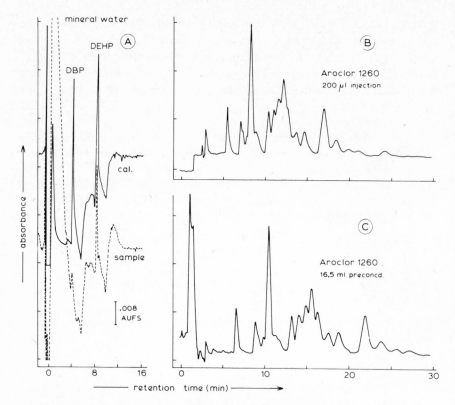

Fig. 4 – HPLC chromatograms with a very short pre-column. *From [19], by permission*.

(a) A standard solution (---) of *di-n*-butyl and diethylhexylphthalate, and a sample (—) of mineral water. Conditions: sample solution, 800 ml; 2 × 4.6 mm 5-μm LiChrosorb RP-18 pre-column; loading at 25 ml/min pumping speed; analysis at 2 ml/min on a 125 × 4.6 mm i.d. LiChrosorb RP-18 separation column with 75% (60 sec), 85% (300 sec) and 95% (270 sec) methanol in water as mobile phase; detection at 233 nm (0.08 AUFS).

(b), (c) Comparison of HPLC chromatograms recorded for equal amounts (about 20 μg) of the PCB mixture Aroclor 1260, using (b) normal 200-μl syringe injection, or (c) trace enrichment from a 16.5 ml aqueous sample. Mobile phase, 85% methanol in water; detection at 205 nm. Pre-column and separation column, as for (a).

FURTHER TRENDS AND FUTURE DEVELOPMENTS

One of the inherent disadvantages of trace enrichment on surfaces of low selectivity is that many sample constituents will be concentrated on the pre-column. Consequently, even putting aside overloading problems, clean-up may well be insufficient, and HPLC separation of the compound of interest will be rendered difficult. There are two chief ways to eliminate this problem: viz.

via an increase of selectivity effected by post-column (and also pre- or on-column) derivative formation [1], or *via* 'multidimensional' HPLC.

An example of on-column derivatization, viz. the analysis of gentamicin as its *o*-phthaldehyde derivative [12], has already been quoted; for peptides a post-column reaction with flurescamine has been performed [1, 8]. The combined use of trace enrichment and post-column derivatization has also been applied successfully in the analysis of chloropheniramine in urine [20]. This tertiary amine drug was pre-concentrated from 2 ml of a 1 : 20 diluted untreated urine sample on a LiChrosorb RP-18 pre-column. Subsequently, analysis was done on a Micro Pak CN-10 column using methanol–water (20 : 80) containing a phosphate buffer as mobile phase. Detection was based on ion-pair formation between the amine cation and the fluorogenic dimethoxyanthracene sulphonate (DAS) and extraction of the ion pair into tetrachloroethane in a micro extraction reactor, and monitoring of the organic solvent stream in a fluorimeter. The pre-column was shown to act as an efficient clean-up device; from the chromatograms shown in Fig. 5 one can see the distinct advantages of ion-pair formation

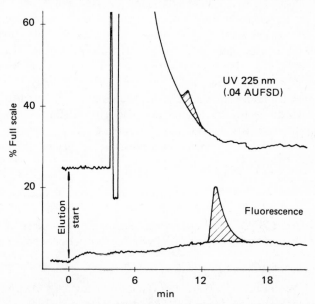

Fig. 5 – Comparison of ion-pair fluorescence and UV (225 nm) chromatograms of a urine sample containing 1.0 μg/ml chloropheniramine, after pre-concentration on a LiChrosorb RP-18 pre-column (for details, *see text*). The cross-hatched peaks illustrate the response due to the presence of chloropheniramine. Chromatographic conditions–Column: Micro Pak CN-10, 300 x 4.6 mm i.d. Mobile phase: 20% methanol in water, pH 3.5. Flow-rate, 0.42 ml/min.
Auto-Analyzer arrangement for post-column ion-pair formation and extraction. Flow-rates: reagent solution 0.32 ml/min; chloroform 0.43 ml/min; chloroform after phase separation, 0.34 ml/min. A 20-turn extraction coil was used. *From [20], by permission.*

(fluorescence trace) over direct analysis of the drug (UV trace) as regards selectivity and sensitivity. In a related study [21] hydroxyatrazine (HA), which is the major metabolite of the well-known herbicide atrazine, was taken as the model compound. With this relatively polar compound, pre-concentration from 10-ml samples on a 2-mm LiChrosorb RP-18 layer was found to be quite critical with regard to recovery due to breakthrough of HA. Reducing the polarity by ion-pair formation with DAS and trace enrichment of HA as its HA-DAS ion pair effectively remedied this situation.

An alternative means [22] to solve the selectivity problem, as applied to the isolation of limonin from grapefuit peel [23], is the multicolumn approach with on-line coupling of one column system to another. In order for the transfer of zones to occur with a minimum of band broadening it is necessary to have a trace enrichment on the second, and subsequent, column(s). To achieve this end, in the quoted papers gel-permeation chromatography (GPC) as a first separation step was combined with an RP system. Combinations such as GPC with organic solvents as mobile phase followed by adsorption chromatography, or ion-exchange separations coupled with RP chromatography, should also be feasible.

Future research will no doubt be directed at the use of more selective pre-column packing materials, such as metal-loaded solid surfaces and immobilized-enzyme-containing phases. Also, different pre-columns will be used in series to isolate different groups of compounds for final analysis. Attention should further be focussed on the use of pre-columns for field sampling. In this way, the samples can be highly reduced in size and stabilized for storage [cf.19] and transportation to a central laboratory. The interest of analytical chemists in these and other trace-enrichment techniques will no doubt be stimulated by the automation potential of the pre-column and 'multidimensional'-HPLC concepts.

References

[1] Frei, R. W. (1978) In *Blood Drugs and other Analytical Challenges* [*Methodological Surveys (A),* Vol. 7] (Reid, E., ed.), Horwood, Chichester, pp. 243–255.
[2] Strubert, W. (1973) *Chromatographia,* **6**, 205–206.
[3] Suffet, I. & Sowinski, E. J. Jr. (1975) in *Chromatographic Analysis of the Environment* (Grob, R. L., ed.), Dekker, New York, p. 435.
[4] Grob, K. (1973) *J. Chromatog.,* **84**, 255–273.
[5] Junk, G. A., Richard, J. J., Fritz, J. S. & Svec, H. J. (1976) in *Identification and Analysis of Organic Pollutants in Water* (Keith, L. H., ed.), Ann Arbor Science Publishers, Ann Arbor, p. 135.
[6] Kirkland, J. (1974) *Analyst,* **99**, 859–885.
[7] Huber, J. F. K. & Becker, R. R. (1977) *J. Chromatog.,* **142**, 765–176.
[8] Frei, R. W. (1978) *Internat. J. Environ. Anal. Chem.,* **5**, 143–155.

[9] Creed, C. G. (1976) *Res. & Develop.*, Sept. issue, p. 40.

[10] Otsuki, A. (1977) *J. Chromatog.*, **133**, 402–407.

[11] Bye, A. (1978) *As for* **1**, 269–280.

[12] Maitra, S. K., Yoshikawa, T. T., Hansen, J. L., Nilsson-Ehle, I., Palin, W. J., Schotz, M. C. & Gruze (1977), L. D., *Clin. Chem.*, **23**, 2275–2278.

[13] Lankelma, J. & Poppe, H. (1978) *J. Chromatog.*, **149**, 587–598.

[14] Ishii, D., Hibi, K., Assi, K. & Nagaya, M. (1978) *J. Chromatog.*, **152**, 341–348.

[15] Ishii, D., Asai, K., Hibi, K., Jonokuchi, T. & Nagaya, M. (1977) *J. Chromatog.*, **144**, 157–168.

[16] Ishii, D., Hibi, K., Asai, K., Nagaya, M., Mochizuki, K. & Mochida, Y. (1978) *J. Chromatog.*, **156**, 173–180.

[17] May, W. E., Chester, S. N., Cram. S. P., Cump, B. H., Hertz, H. S., Enagonio, D. P. & Dyszel, S. M. (1975) *J. Chromatog. Sci.*, **13**, 535–540.

[18] Eisenbeiss, F., Hein, H., Joester, R & Naundorf, G. (1978) *Chromatog., Newsletter*, **6**, 8–11.

[19] van Vliet, H. P. M., Bootsma, Th. C., Frei, R. W. & Brinkman, U. A. Th. (1979) *J. Chromatog.*, **185**, 483–495.

[20] Frei, R. W., Lawrence, J. F., Brinkman, U. A. Th. & Honigberg, I. L. (1979) *J. High Res. Chromatog. Chromatog. Commun.*, **2**, 11–14.

[21] Lawrence, J. F., van Buuren, C., Brinkman, U. A. Th., Honigberg, I. L. & Frei, R. W. (1980) *Anal. Chem.*, **52**, 700–704.

[22] Erni, F. & Frei, R. W. (1978) *J. Chromatog.*, **149**, 561–569.

[23] Johnson, E. L., Gloor, R. & Majors, R. E. (1978) *J. Chromatog.*, **149**, 571–585.

#NC(B) Notes and Comments related to Water and Effluents

#NC(B)–1 *A Note on*
SPARGING, TRAPPING AND CONCENTRATING TRACE-ORGANICS IN WATER SAMPLES

C. E. ROLAND JONES, Chromsultants, Redhill, Surrey RH1 2DF, U.K.

All environmental samples present the analyst with the problem of isolating the species of interest from the matrix by extraction, distillation or some other effective technique. Frequently such methods demand tedious clean-up procedures and thus are fraught with the fallibilities common to multi-stage processes. Three of the most widely used methods to trap and concentrate organics from aqueous matrices are:

(i) direct sorption of organics from water onto charcoal, followed by solvent extraction and subsequent concentration;

(ii) direct liquid–liquid extraction followed by solvent evaporation; and

(iii) inert gas stripping onto a suitable adsorbent followed by thermal desorption as originally proposed by Bellar & Lichtenberg [1] and subsequently developed by Novotny *et al.* [2], May *et al.* [3] and Zlatkis [4].

It would seem that the third option, despite its limitations, affords the best approach to the analysis of trace volatiles in an aqueous environment. The method avoids the complication of long pre-concentration times, extensive sample handling and the use of significantly large (and hence expensive!) amounts of ultra-pure organic solvents.

Typically, present practice requires the use of a sparging vessel of 100–400 ml capacity which has a dense glass frit sealed into the gas entry at its base. The vessel itself is contained within an isothermal jacket and is sparged by bubbling an inert gas, preferably helium, through an aliquot of the sample. The effugate is passed into a standard 3 × ¼ in. o.d. packed adsorption tube and then desorbed in the back-flush mode by a fast thermal pulse, the analytical 'finish' normally being GC.

An oft-voiced objection to the method has been the apparently complex and extensive hardware required to desorb a sample concentrate from a suitable trap by means of a fast thermal pulse. This requires a heated switching valve of low dead volume, a desorption oven of quite unusual characteristics, and heated transfer lines, again of minimum volume. Clumsy methods of overcoming this transfer problem have included removal of the GC column head and direct insertion of the trap into the injection port, replacing the head and re-establishing the gas flow; alternatively the adsorption tube may be sealed and broken mechanically after insertion into the injection port and completion of the carrier gas circuit. Back-flushing of the volatiles and the limited desorption efficiency of the carrier stream militate against these primitive approaches.

Although the method was originally designed to accommodate aqueous samples containing, e.g., halocarbons or light hydrocarbons, the range has been extended by Voznakova et al. [5] to include organic compounds having a b.p. of some 330/340° (e.g. alkylbenzenes and methyl naphthalenes).

Obviously the success of the procedure is in part a function of the quality of the adsorbent. Carbo-Pack C-HT, Tenax GC and Porapaks Q and T have all found favour and indeed have been the subject of an evaluation by Vidal-Madjar et al. [6]. Tenax GC is offered by many advocates as the preferred adsorption medium; but Porapak Q has an infinitely greater capacity and therefore, in the back-flush mode, commensurately greater concentrating power. Unfortunately, the thermal-shock desorption can lead to pyrolysis of the adsorbent, and in the presence of e.g. a halocarbon the sample can be lost as it effectively acts as a 'chain-stopper' for degradation products; the phenomenon is well-known in polymer chemistry and is effectively used as a means of limiting mol. wt. during a polymerization.

Carbo-pack C-HT has obvious virtues; but the low specific surface area, about 13.5 m^2/g for 80-100 BSS mesh, diminishes its efficacy. Tenax GC is not much better in this respect, being only about 20 m^2/g for 60-80 BSS mesh as against about 500 m^2/g for Porapak Q; Porapak T, although of comparable capacity, is a far more selective adsorbent and therefore perhaps loses a little on that score in some contexts.

Thus one has a limited choice that is even narrower for a specific application and it is obvious that selection of an adsorbent must be made with a certain informed prudence, and the material eventually selected tested with great care before finalizing the details of methodology for a particular determination.

References

[1] Bellar, T. A. & Lichtenberg, J. J. (1974) *J. Am. Water Works Ass.* **66**, 739-744.

[2] Novotny, M. V., Lee, M. L. & Bartle, K. D. (1974) *Chromatographia*, 7, 333-338.

[3] May, W. E., Chester, S. W., Cram, S. P., Gump, B. H., Hertz, H. S., Enagonio, D. P. & Dyszel, S. M. (1975) *J. Chromatog. Sci.*, **13**, 535-540.

[4] Zlatkis, A. U.S. Pat. Appl. 450-500, 12 March 1974.

[5] Voznakova, Z., Popl, M. & Berks, M. (1978) *J. Chromatog. Sci.*, **16**, 123-127.

[6] Vidal-Madjar, C., Gonnord, M.-F., Benchah, F. & Guiochon, G. (1978) *J. Chromatog. Sci.*, **16**, 190-196.

―――――――

#NC(B)-2 *A Note on*

METHODOLOGY FOR POLLUTANTS, AS IN PLANT OUTFALLS*

JAMES S. SMITH, Allied Chemical Corporate Headquarters, P.O. Box 1021R, Morristown, NJ 07960, U.S.A.

The analytical procedures now outlined relate to compliance, in respect of industrial waste streams, with standards laid down by the U.S. Environmental Protection Agency (EPA), in accordance with the Clean Water Act of 1977. The 'priority' (toxic) pollutants concerned, amounting to 129 as of April 1977, are classified from the analytical viewpoint as (a) volatile organics, (b) basic or neutral ('base neutral') extractables, (c) acid(ic) extractables, and (d) pesticides; also (e) 'elements', not considered here. Excluded from the classification are polar compounds of low mol. wt., surfactants, naturally occurring compounds of high mol. wt., and certain other materials that are difficult to identify and quantitate.

Volatile organics, by purge-and-trap method

Examples	Bromochloromethane; 2-bromo-1-chloropropane; 1,4-dichlorobutane (used as standards, at 20 μg/ml). GC/MS checked by decafluorotriphenylphosphine.
End-step	GC-MS(EI), on 0.2% Carbowax 1500/Carbopack C packed below 3% Carbowax 1500/Chromosorb W, 8 ft. and 1 ft. respectively (2 mm i.d.).
Sample handling	Sample (5 ml) helium-purged within 24 h onto Tenax-GC + Chromosorb 102 (4 in. + 2 in.). Heat-desorption into GC performed within 2 weeks.
Comments	Detection limit 10 μg/l. The characteristic ion intensities for each compound should agree within ±20% with those expected: thus for $CHCl_2Br$, m/e 127 (used for quantitation; $CHClBr^+$) = 13% of 83 ($CHCl_2^+$).

* Editor's condensation of a survey intended for presentation at the Forum, which the author was unable to attend, and based on a protocol prepared by Environmental Monitoring and Support Laboratory, at the request of the Effluent Guidelines Division, Office of Water and Hazardous Wastes, and with the cooperation of Environmental Research Laboratory, Athens, GA, March, 1977; revised April, 1977. The preceding articles serve as perspective for this condensed outline, which lacks author's endorsement.

Base-neutral extractables, by liquid–liquid extraction

Examples| Anthracene (d_{10} used as internal standard, 10 μl of 2 $\mu g/\mu l$ solution);
| dichlorobenzenes (1,3; 1,4; 1,2); *N*-nitrosodiphenylamine; benzidine.
End-step |GC-FID, then GC-MS(CI), on 1% SP-2250 DB/Supelcoport, 6 ft.
Sample | Sample (2 l; + a blank water sample) continuously extracted for 24 h
handling | with dichloromethane (700–800 ml) after adding 6 M NaOH to bring
| pH above 11. Extract concentrated to 1.0 ml by Kuderna-Danish
| evaporation [*see* Note by M. J. Saxby, *this vol.—Ed.*] with blanks to
| ensure clean glassware.

Acid extractables, by liquid–liquid extraction

Examples| Phenolic compounds only, e.g. pentachlorophenol (100 ng should be
| detectable).
End-step |As for base-neutrals, but 1% SP-1240 DA coating.
Sample | As for base-neutrals, but 6 M HCl added to bring pH below 2.
handling |

Pesticides, by liquid–liquid extraction

Examples| BHC (α, γ, β); dieldrin; dicapthon (0.06 ng should be detectable).
End-step |GC-ECD, then GC-MS(CI), on 1.5% SP-2250 & 1.95% SP-2401.
Sample | Sample (1 l) serially extracted with 3 × 60 ml of 15% (v/v) dichloro-
handling |methane in hexane in separating funnel. Extract concentrated by
| Kuderna-Danish evaporation.

SAMPLING AND STANDARDS

Guidelines have been drawn up to ensure that samples are representative and do not become contaminated. Thus, for automatic sampling by peristaltic pump, at each location new unused tubing should be used, suitably Teflon or possibly silicone but not PVC. For organics the glass containers should be narrow-mouth, sealed with Teflon-lined caps (pre-rinsed with solvent and dried), and should have been rinsed with dichloromethane (air dry at room temp. protected from atmospheric or other sources of contamination) or, better, washed with hot detergent solution and thoroughly rinsed with tap-water and 'blank water' (dry in oven). Phenol-containing samples should be taken to pH 4 by H_3PO_4 or H_2SO_4, and protected from direct light. For volatile organics the container should be filled to overflowing, with no bubbles.

From primary stock solutions of standards (2 $\mu g/ml$ in methanol), working standards are prepared by diluting 20 μg to 100 ml with organic-free water; alternatively they may be purchased (Supelco or Radian Corporation).

Comments on #B-1, J. S. Fritz — SOLUTES IN WATER

J. S. Fritz, *replying to* R. Tanner. — Preparation of the XAD-4 resin entails sieving (225–300), suspending to get rid of fines, then heating (*see text*). *Reply to* P. Hendley: poor trapping rather than poor desorption seems to be the cause of low recoveries with XAD-2 for certain solutes; such compounds were better analyzed by trapping with anion-exchange resins. Whether the latter might give falsely high values for acids by hydrolyzing amides or esters is not known; but such compounds were thought not to be present in the water samples studied (*reply to* B. Scales). *Query by* U. A. Th. Brinkman: for acids (sulphonic), might it be advantageous to perform ion-pair solvent extraction with, say, methylene blue or acridine?

Comments on #B-3, U. A. Th. Brinkman — TRACE ENRICHMENT

Reply to A. P. Woodbridge. — Pre-column packing technique and efficiency are important factors beyond 5 mm length; but up to 5 mm there is no band broadening and there can be the benefit of some migration during the adsorption step where the polarity of the analytes permits this. *Comment by* K.-G. Wahlund: taking into account that micro-particles are used, evidently the pre-columns really are well packed, by a very simple technique.

Comments on #NC(B)-1, C. E. Roland Jones — SPARGING etc.

Porapak Q becomes degraded above about 160° (*reply to* R. J. Tanner). *Remark by* P. Hollingdale Smith: some years ago we found that Porasil (porous silica beads) and Bioglas (porous glass beads) had good adsorptive capacity, but we had difficulty in getting good recoveries. *Comments by* C. E. Roland Jones. — Other workers too have had difficulty in desorbing active species from such materials, the surface area of which is good (about $150 \, m^2/g$). For nitrosamines, which are causing the WHO increasing concern even at sub-ppb levels, a notably effective technique has been developed (Drescher & Frank, at Iowa State Univ.) for extracting from water, with dichloromethane.

#C Samples for Residue and Other Non-Pharmacological Studies

#C-1 ENZYMIC LIBERATION OF CHEMICALLY LABILE OR PROTEIN-BOUND DRUGS FROM TISSUES*

M. D. OSSELTON†, Home Office Central Research Establishment, Aldermaston, Reading, U.K.

Procedures commonly used for the isolation of drugs from biological samples involve either direct solvent extraction or tissue denaturation followed by solvent extraction. Such procedures may work poorly for compounds that are bound tightly to tissue proteins, as with benzodiazepines, phenothiazines and phenylbutazone, or are degraded by the denaturing reagent, as with benzodiazepines and pentazocine.

Digestion of biological materials with the potent proteolytic enzyme subtilisin Carlsberg releases high levels of drugs belonging to all of the major drug classes (acids, bases, neutrals) under mild physical and chemical conditions. The enzymic digestion procedure is simple, usually complete within 60 min, and may be used for the liberation of drugs from liver and other solid tissues, blood, blood stains and injection sites. The procedure for tissues is described, and the results compared with those from conventional extraction procedures.

For isolating and identifying drugs in tissue, its structure usually has to be destroyed in order to liberate them into an environment from which they may be extracted, purified and subsequently identified. The degradation has hitherto commonly involved an initial mechanical disruption of the tissue structure, such as blending or homogenization, followed by hydrolysis of tissue proteins using a combination of heat and strong acid or alkali [1], or precipitation of proteins by addition of a protein precipitant such as tungstic acid [2], ammonium sulphate or aluminium chloride [3]. Such techniques were generally suitable for analyzing a wide range of drugs in the context of overdoses or of relatively high levels as needed to attain therapeutic effects.

† Present address: Home Office Forensic Science Laboratory, Shakespeare Street, Nottingham, NG1 4FR.

Recent trends in drug design have led to many potent and highly active drugs that bind tightly to tissue proteins and are administered in low doses with consequently low levels in blood and other tissues, as exemplified by some of the newer benzodiazepines. It was the introduction and widespread use of benzodiazepines that prompted us to re-examine the problem of drug extraction, for here was a class of compound which confronted analysts with a problem owing to a high capacity for protein binding and low stability under conditions of acid hydrolysis. What seemed to be required was a new, low-cost, drug-extraction procedure which would rapidly destroy tissue proteins under mild physical and chemical conditions to release a wide range of drugs or other toxic substances. One theoretical means of satisfying these requirements was enzymic hydrolysis if the right enzyme could be found. [Cf. N. T. Crosby's enzymic approach (*this vol.*) to liberating bound constituents from foods.—*Ed.*] The only commercially available enzyme that offered promise was subtilisin Carlsberg. This is a proteolytic enzyme, not group-specific, which will hydrolyze any peptide bond in a protein chain unlike most other proteases which hydrolyze bonds involving specific amino acids. It is also unusual in that it acts well over a wide pH range (viz. 7.0 to 11.0) with optimal activity at pH 9.5 [4]. The enzyme is also highly active and stable at temperatures between 50° and 60°, with optimum activity at 55° [4]. It proved satisfactory for various tissues (liver, brain, injection sites and blood), efficiently releasing acid-labile or protein-bound drugs and their metabolites into an aqueous medium from which they were extractable [5,6].

PROCEDURES FOR TISSUE DIGESTION WITH SUBTILISIN

The procedure for solid or semi-solid tissues such as liver is outlined in Scheme 1. The resulting aqueous solution may be analyzed for the presence of drugs from different chemical classes following Scheme 2. [Cf. de Silva's Fig. 4 in Vol. 7 [7].—*Ed.*].

For blood or plasma the procedure (Scheme 3) is most conveniently performed in a small glass tube (6 mm × 50 mm). This procedure, modified from that of Rutherford [8] for diazepam, provides in conjunction with the use of subtilisin a method of screening for a wide range of basic drugs and metabolites. Controls were prepared by addition of 'Trizma' base (i.e. purified 'Tris'; Sigma Chemical Co.) to the blood sample which did not contain subtilisin. The phase interface formed after centrifugation is well defined with enzyme-treated samples, but less so if enzyme treatment is omitted.

ILLUSTRATIVE RESULTS

The theory underlying the use of subtilisin to release drugs from tissue entails two assumptions: (1) that the conditions used for tissue degradation are sufficiently mild for non-peptide drugs and their metabolites to survive tissue digestion, and

LIVER OR OTHER TISSUE (10 g)

Add tris base (1.0 M, pH 10.5
(no adjustment); 35 ml) and blend
or homogenize

Add subtilisin Carlsberg (10 mg)
and incubate (55°, 60 min)

AQUEOUS DIGEST
for solvent extraction and fractionation (*Scheme 2*)

Scheme 1   Procedure for enzymic digestion of tissues.

BLOOD OR PLASMA (200 μl)

Add tris buffer (1.0 M, pH up to
10.5; 50 μl) and subtilisin (1 mg),
mix, and incubate (55°, 60 min)

AQUEOUS DIGEST

Add butyl acetate (50 μl), mix, and
centrifuge (2 min, 10,000g)

ORGANIC PHASE

Assay by GC-MS

Scheme 3   Isolation of drugs from blood.

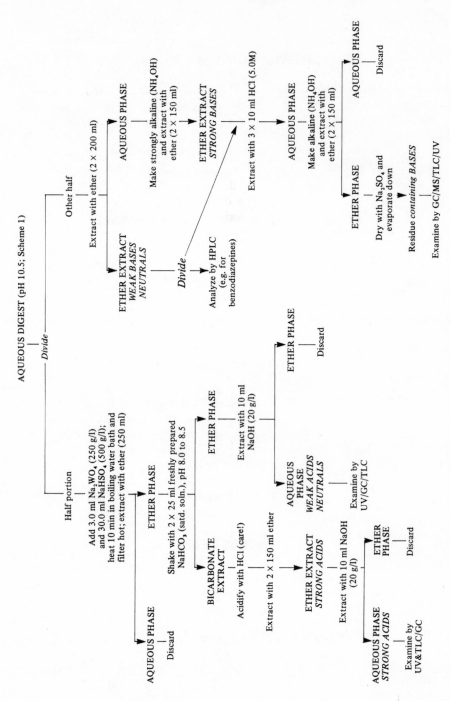

Scheme 2   Separation of different drug classes

(2) that if tissue hydrolysis is taken to completion the drug will be maximally liberated, and subsequently extracted subject to use of the optimum combination of solvent and pH. Hence, enzyme digests may give higher yields of many drugs than with more conventional techniques, as now illustrated for benzodiazepines and pentazocine.

Benzodiazepines
Members of this class of compounds bind tightly to tissue proteins [9,10] and are broken down during heating with strong acids. Prior to the introduction of subtilisin into toxicological analyses the favoured method for analyzing these compounds involved acid hydrolysis of tissues with conversion [7] into the corresponding benzophenones, a drawback being that many benzodiazepines are converted into a common benzophenone, making it difficult to ascertain which parent drug had initially been present. The use of subtilisin followed by reversed-phase HPLC enabled benzodiazepines and their metabolites to be analyzed unchanged without lengthy clean-up procedures [5]. Examples from poisoning cases involving benzodiazepines are given in Table 1.

Table 1 Concentrations of benzodiazepines in liver in cases of human self-poisoning. Concentrations measured using subtilisin are of unchanged drugs. Those for acid hydrolysis column are benzophenone recoveries.

Case	Drug	Drug level, $\mu g/g$	
		Subtilisin	Acid hydrolysis
A	Chlordiazepoxide	3.6	0.2
B	Demoxepam	1.6	0.1
	Desmethylchlordiazepoxide	6.0	–
C	Diazepam	21.5	1.2
	Desmethyldiazepam	1.2	–
D	Oxazepam	9.7	0.4
E	Oxazepam	13.2	2.0
	Desmethyldiazepam	80.6	–
	Flurazepam	0.9	–

Pentazocine

Pentazocine is another acid-labile drug which for a long time presented analysts with difficulties. It was first observed to break down under the mild acid hydrolysis conditions used to cleave glucuronide conjugates [11], to form a hydrated derivative, 2^2-hydroxy-2-(3-methyl, 3-hydroxy-butyl)-5,9-dimethyl-6,7-benzmorphan [12]. Whilst isolating pentazocine from a liver using the acid hydrolysis procedure, in low yield, we observed in the extracts by TLC and GC a second compound which was subsequently identified as the hydrated breakdown product. No breakdown product was formed using enzymic tissue digestion and an increased yield of pentazocine was obtained (Fig. 1).

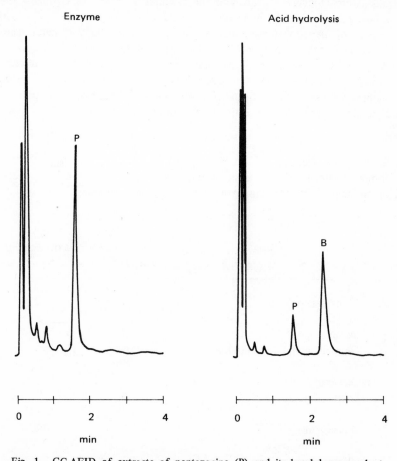

Fig. 1 – GC-AFID of extracts of pentazocine (P) and its breakdown product 2-hydroxy-2-(3-methyl, 3-hydroxy-butyl)-5, 9-dimethyl-6, 7-benzmorphan (B) in liver exracts after subtilisin digestion and acid hydrolysis.
Column: 3% OV-17 on Chromosorb G-HP, 100–200 mesh, 1 m, run at 230° (injection temp. 275°) with N_2 as carrier gas, in a Perkin-Elmer F-33 assembly.

Some typical drug recoveries obtained from liver tissue comparing enzymic and acid hydrolysis are presented in Table 2. The ratio between the results of conventional and enzymic procedures may vary for any given drug depending on the state of the tissue and on the presence or absence of other drugs. Thus, if the tissue is partially or totally putrefied, bacterial and intracellular proteolytic enzymes may have produced autolysis of the tissue, so rendering the conventional procedure relatively efficient.

Table 2 Comparison of drug recoveries obtained from liver after enzymic hydrolysis or acid hydrolysis in cases of self-poisoning.

Drug	Drug level, $\mu g/g$	
	Subtilisin	Acid hydrolysis
Amitriptyline	5.0	2.8
Chlormethiazole	230.0	106.0
Chlorpromazine	29.7	18.0
Pentazocine	15.0	7.0
Pethidine	16.2	11.0
Salicyclic acid	54.3	24.0
Trimipramine	66.8	66.3

ANALYSIS OF DRUGS IN BLOOD

The difference in drug levels measured by direct solvent extraction and enzymic hydrolysis in whole blood or blood plasma (Tables 3 and 4) are less marked than those obtained with liver. The largest differences are associated with compounds which are tightly protein-bound, e.g. phenylbutazone and benzodiazepines. This might be due to the relative lack of cellular compartmentalization in whole blood and plasma.

The tabulated results were obtained using butyl acetate as extracting solvent since this appears to be good for a wide range of compounds, is less dense than the aqueous blood phase thus minimizing contamination of GC syringe needles, and is less volatile than many other solvents commonly used for drug extraction,

Table 3 Drug recoveries obtained from whole blood after enzyme digestion (Scheme 3) or direct solvent extraction (without any hydrolysis).

Compound	Drug level, μg/ml	
	Enzyme	Direct
Amitryptyline	12.3	11.1
Desmethyldiazepam	0.41	0.1
Diphenhydramine	1.8	1.8
Phenylbutazone	5.5	1.0
Salicyclic acid	40.0	20.0
Trimipramine	2.6	2.1
Cholesterol	460.0	155.0

Table 4 Drug recoveries obtained from blood plasma after enzyme digestion or direct solvent extraction.

Compound	Drug level, μg/ml	
	Enzyme	Direct
Desmethyldiazepam	1.0	0.5
Desmethyldiazepam	0.25	0.12
Diazepam	0.53	0.38
Phenytoin	11.8	2.8
Phenytoin	3.8	1.6
Salicylic acid	90.0	61.0
Cholesterol (2 expts.)	537.5, 137.5	140.0, 57.8

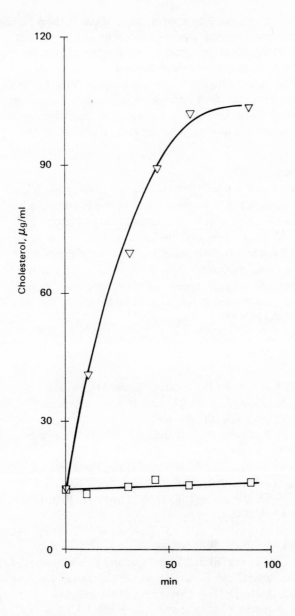

Fig. 2 – Extraction of endogenous cholesterol (free + esterified) from human blood plasma during enzyme incubation (△) and by direct extraction (□) into butyl acetate.
Assays were performed by GC-FID (column as in Fig. 1; 275°).

thereby reducing errors caused by solvent evaporation. In these experiments the effect of enzyme treatment on drug yields may be attributed only to enzymic hydrolysis of proteins, since the presence or absence of enzyme was the only difference between the enzyme tests and controls.

Analysts who employ highly selective solvent systems for optimizing the recovery of one particular drug or class of drugs may not obtain significant increases when subtilisin digestion is incorporated into their existing assays, since either procedure will then yield optimum drug recoveries.

Cholesterol release

We have always noted a significant increase in choloesterol release following enzyme digestion, as in Fig. 2 where in non-enzyme controls the level remained constant with time. This consistent increase in extracted cholesterol after enzyme treatment may serve two useful functions to the drug analyst. It provides a ready means of ascertaining that the enzyme was active during the experiment, this being particularly useful when no significant increase in drug levels is seen between enzyme digests and direct extracts. Secondly, an observable cholesterol peak with our GC conditions connotes that most drugs will have eluted from the column.

References

[1] Dubost, P. & Pascal, S. (1953), *Ann. Pharm. Fr.*, **11**, 615.
[2] Curry, A. S. (1976) *Poison Detection in Human Organs,* 3rd edn, Charles C. Thomas, Springfield, Ill., pp. 106-107.
[3] Stevens, H. M., Owen, P. & Bunker, V. W. (1977) *J. Forens. Sci. Soc.,* **17**, 169-176.
[4] Markland, F. S. Jr. & Smith, E. L. (1975) in *The Enzymes,* Vol. 3 (Boyer, P. D., ed.), Academic Press, New York, pp. 561-608.
[5] Osselton, M. D., Hammond, M. D. & Twitchett, P. J. (1977) *J. Pharm. Pharmac.,* **29**, 460-462.
[6] Osselton, M. D., Shaw, I. C. & Stevens, H. M. (1978) *Analyst,* **103**, 1160-1164.
[7] de Zilva, J. A. F. (1978) in *Blood drugs and Other Analytical Challenges* (Vol. 7, *this series*; Reid, E. *ed.*) Horwood, Chichester, pp. 7-28.
[8] Rutherford, D. M. (1977) *J. Chromatog.,* **137**, 439-448.
[9] Kaplan, S. A., Jack, M. L., Alexander, K. & Weinfeld, R. E. (1973) *J. Pharm. Sci.,* **62**, 1789-1796.
[10] Zingales, I. Z. (1973) *J. Chromatog.,* **75**, 55-78.
[11] El-mazati, A. M. & Way, E. L. (1971) *J. Pharmacol. Exp. Ther.,* **177**, 332-341.
[12] Vaughan, D. P. & Beckett, A. H. (1973) *J. Pharm. Pharmac.,* **25**, 993-995.

#C-2   EXTRACTION AND CLEAN-UP PROCEDURES FOR
        FOODS

N. T. CROSBY, Laboratory of the Government Chemist, London,
SE1 9NQ, U.K.

*Through advanced analytical techniques such as HPLC, GC and GC-MS many
trace constituents in food such as additives and contaminants are now identifiable
down to µg/kg or even below. However, advances in end-detection methods have
not been matched by equivalent advances in sample pre-treatment. Solvent
extraction and distillation are commonly used for the initial separation step,
followed by further purification procedures based on column chromatography,
solvent partition or TLC. For very low concentrations, if the final identification
is to be unequivocal, the clean-up techniques must be elaborate and extensive,
and are often tedious, particularly when many samples have to be analyzed.
Moreover, in multi-stage methods losses will occur at each stage, so that the
overall recovery through the method is often only 50–80%. Further problems in
quantitation arise because some trace contaminants are bound to constituents
of the food and are not readily released into simple extracting solvents.*

*Some progress related to these problems is now described, especially for the
following types of compound:*

Requirement:	Colouring matter in foods, down to 1 mg/kg	Mycotoxins in food down to 0.1 µg/kg	Polynuclear aromatic hydrocarbons (PAH) in foods, down to 1 µg/kg
End-step:	HPLC with 5 µm SAS-Hypersil; methanol/water/ cetrimide; visible range detector	HPLC with 5 µm Lichrosorb SI60; chloroform/water/ acetic acid; UV + fluorescence detector	HPLC with phthalimido-propyltrichlorosilane; fluorescence detector; 10% (v/v) toluene in hexane
Sample preparation:	enzymic digestion release; anionic resin extraction; polyamide column	enzymic digestion release; acetonitrile extraction; solvent partition: TLC	enzymic digestion release; acetonitrile extraction; solvent partition, saponifi- cation; silica gel chromatography

SYNTHETIC FOOD COLOURS

There is growing concern about the safety-in-use of certain synthetic compounds used to colour foodstuffs. This has led to a gradual reduction in the number of compounds permitted in foods in most countries of the world. Hence, methods of analysis for the identification of colours added to foods are well established and many are based on paper [1] or thin layer [2] chromatography. More recently, there has been an interest not only in the particular colours present in food but also in the levels present, and in the presence of degradation products that may be formed by the action of heat or light during processing and storage of the food. Such compounds are generally non-volatile and readily separated, and are measured by HPLC using a spectrophotometric detector. A suitable system for the separation of synthetic water-soluble colours by HPLC has been described by Chudy et al. [3] .

The most difficult step in the measurement of colours in foods is undoubt-edly the initial separation of the colour from the food matrix. Investigation of a number of different extraction procedures has shown that recovery of added dyes is seldom quantitative from proteinaceous or bakery products. For qualitative analysis, small losses in recovery are not serious since the result is merely a slight worsening in the detection limit attainable. However, for quantitative work, such losses must be reduced to a minimum.

In our initial studies, the method of Graichen [4] , in which an anionic ion-exchange resin dissolved in an organic solvent is used to remove the colour from foods, proved to be the most reliable extraction technique, applicable to a wide range of foods. Further clean-up using a polyamide column to remove any natural colours and to concentrate the purified extract was employed. However, the extraction was not complete where irreversible binding between the colouring matter and a food constituent occurred. A supplementary approach involving pre-treatment of the sample with an enzyme or a mixture of enzymes [cf. M. D. Osselton, this vol. –Ed.] , has been developed to overcome this problem [5]. The function of the enzymic digestion step is to break down the food matrix to secure a more rapid and complete release of any bound colouring matter present in the food. Table 1 shows some recovery values for dyes added to foods and analyzed subsequently with, and without, enzyme pre-treatment.

MYCOTOXINS

Toxic mould metabolites (mycotoxins) are produced by the growth of spoilage organisms on various foods and animal feedingstuffs under conditions of high temperature and humidity. Many compounds have already been characterized, although most attention has been paid so far to the aflatoxin group. The chemistry of these compounds has been discussed by Rodricks [6] , and by Heathcote & Hibbert [7]. Mycotoxins are complex molecules containing one or more oxygenated alicyclic rings. In view of the potent carcinogenicity of some

Table 1   Enhanced recovery of colours from food with preliminary enzyme treatment. The enzymic treatment was applied to aqueous dispersions.

Food	Enzyme	Colour material	Graichen method [4]	After digestion with enzyme
			Amount recovered, mg/kg	
Kippers	Papain + lipase	Brown FK	9.4	14.7
Kippers	Papain + lipase	Brown FK	12.7	28.9
Kippers	Papain + lipase	Brown FK	5.4	6.0
Plain cake	Papain	Ponceau 4R	13.8	19.9
		Tartrazine	8.2	8.8
Chocolate sponge	Papain + phospholipase	Chocolate brown HT	5.7	112.0
		Carmoisine	8.7	15.0

aflatoxins in particular, very sensitive analytical methods for their identification and determination are required. The natural fluorescence possessed by aflatoxins and some similar compounds forms the basis of most methods of detection following separation from other naturally occurring fluorescent compounds by TLC or HPLC. Nevertheless, the measurement of minute quantities of such compounds in a complex matrix such as food presents a most difficult analytical challenge.

The application of HPLC with fluorimetric and UV detectors has been described recently by workers at this Laboratory [8,9]. For the aflatoxin group and ochratoxin, the food is extracted with aqueous acetonitrile and this extract is then washed with 2,2,4-trimethylpentane to remove any lipid material. Mycotoxins are then re-extracted into chloroform. Further purification is effected using grooved TLC plates coated with silica gel. After development, fluorescent bands are removed and examined by HPLC. The separation of the aflatoxins and ochratoxin is illustrated in Fig. 1. Detection limits of individual compounds varied from 0.3 to 12.5 μg/kg, with recoveries from foods in the range 50–90% at levels only twice the limit of detection.

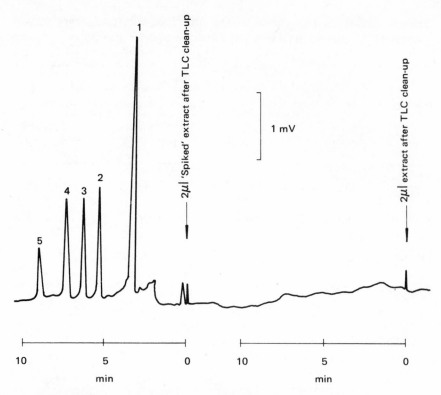

Fig. 1 – HPLC chromatograms of a total diet fat sample before and after addition of mycotoxins.
Fluorescence detector at $\frac{1}{3}$ full sensitivity.
Peak identification: 1. Ochratoxin A (25 ng); 2. aflatoxin B_1 (4.8 ng); 3. aflatoxin B_2 (4.7 ng); 4. aflatoxin G_1 (0.6 ng); 5. aflatoxin G_2 (0.2 ng).

Other mycotoxins such as zearalenone, patulin and penicillic acid that exhibit no, or only weak, fluorescence are more difficult to determine at these very low levels. The loss of the discriminating power of the fluorescence phenomenon requires that an even more rigorous clean-up technique be employed. In a second paper [9], the same authors describe a clean-up technique of the same chloroform extract by chromatography on a silica gel column using (a) diethyl ether and (b) chloroform/methanol, to prepare two separate fractions. Further purification by chromatography on a semi-preparative HPLC alumina column is described, followed by examination and determination on an analytical column by HPLC.

More recent studies [10] have been concerned with the determination of ochratoxin A in pig kidney tissue. Sample pre-treatment consists of enzymic digestion concomitant with dialysis under controlled pH conditions. This technique allows co-extracted substances to be removed without the need for column chromatography. HPLC is used for the subsequent determination stage, and a novel feature of the method is that a post-column derivatization procedure using ammonia is performed to increase the sensitivity of the method 10-fold. Thus, as little as 0.25 ng of ochratoxin A can be detected, equivalent to 1 μg/kg in the original sample. By repetition of the chromatographic run without the post-column ammonia pump in operation, disappearance or diminution of the peak serves as a confirmation of the true identity of ochratoxin A (Fig. 2). This is of particular importance with this type of substrate when only minute levels of the compound are present. A further method of confirmation by chemical derivatization to the methyl ester with BF_3 has been developed [11]. In this case, the disappearance of the ochratoxin A peak following esterification is matched by a parallel appearance of a new peak of similar size at a longer retention time (Fig. 3).

POLYNUCLEAR AROMATIC HYDROCARBONS (PAH)

Contamination of foods by PAH compounds may arise from environmental sources or through treatment or processing at high temperatures, e.g. roasting grilling or smoke-curing. PAH compounds produced during the incomplete combustion of organic matter are numerous, and only minute quantities (if any) can be detected in foods. As not all such compounds are carcinogenic, there is a need for reliable and quantitative methods of analysis that are applicable to foods at the μg/kg level. General methods of analysis involve solubilization of the food with alcoholic KOH followed by solvent extraction, and partition between dimethyl sulphoxide and an aliphatic hydrocarbon. Further clean-up can be achieved by column chromatography on Florisil and by TLC. Solvents used must be purified by re-distillation; hence the complete method is tedious and time-consuming.

Work at this Laboratory has been directed towards the application of HPLC techniques to shorten the analysis time and improve the sensitivity and specificity of the method. A new column-packing material (phthalimidopropylsilane— PPS) has been developed [12] and tested [13]. The performance of this material chemically bonded to silica gel has been compared with ODS-Partisil and some other commercially available column-packing materials. Many of the differences in retention times observed are useful analytically for the identification of PAH compounds in foods. Undoubtedly, the greatest problem again lies in the extraction and clean-up procedures required prior to separation and detection. The method devised by Guerrero *et al.* [14] for shellfish has been applied successfully to other types of food. PAH compounds are extracted by maceration in acetonitrile followed by partition into *iso*-octane to remove lipid material. The residue containing PAH compounds is then saponified and further purified by chromatography on a silica gel column prior to examination by HPLC using a fluorescence detector. By this approach it is possible to identify and measure a number of different PAH compounds down to levels as low as 0.1 μg/kg in foods.

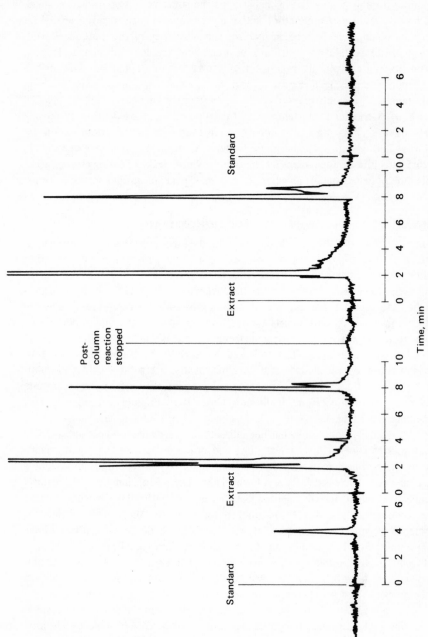

Fig. 2 – HPLC of an ochratoxin A standard (2.5 ng) and a naturally contaminated kidney extract.

C$_{22}$ column: with the post-column reaction stopped working (*left*) or stopped (*right*).

For other conditions see text.

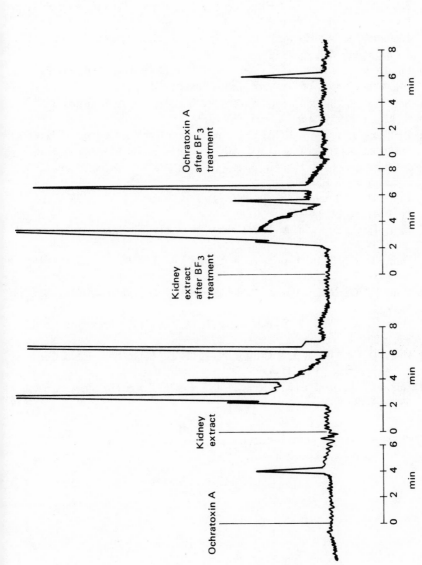

Fig. 3 — HPLC runs as in Fig. 2, but to show effect of esterification (*on right; see text*).

References

[1] Pearson, D. (1976) *The Chemical Analysis of Foods,* 7th edn., Churchill-Livingstone, Edinburgh, p. 53.

[2] Gilhooley, R. A., Hoodless, R. A., Pitman, K. G. & Thomson, J. (1972) *J. Chromatog.,* **72**, 325–332.

[3] Chudy, J., Crosby, N. T. & Patel, I. (1978) *J. Chromatog.,* **154**, 306–312.

[4] Graichen, C. (1975) *J. Ass. Off. Anal. Chem.,* **58**, 278–282.

[5] Boley, N. P., Bunton, N. G., Crosby, N. T., Johnson, A. E. Roper, P., & Somers, L. (1979) *Analyst,* **105**, 589–609.

[6] Rodricks, J. V. (ed.) (1976) *Mycotoxins and other Fungal Related Food Problems,* American Chem. Soc. Adv. in Chem. Series, Washington, D. C.

[7] Heathcote, J. G. & Hibbert, J. R. (1978) *Aflatoxins: Chemical and Biological Aspects,* Elsevier, Amsterdam.

[8] Hunt, D. C., Bourdon, A. T., Wild, P. J. & Crosby, N. T. (1978) *J. Sci. Food Agric.,* **29**, 234–238.

[9] Hunt, D. C., Bourdon, A. T. & Crosby, N. T. (1978) *J. Sci. Food Agric.,* **29**, 239–244.

[10] Hunt, D. C., Philp, Lesley, A. & Crosby, N. T. (1979) *Analyst,* **104**, 1171–1175.

[11] Hunt, C. C., McConnie, B. R. & Crosby, N. T. (1980) *Analyst,* **105**, 89–90.

[12] Hunt, D. C., Wild, P. J. & Crosby, N. T. (1977) *J. Chromatog.,* **130**, 320–323.

[13] Hunt, D. C., Wild, P. J. & Crosby, N. T. (1977) *Rapp. P.-V. Réun. Cons. Int. Explor. Mer.,* **171**, 41–48.

[14] Guerrero, H., Biehl, E. R. & Kenner, C. T. (1976) *J. Ass. Off. Anal. Chem.,* **59**, 989–992.

#C-3 DETECTION OF ADVENTITIOUS NITROGENOUS ORGANICS IN FOOD SAMPLES

C. L. WALTERS, Leatherhead Food Research Association,
Randall's Road, Leatherhead KT22 7RY, U.K.

Early attempts to isolate neutral nitrogenous compounds such as the simple dialkyl, alkaryl or heterocyclic nitrosamines involved the prior separation of acidic and basic components of extracts of biological materials by extraction into alkali and acid respectively. The nitrosamines themselves were then rendered basic by reduction to a hydrazine derivative or by hydrolysis to the parent amine, either of which could be derivatized for separation, detection and quantitation. With the advent of procedures such as GC-MS (high resolution), or GC or HPLC combined with the thermal energy analyzer, that offer specificity for individual N-nitroso compounds, most clean-up procedures take advantage of the volatility of the simpler nitrosamines, usually as azeotropes with water. There is now emphasis on maximum recovery, as distinct from the removal of as many extraneous compounds as possible.

Difficulties persist with the more complex non-volatile nitrosamines and the labile nitrosamides. Derivatization of some nitrosamino acids may allow of GC after complex separation procedures incurring considerable losses. A procedure has been proposed for the group determination of such contaminants directly on a food or other matrix without the necessity for clean-up procedures other than water removal, e.g. by freeze-drying.

Nitrogenous organic compounds with a basic function can generally be separated as a group from extracts of biological matrices in an immiscible solvent by virtue of their preferential solubility in an acid medium. Furthermore, many derivatives of such compounds are available for their individual separation and characterization by such procedures as GC, TLC, HPLC and MS. No similar procedures are, however, common to all neutral nitrogenous contaminants, and hence consideration is now given to the N-nitroso compounds as representatives of organic compounds of nitrogen without a basic function.

IMPORTANCE OF N-NITROSO COMPOUNDS

N-Nitroso derivatives of secondary amines and amides are considered to be of considerable potential importance in foods not only because the majority are carcinogenic—some in as many as twelve different species—but also for the following reasons:

1.   They are most active when administered in repeated small doses as compared with larger, less frequent applications. The former regimen simulates human exposure to such carcinogens.
2.   Small changes in their chemical structures can lead to marked variations in their potency and site of action. N-nitrosocyclohexamethyleneimine is primarily a liver carcinogen; yet N-nitrosocycloheptamethyleneimine, for instance, produces tumours in the lung even when it is administered orally [1].
3.   Both nitrosating and amine or amide precursors are widely distributed in the environment.
4.   Conditions found in foods, etc., match those under which synthesis of N-nitroso compounds can occur, particularly in the presence of naturally occurring stimulators such as nitrate-reducing bacteria, phenols or simple aldehydes.

SEPARATION AND DETECTION OF VOLATILE NITROSAMINES

Initially emphasis was placed upon the removal of as many nitrogenous compounds other than nitrosamines as possible, because of the limited specificity of the detection procedures then available, which included GC-FID, polarography, TLC and photolysis to inorganic nitrite.

   Thus, basic nitrogenous compounds were removed from extracts in immiscible solvents by washing with mineral acid. Care had to be taken that the extracts did not contain nitrite or any compound breaking down to nitrite. Prior use of alkali would have led to complete loss of nitrosamides but should result in removal of oxides of nitrogen. As a result, nitrosamines present without a basic function would remain in solution in the immiscible solvent. Subsequent separation and characterization of these compounds has often been achieved after either:
(i)   reduction with, for instance, lithium aluminium hydride, to asymetrically substituted hydrazines, or
(ii) hydrolysis with, for instance, HBr to the parent amine.

   These procedures were followed by derivatization with, for instance, dansyl chloride, HFB chloride or condensation with 5-nitro-2-hydroxybenzaldehyde or other similar reagent. It is therefore vital to remove all other amines present in the food or other matrix before detection or determination of nitrosamines as amines or hydrazines. The presence of amines in the matrix was controlled in the method of Sen [2] who made use of ninhydrin on TLC with and without UV irradiation. No reaction with ninhydrin occurs with nitrosamines until they have been cleaved by irradiation. It is therefore possible to discriminate ninhydrin-positive spots appearing only after irradiation.

   At one stage, a procedure for the separation and concentration of all N-nitroso compounds as a group was considered, namely the quantitative adsorption of such compounds on activated carbon [3]. However, the efficiency of desorption proved to be very variable, with complete retention of those containing aromatic rings. Workers at the FDA in the U.S.A. proposed the complete digestion of foods in methanolic KOH prior to the extraction of nitrosamines into dichloromethane [4].

With improvements in the specificity of the detection procedures available the necessity for the removal of other compounds became less acute. Concurrently, attention was directed principally towards the simpler volatile dialkyl, alkaryl and heterocyclic nitrosamines which could more readily be separated and characterized and whose biological activity was better known. These potent carcinogens had to be not merely separated but also concentrated so that volatile nitrosamines could be detected at the very low levels considered to be of potential epidemiological importance. One method of concentration developed for such compounds was that of the fractional distillation of an aqueous slurry of homogenate to which was added a small proportion of a hydroxylic solvent such as methanol. With a high reflux ratio, and an efficient spinning band column, volatile nitrosamines present can be concentrated in a narrow-boiling fraction between the methanol and the water, as is illustrated for N-nitrosodiethylamine in Fig. 1.

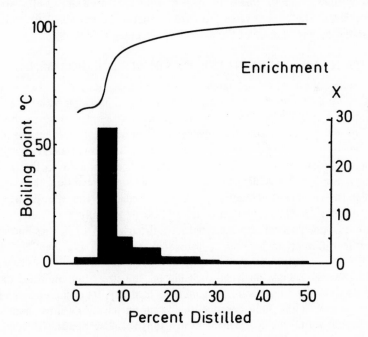

Fig. 1 – Boiling profile in spinning band column of fraction obtained from solution of N-nitrosodiethylamine in water containing 5.0% (v/v) methanol, and enrichments of nitrosamine concentrations obtained.

With the advent of GC linked to high-resolution MS, the need for the selective separation of volatile nitrosamines became even less. Advantage has therefore been taken of the volatility of such compounds in steam during the distillation of aqueous homogenates of foods, etc. Recently, the stratagem of having a layer of mineral oil over the surface of the homogenate has been adopted to facilitate smooth and efficient distillation. In some cases, alkali was added to suppress artifactual nitrosation and to restrict the volatility of any nitrite present as nitrous gases. The practice has been generally discontinued since the finding that nitrosation can take place quite readily under alkaline conditions, particularly from oxides of nitrogen. Where nitrite is present, this can be removed selectively by prior treatment with sulphamic acid.

The extraction of volatile nitrosamines into dichloromethane from the distillates has been facilitated after the addition of NaCl to 10%. The extracts have been dried and concentrated very carefully in a Kuderna-Danish evaporator [#NC(C)-2, *this vol.—Ed.*] placed in a thermostatically controlled water bath and shielded from direct sunlight. Finally, the usual practice has been to add spectroscopically pure hexane to the residue in dichloromethane, so leaving behind the volatile nitrosamines in a small volume of hexane. Thereby an approximately 1000-fold concentration can be achieved, i.e. μg/kg concentrations in foods can be raised to mg/kg, with recoveries of 70–80% at the 10 μg/kg level. The concentrates thus produced can be examined for volatile nitrosamines by one or more of the procedures outlined below.

DETECTION AND DETERMINATION OF VOLATILE NITROSAMINES

For screening purposes, concentrated extracts of foods, etc. can be examined by GC with a Coulson detector operating in the pyrolytic mode or with ECD after oxidation to N-nitramines with peroxytrifluoroacetic acid. GC is the method of choice for the detection and determination of individual volatile nitrosamines in combination either with high-resolution MS or a thermal energy analyzer, both of which can be highly selective towards such compounds. In the former case, each nitrosamine is detected at its characteristic retention time by its molecular (or another characteristic) ion, as recorded by the MS operating at high resolution [6–8]. A resolution of 10,000 is generally considered to be adequate for unequivocal characterization, although even higher resolution is required to differentiate between N-nitrosodimethylamine and the $^{29}Si(CH_3)_3$ ion.

The thermal energy analyzer depends on the detection of nitric oxide cleaved catalytically from each nitrosamine in turn [9]. Its chemiluminescence analyzer module (more fully described below) is extremely sensitive, and N–N bonds are split within the detector far more readily than C–N bonds. Nevertheless, the analyzer is recognized to respond to other types of nitrogenous compounds such as C-nitroso compounds [10] and N-nitramines [11], and probably detects as NO many others derived from nitrite.

SEPARATION AND DETECTION OF NON-VOLATILE
N-NITROSO COMPOUNDS

Most nitrosatable precursors in a food or other matrix, e.g. proline or sarcosine, would give rise to non-volatile N-nitroso compounds. Thus, the polyamine spermidine, which is widely distributed in foods, etc., can form on reaction with nitrite a number of hydroxy N-nitroso derivatives [11a] which are not volatile in steam. Furthermore, many other types of compounds, e.g. the pseudonitrosites of unsaturated lipids, can be potentially formed from nitrite in a biological matrix and would be likely to break down to nitric oxide with thermal catalysis. Nitrogen from nitrite 'labelled' with ^{15}N and used as a food preservative can be located in part in the lipid portion where it has probably reacted to form pseudonitrosites of unsaturated glycerides. Such compounds are nitrosating agents and may be responsible for the lipid localization of most of the N-nitrosopyrrolidine formed during the frying of bacon.

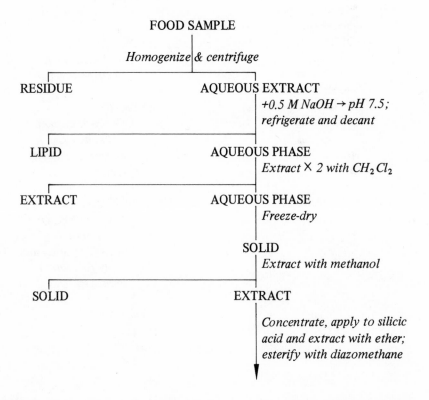

Scheme 1   Separation procedure for N-nitrosamino acids according to Kushnir et al. [12].

Emphasis has hitherto been placed upon the nitrosamino acids insofar as individual non-volatile nitrosamines are concerned. Elaborate procedures have been developed for their separation and detection as methyl esters, e.g. by GC-MS, in the manner of Kushnir *et al.* [12; Scheme 1], Dhont & Van Ingen [13; Scheme 2] and Sen [14; Scheme 3]. Not suprisingly in view of the complexity of the procedures, the recoveries of the N-nitrosamino acids are usually far below theoretical.

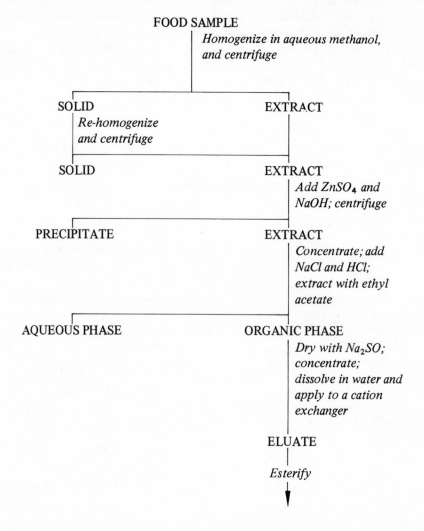

Scheme 2 Separation procedure for N-nitrosamino acids according to Dhont & Van Ingen [13].

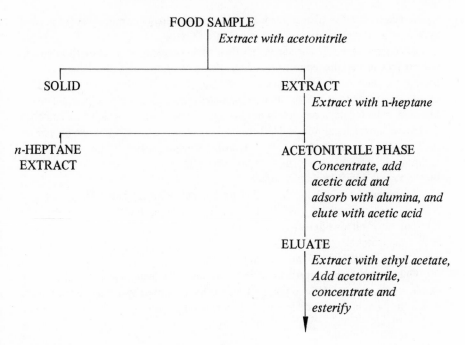

FOOD SAMPLE
Extract with acetonitrile

SOLID   EXTRACT
Extract with n-heptane

n-HEPTANE
EXTRACT

ACETONITRILE PHASE
Concentrate, add
acetic acid and
adsorb with alumina, and
elute with acetic acid

ELUATE
Extract with ethyl acetate,
Add acetonitrile,
concentrate and
esterify

Scheme 3   Separation procedure for *N*-nitrosamino acids according to Sen *et al.* [14].

A procedure has been developed for the determination of non-volatile *N*-nitroso compounds as a group, including those not extractable into an organic solvent, in a food or other matrix [15]. It depends on the determination of nitric oxide formed from the denitrosation of *N*-nitroso compounds which is reacted with ozone to form activated nitrogen dioxide. As this decays to its ground state, light is emitted in the far-visible and near-infrared regions. The light from most other similar reactions peaks at much lower wavelengths and is discriminated through the use of a red filter.

Thus, a freeze-dried-matrix or an extract thereof is suspended or dissolved in a refluxing solvent under an atmosphere of nitrogen. The addition of acetic acid produces NO from any nitrite present, which can thereby be determined down to much lower concentrations than by the more conventional spectrophotometric procedures. The subsequent addition of HBr leads to a further evolution of NO from any *N*-nitroso compounds present. Finally, the inclusion of titanous chloride reduces nitrate present to a form, probably nitrite, that yields NO on acidification. Thus, the method permits the determination, in sequence, of nitrite, *N*-nitroso compounds as a group and nitrate directly on many food matrices. Its application to a sample of codfish for instance gave

μg/kg values 2.0 for nitrite as NO_2^-, 50 for N-nitroso compounds (expressed as N-nitroso-pyrrolidine), and 530 for nitrate as NO_3^-.

Attempts have been made to confirm as N-nitroso compounds those giving rise to NO under the conditions of the test procedure. This has been done by noting an increase in the response resulting from the deliberate nitrosation of the food or other matrix with, e.g., alkyl nitrite. Secondly, transnitrosation has been affected to a secondary amine, e.g. morpholine, of which the N-nitroso derivative can be separated and determined selectively by an alternative procedure such as GC-MS (high resolution). Thirdly, the properties of the many other compounds potentially formed from nitrite in a biological matrix are such that the great majority would either:

(i) decompose to NO in the refluxing solvent without the addition of any reagent, this being the case with, e.g., the pseudonitrosites of unsaturated lipids or the S-nitroso-derivatives of thiols; or

(ii) remain stable so far as NO evolution is concerned throughout the determinations of nitrite and N-nitroso compounds.

Only nitrolic acids and S-nitrothiols, neither of which have been reported in edible foods, require HBr to yield NO as for nitrosamines and nitrosamides from which they could not be differentiated by this procedure.

References

[1] Lijinski, W. & Taylor, H. W. (1977) *Nitrosamines and their Precursors in Food, the Origins of Human Cancer*, Cold Spring Harbor Laboratory, pp. 1579-1590.

[2] Sen, N. P. & Dalpe, C. (1972) *Analyst*, **97**, 216-220.

[3] Walters, C. L., Johnson, E. M. & Ray, N. (1970) *Analyst*, **95**, 485-489.

[4] Fazio, T., Howard, J. W. & White, R. (1972) *N-Nitroso Compounds— Analysis and Formation*, IARC Scientific Publications No. 3, IARC, Lyon, France, pp. 16-24.

[5] Walters, C. L. (1971) *Lab. Practice*, **20**, 574-578.

[6] Gough, T. A. & Webb, K. S. (1972) *J. Chromatog.*, **64**, 201-210.

[7] Gough, T. A. & Webb, K. S. (1973) *J. Chromatog.*, **79**, 57-63.

[8] Crathorne, B., Edwards, M. W., Jones, N. R., Walters, C. L. & Woolford, G. (1975) *J. Chromatog.*, **115**, 213-217.

[9] Fine, D. H., Rufeh, F., Lieb, D. & Rounbehler, D. P. (1975) *Anal. Chem.*, **47**, 1188-1191.

[10] Stephany, R. W. & Schuller, P. L. (1976) *Proc. 2nd Int. Symp. Nitrite Meat Prod.*, Zeist, Pudoc, Wageningen, pp. 249-255.

[11] Hotchkiss, J. H., Barbour, J. F., Libbey, L. M. & Scanlan, R. A. (1978) *J. Agric. Food Chem.*, **26**, 884-887.

[11a] Scanlan, R. A. (1974) *C.R.C. Critical Reviews in Food Technology*, **5**, 357-402.

[12] Kushnir, L., Feinberg, J. J., Pensabene, J. W., Piotrowski, E. G., Fiddler, W. & Wasserman, A. E. (1975) *J. Food Sci.,* **40**, 427–428.

[13] Dhont, J. H. & Van Ingen, C. (1976) in *Environmental N-nitroso Compounds —Analysis and Formation,* IARC Scienific Publications No. 14, IARC, Lyon, France, pp. 355–360.

[14] Sen, N. P., Donaldson, B. A., Seamen, S., Iyengar, J. R. & Miles, W. F. (1978) in *Environmental Aspects of N-nitroso Compounds,* IARC Scientific Publications No. 19, IARC, Lyon, France, pp. 373–393.

[15] Walters, C. L., Downes, M. J., Edwards, M. W. & Smith, P. L. R. (1978) *Analyst,* **103**, 1127–1133.

#C-4   WORK-UP OF CROPS AND SOILS FOR RESIDUES OF
        PESTICIDES AND THEIR METABOLITES

A. P. WOODBRIDGE and E. H. McKERRELL, Shell Research Ltd.,
Shell Biosciences Laboratory, Sittingbourne Research Centre,
Sittingbourne, Kent ME9 8AG, U.K.

The procedures for sample handling, extraction and clean-up must ensure minimal loss of residues from the time of sampling to the time of analysis. Samples are usually deep-frozen for transit from remote field locations and are stored deep-frozen before extraction. The method of extraction is determined by the type of sample and can include homogenization, tumbling and steam-distillation. The efficiency of the chosen procedure, which must be as near quantitative as possible, is confirmed by extractability studies using radiolabelled pesticides and field-aged samples.

Clean-up of extracts is almost always necessary, and techniques based on HPLC using microparticulate bonded-phase packings are now supplementing and replacing traditional techniques such as classical column chromatography and solvent-solvent partition.

Pesticides are introduced into the environment for a variety of purposes, notably for crop protection. By definition, as insecticides, herbicides and fungicides, they are materials with high biological activity. It is therefore necessary to monitor the fate of these materials and their breakdown products in the environment following their agricultural use.

Such residue and metabolism studies are carried out from the early development stages of a pesticide, and at Shell Biosciences Laboratory (SBL) the fate of materials in a wide variety of crops, crop products, soil and water is examined. The analytical data from these studies and the interpretation of their significance are required for registration of the pesticide before marketing of the product can begin. The analytical techniques needed to obtain the data stem from the different studies carried out.

The nature of the products formed as the pesticide breaks down is ascertained by metabolism experiments, carried out in the laboratory, in glasshouses and in outdoor enclosures. Extensive use is made of radiolabelled compounds to allow the small quantities of metabolites to be detected for isolation and identification. Fig. 1 shows the metabolic changes that could be undergone by cypermethrin [1].

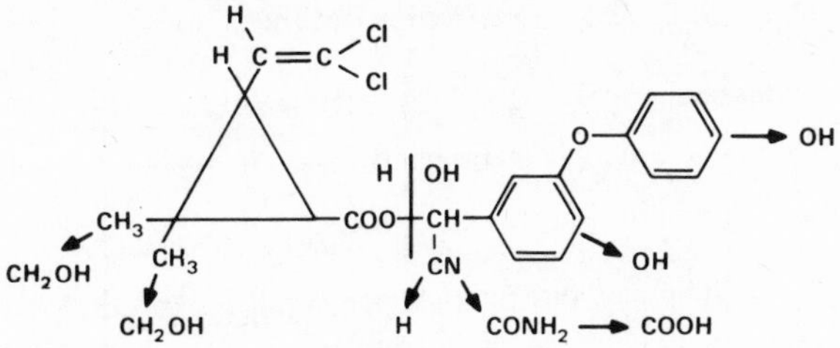

Fig. 1 – Structure of cypermethrin (WL 43467; NRDC 149), with metabolic changes indicated.

Residue experiments are designed to give such information as: (a) the rate of breakdown and distribution of the pesticide in the environment; (b) the concentration of material, including key metabolites, to be found at harvest and in soil following application; (c) the fate of residues in food during processing. Trials are laid down to examine these various factors under the different conditions to be found in normal agricultural practice. The trial plots range in size from as little as $100 \, m^2$ to 1 or 2 ha. Modern pesticides are applied at dosage rates normally in the range 0.1 to 1.0 ai/ha but sometimes at rates < 0.1 kg ai/ha [ai = active ingredient]. The activity of the compounds is so high that they are applied sparsely.

From all these studies representative samples are taken which usually range from a few hundred grams in some metabolism studies to several kg in residue experiments. These samples contain the very small quantity of material to be identified and determined, often in sub-mg/kg (ppm) concentrations. This concentration is the residue concentration for the material to be determined, and defines the magnitude of the problem facing the residue analyst.

Residue analytical methods
The development of a method frequently requires several stages (Scheme 1). Almost all samples, except possibly some aqueous ones such as natural water, require sample preparation and extraction of the residue. This must be done as quantitatively as possible, with minimum extraction of other compounds that may interfere with the determination step. For pesticide residues, samples are normally extracted with the least polar solvent capable of extracting the residue efficiently, as verified by extractability experiments using field-aged samples and samples fortified with radiolabelled compounds.

With some residues direct determination using the initial extract is possible. Thus, for organophosphorus insecticides in crops, GC with phosphorus-selective flame photometric detection can often be used without clean-up of the initial extract. Such cases are the exception rather than the rule; the quantity of co-extractives is almost always too high at this stage and most of the

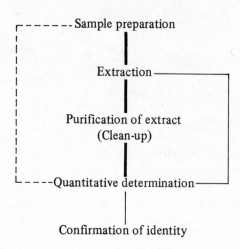

Scheme 1. Key steps in a residue analytical method. The purification step is not always obligatory. [*Editor's Note*: Commonly, as in other contributions to this vol., term 'sample preparation' also embraces extraction and purification.]

available detectors are insufficiently selective. In general, therefore, clean-up of the extract is required, and this must achieve the necessary balance between the selectivity of the detector and the degree of clean-up needed, while keeping the procedure as short and as quantitative as possible.

Preparation and extraction of samples for residue analysis are carried out using a variety of techniques. In the difficult area of clean-up where such methods as solvent partition and adsorption column chromatography are traditionally used, the modern technique of HPLC is now making an impact. Our experience in these areas of residue analysis will now be discussed in more detail.

SAMPLE TREATMENT AND EXTRACTION

It is essential that representative residues samples from the field be handled, from the time they are collected to the time the residue is determined, in such a way that the results of the analysis relate quantitatively to the residue concentration in the field. In this respect, sample handling, transport, storage, preparation and extraction form as important a part of the analytical procedure as the subsequent clean-up and determination stages. Residue method development studies must examine these initial steps in the procedure with equal importance, to answer the following questions.

(a) How should the sample be transported following collection and be stored before analysis?

(b) What processing is required to provide a suitable sub-sample for extraction?

(c) How should this sub-sample be extracted to ensure quantitative removal of the residue?

Sample handling and storage

Following collection of the crop or soil from the field it is most important to prevent any further loss or degradation of the residue in the sample. When the trial site is remote from the laboratory or the interval between treatment and harvest is short compared with the transit time, loss of residues is normally prevented by deep-freezing the sample with solid carbon dioxide for transit to the laboratory. On arrival, samples are stored deep frozen at about $-20°$.

To confirm that no loss of residue occurs even when the sample is deep-frozen, untreated samples of selected crops and soils are fortified by syringe with pesticide or metabolite to give concentrations similar to those expected for field-treated samples, usually in the range 0.1–1.0 mg/kg. These samples are also stored deep-frozen, usually for periods up to about 1 year, before analysis to check the stability of the residue. To help ensure quantitative recovery of the pesticide, samples are fortified and stored individually in glass containers that can be washed out at the time of analysis. Usually there is no loss of residues from deep-frozen samples, as exemplified in Table 1. However, in rare cases, where loss of pesticide occurs even under these conditions, then samples must be extracted within the appropriate time period. This is the case for tobacco if it is dried following treatment with cypermethrin.

When the interval between treatment and harvest is long (several months in the case of some herbicides), samples can be transported to the laboratory at ambient temperature if the transit time is only a few days. Some storage-stability experiments at ambient temperature should be carried out to confirm that no loss of residue occurs.

Sample preparation

[*Editor's note*; this term is more commonly used to signify *all* pre-measurement steps; the book title 'Sample Handling' likewise has this connotation.]

A representative sub-sample, often in the range 10–100 g, has to be taken from the field sample for extraction of the residue. Sub-sampling and extraction are facilitated if the sample is first finely divided or made homogenous. With the procedures currently used in our laboratory it is convenient to divide different sample types into groups requiring different treatment. One classification is as follows:

Cereal grain and rice: deep-freeze by mixing with solid carbon dioxide and grind to powder.

Soft fruits, tomatoes: Homogenize by mincing frozen or semi-frozen crop.

Vegetables, top fruit, citrus and root crops: dice or chop with minimum bruising and juice separation.

Table 1 Deep-freeze storage stability of cypermethrin on fruit and vegetables

Matrix	Variety	WL 43467 added, mg/kg	Time of storage at $-18°$ weeks	Recovery, %
Apple	Golden Delicious	0.5*	20	100
	Cox	0.2	4	110
			52	105
	Anne Elizabeth	0.2	4	100
			52	100
Bean	Findor	1.0	1	100
Grapes	Barbera (black)	0.5	9	100
	Carignan (black)	0.5	9	90
	Carignan (white)	0.5	9	90
	Carignan (white)	0.2	34	105
Lettuce	Val D'Orge	1.0	3	105
Maize grain	—	0.2	52	110
Maize silage	—	0.2	4	95
Peach	Royal Gold	1.0*	1	105
Pear	Cedrata Romana	0.5*	22	105
	Packham	0.5	14	105
Phaseolus bean	ICA/PJAO	0.6	16	90
Savoy	Vorbote	1.0	17	95
Sprouts	Jade Cross	1.0	7	85
Tobacco	Garcia (dry)	0.6	26	45
	Siroga (green)	0.5	54	95

* Part of the applied pyrethroid was injected up to 0.5 cm into the surface of the fruit.

Cotton seed and oil seed crops: deep-freeze with solid carbon dioxide/liquid nitrogen and grind to powder.
Straw: chop into short lengths and mill to a coarse powder.
Soil: separate stones and residual plant material, break up lumps and riffle or cone and quarter.

Some crops require additional processing; thus oranges may be peeled for separate analysis of the pulp and peel, and cherries are de-stoned.

The most important factors in the processing stage are keeping samples cold, particularly when grinding grain samples, to prevent loss of residues by local overheating, and avoiding the bruising of aqueous crop samples to prevent separation of juice which leads to sample inhomogeneity.

Sample extraction

Solvent extraction is used to remove the pesticide residue from the crop or soil matrix. The key to this step is quantitative extraction of the residue with reasonable speed. If the method used removes the minimum amount of co-extracted material from the matrix the subsequent stages of clean-up and determination are correspondingly easier. Some of the more widely used modes of extraction for crops and soils are given in Table 2. Most samples, however, can be extracted by homogenization or tumbling.

Table 2   Techniques used in the extraction of crop and soil samples

Extraction technique	Sample type	Time
Blending/homogenization (maceration)	Aqueous crops incl. fruit and vegetables, oil seeds	5–10 min
Tumbling in a closed jar	Soil, grain, straw	2–5 h
Soxhlet extraction	Dry crops, grain, straw, soil	Up to 24 h
Steam distillation	General	Up to 2 h

To confirm that the selected solvent extracts the residue efficiently, extractability experiments have to be carried out. The most conclusive and in our opinion important of these are studies employing radiolabelled compounds, although supporting information is obtained by examination of field-aged samples.

In our laboratory two types of radiolabelled experiments are normally carried out:

(a) The crop is grown in the glasshouse and is treated, at intervals during several weeks up to the time of harvest, with the radiolabelled pesticide or metabolite of interest. After harvest the sample is processed and extracted using the selected technique. The extract is examined for unchanged compound, and a portion of the undissolved matrix is combusted, the released CO_2 being trapped for scintillation counting to check for the absence of radiolabel. Should any radioactivity be detected than the remaining residuum is re-extracted using more vigorous conditions and the new extract and residuum are analyzed. In general, the solvent initially chosen is the least polar one expected to extract the residue, progressing to more polar solvents if the extraction is inefficient. The results of an extractability study carried out with cypermethrin, given in Table 3 show that acetone + petroleum spirit is a good choice of solvent for efficient extraction of the pesticide from cotton seed.

(b) Samples fortified with radiolabelled compound, using the procedures described for deep-freeze storage stability experiments, are stored at ambient temperature for several weeks. They are then treated as described for glasshouse crops. Soil extractability experiments are usually carried out in this way.

In this type of study it is important that the pesticide be stored in contact with the crop or soil for a minimum of several days. This simulates field conditions and ensures that any 'ageing' of the residue on the surface does not prevent efficient extraction with the solvent. Addition of the pesticide to the matrix for only a few hours followed by extraction *does not constitute a real extractability experiment*; it provides only a recovery experiment to check subsequent steps in the analytical procedure.

Useful supporting data can be obtained by extracting field-trial samples, where the residue (not radiolablelled) has been aged under natural conditions, using similar procedures to those for extractabilities with radiolabelled compounds. For cypermethrin in field-aged soil, acetone + petroleum spirit (1:1 by vol.) was found to be as efficient as water + acetonitrile (3:7 by vol.) for extraction of the residue; but the former solvent mixture is preferred because of its lower polarity.

Many pesticides are applied to a range of crops, and it is useful to classify samples for extractability studies into groups depending on matrix type:

Aqueous crops: fruit, vegetables, root crops.
Dry crops: cereal grain and straw, rice, tea.
Oily crops: cotton, soya and rape seed.
Soil: loam, sandy loam, clay loam and chalky loam.

Table 3 Extractability results for cypermethrin on cotton seed grown in the glasshouse. Analysis of a control sample grown at the same time as the treated material showed no detectable residue.

Extraction stage	Cases + linters			Kernels		
	Extract		Residuum	Extract		Residuum
	[14C]-label	WL 43467	[14C]-label	[14C]-label	WL 43467	[14C]-label
Before extraction	–	–	0.47	–	–	0.013
Acetone + petroleum spirit 1 + 1 (v/v) extraction (homogenization, 10 min)	0.39	0.39*	0.026	0.01	0.002	0.005
Water + acetonitrile 3 + 7 (v/v) extraction (homogenization, 10 min)	0.019	0.015	–	0.012	<0.001	–
Methanol + chloroform 7 + 3 (v/v) extraction (Soxhlet, 16 hrs)	0.001	–	0.05	<0.001	–	0.002

Residue mg/kg

* Result confirmed by GC/ECD determination.

Normally extractability studies would be carried out using representative sample types from each of the groups, and cotton seed, lettuce, wheat or barley and 2 or 3 soil types are good examples.

Two important practical points to remember at the extraction stage are the use of desiccants and the surface activity of the matrix. When aqueous crops are extracted with non-aqueous solvents such as petroleum spirit, petroleum spirit + acetone or isopropanol mixtures, dichloromethane or ethyl acetate, two immiscible layers will be fomed unless an excess of a drying agent such as anhydrous sodium sulphate is added to absorb the water. With dry materials such as tea, cured tobacco, or soil of low water content ($<10\%$ m/m), the surface activity can be high, and dampening of the sample with water 1-2 h before extraction may be essential before the residue can be removed efficiently with a non-aqueous solvent. The effect of dampening tea samples treated with BIDRIN (dicrotophos) on the extractability of the residue using chloroform is illustrated in Table 4. In the case of dry tea, less than 5% of the residue was extracted and a totally misleading set of data was obtained. Such data are worse than no data at all.

[*Editor's note:* later in this vol, M. J. Saxby summarizes procedures for extraction, and also for clean-up (cf. below).]

Table 4 Extraction of dry and dampened tea with chloroform for BIDRIN (dicrotophos) residues. Samples were mixed with anhydrous sodium sulphate and were tumbled with chloroform. Dampened samples had an equal mass of water added 1-2 h before extraction. Recoveries of BIDRIN added immediately before extraction were in the range 80-100% for both dry and dampened tea.

Sample code	Residues of BIDRIN found, mg/kg	
	Dry tea	Dampened tea
B1	0.3	8
B2	0.6	23
B3	0.1	4
B4	0.6	19

CLEAN-UP OF PESTICIDE RESIDUES IN EXTRACTS

Amongst the several key steps in residue analytical procedures (Scheme 1), clean-up is especially critical. Of the several techniques, available solvent partition and column chromatography are the most widely used. The technique of sweep co-distillation for clean-up of extracts is mentioned by M. J. Saxby #NC(C)-2, [*this vol.*]. Partition, when required, is normally used for an initial bulk clean-up of the extract, as in removing fats and oils during the analysis of oil-seed crops. Column clean-up can be used when the total concentration of solutes in the extract is not so high that the capacity of the column packing is exceeded. Such clean-up gives further general purification but is most important for the more difficult separation of the residue from co-extractives of a similar nature.

Partition procedures

Partition procedures are well known and are discussed by K. H. Dudley [#D-3, *this vol.*]. Those based on petroleum spirit find wide application. If radiolabelled compound is available, a valuable technique to rapidly calibrate a range of partition systems is to add it to equal volumes of two phases, to shake thoroughly and to radio-count an aliquot of each phase after partition. If calibrated tubes are used for the partition, relative changes in the volumes of the phases, due to solubility, can be taken into account. An additional advantage of this method is that any effect of crop co-extractives on the partition coefficient can be monitored before subsequent clean-up and determination steps have been worked out.

Chromatographic clean-up techniques

Conventional column clean-up techniques are based on liquid-solid (adsorption) and liquid–liquid (partition) chromatography. A third technique known as size exclusion chromatography (sometimes referred to as gel-permeation chroma-tography) is also employed, particularly for the removal of high mol. wt. species.

In adsorption chromatography such packings as aluminas of various pH ranges, silicas and Florisil are most commonly employed, whereas in partition chromatography, polar (normal-phase) and non-polar (reversed-phase) liquids are used as stationary phases on suitable supports such as fire-brick or Gas Chrom Q. However, all these materials have disadvantages:

(a) With the adsorbents, the characterization of the surface and the mechanism of interaction with the solute tend to be uncertain, and for the chromatography of pesticides in nanogram quantities these materials can show variations, in particular from batch to batch, in efficiency, performance and recovery of pesticide.

(b) Partition columns of the type described are often difficult to prepare and use, because they rely on the immiscibility of liquids of which few satisfactory and easily used combinations exist.

(c) Clean-up columns prepared with these conventional packings often exhibit relatively low efficiencies that can largely be attributed to the relatively large size and wide distribution of particle size range present.

(d) Such columns are not usually capable of being readily regenerated with the result that they tend to be used only once.

On the other hand, these packings have the advantage of being inexpensive and can be readily packed into columns that are usually made of glass.

Many of the aforementioned disadvantages can be overcome through the use of packings (usually silica-based) onto the surface of which an organic moiety has been chemically bonded. Such materials have several potential advantages:

(a) The surface is characterized by a known bonded-phase resulting in a controlled activity.

(b) The stationary phase is permanently bonded to the surface (often stable within the pH range 2–8), resulting in the safe use with a wide range of solvents.

(c) With experience these packings can be reproducibly prepared, and a limited range is now commercially available.

(d) Columns packed with these materials can be readily regenerated; hence they can be used many times and therefore offer scope for automation.

(e) A variety of functional groups are available for both normal-phase (e.g. -amino, -cyano) and reversed-phase (e.g. -octadecyl, -octyl, -phenyl) modes of operation.

By using these bonded phases one can achieve high selectivity without sacrificing sample capacity.

Improved efficiency is achieved by using microparticulate packings (10 μm) of narrow size distribution; but these are more expensive and can be more difficult to pack into columns. However, their advantages far outweigh the disadvantages. HPLC is making a major contribution to pesticide metabolism studies, to clean-up of residues in extracts, and to the actual determination. We have incorporated clean-up steps based on HPLC techniques in a number of our recent residue analysis methods, as will now be discussed.

Sep-Pak cartridges

Sep-Pak cartridges are made of high-density polyethylene packed with liquid chromatographic separating materials ($\sim$ 50 μm mean particle size) optimized for sample preparation. At present two types of cartridge, Sep-Pak silica and Sep-Pak C18, are commercially available from Waters Associates, who describe their use [2]. Each cartridge is self-contained and has a small bed of packing (10 mm o.d.) 25 mm long in the case of silica and 10 mm long in the case of C18. Sep-Pak cartridges are employed as a pre-column/concentration step in several of our current residue methods. In practice two cartridges are linked together *via* a short length of glass tube in order to obtain increased sample capacity.

The cartridges are first washed with either 5–10 ml each of methanol and water (Sep-Pak C18) or 5–10 ml each of diethyl ether and hexane (Sep-Pak silica) prior to calibration. The cartridges are calibrated in one of two ways:

(a) the compound of interest is eluted at or near the solvent front such that it is contained in the first 5 ml of eluate;

(b) the eluting solvent is chosen such that 4–10 ml of the mobile phase can be discarded, and the compound of interest can then be eluted within the next 10 ml of solvent without recourse to changing the composition of the eluent.

The former approach does not require detailed calibration since there is no discard volume, and this has been most widely applied with the Sep-Pak C18 cartridges in the following manner. The extract, normally after solvent partition, is dissolved in a solvent in which it is very soluble, commonly methanol. A minimum volume of solvent is used, and up to 0.5 ml methanol will completely dissolve most crop extracts. At this stage the concentration of the extract will be very high. If 50 g of crop had been extracted the crop-to-solvent ratio would be 100 g of crop equivalent in 1 ml of solvent. This concentrate is then transferred to the Sep-Pak C18 cartridge by syringe.

Water + acetonitrile mixtures can be selected such that the fraction containing the residue elutes in the region of the solvent front. With crop extracts the bulk of the co-extractives tends to be held back on the cartridge while the residues are quantitatively recovered. The collected fraction, normally 4–6 ml volume, is either reduced in volume before further clean-up by reversed-phase HPLC, or partitioned into an organic solvent for further clean-up by normal-phase HPLC. Occasionally no further clean-up is required before determination.

HPLC clean-up

The HPLC clean-up is carried out using stainless steel columns (20 cm × 10 mm i.d.) packed with 8–10 g of 10 μm packing. The packings most frequently used in our laboratory are LiChrosorb-NH$_2$ and LiChrosorb-RP18 or –RP8. A volume of 0.5 or 1 ml is injected by a syringe loading injection valve, e.g. Rheodyne Model 7120. The crop-to-solvent ratio required for this stage of the procedure is dependent on the sensitivity of the determination stage.

Table 5 gives recovery data, illustrating the effects of crop-to-solvent ratio, co-solvent and volume of injection during the reversed-phase HPLC clean-up of lettuce extracts, for WL 47133, viz. the amide metabolite (cf. Fig. 1) of cyper-methrin, α-amido-3-phenoxybenzyl-2,2-dimethyl-3-(2,2-dichloro-vinyl)cyclo-propanecarboxylate. The column was first calibrated by injecting 1.0 ml of a standard solution of the amide WL 47133 dissolved in the mobile phase. From the resulting UV pattern the 'discard' and 'collect' volumes were calculated. Lettuce extracts fortified with [^{14}C]-WL 47133 were injected onto the column, with variation of the crop-to-solvent ratio, solvent, and volume of injection.

Table 5   Recovery of [^{14}C]-WL 47133 from HPLC clean-up of typical lettuce extracts on LiChrosorb RP18. For conditions, see text; the flow rate was 4 ml/min, with water + acetonitrile (1 : 4 by vol.) as mobile phase.

Whether Sep-Pak C18 cartridge used initially	Crop : solvent ratio (g : ml)	Solvent (injection medium)	Injection vol., μl	% recovery of ^{14}C label
No	1:1	methanol	1000	93
No	5:1	methanol	1000	42
No	5:1	methanol	500	95
No	5:1*	mobile phase	1000	93
Yes	25:1*	mobile phase	1000	95

* Maximum concentration attained without precipitation.

The following conclusions emerged:

(a) The low (40%) recovery obtained, when the crop-to-solvent ratio was increased using methanol as solvent, was due to a shift in the retention time of WL 47133.

(b) When the sample was dissolved in the mobile phase (water + acetonitrile) at the same crop-to-solvent ratio, no such shift in retention time occurred and a good recovery was obtained.

(c) Use of a Sep-Pak cartridge as a pre-HPLC clean-up enabled the crop-to-solvent ratio to be increased.

A wash cycle is often used after each clean-up in order to regenerate the column for re-use. The clean-up sequence shown in Scheme 2 is proving very useful in our current work.

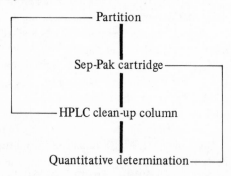

Scheme 2. Key steps in the purification (clean-up) of an extract

Analysis of crops using combined clean-up techniques
Schemes 3 & 4 outline the analysis of crops for WL 47133 and BARNON
(flamprop-isopropyl, a wild oat herbicide) respectively. Fig. 2 shows a chromatogram obtained for WL 47133 recovered from lettuce. This example illustrates

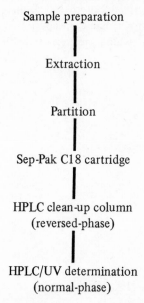

Scheme 3.   Scheme for the analysis of crops for WL 47133.

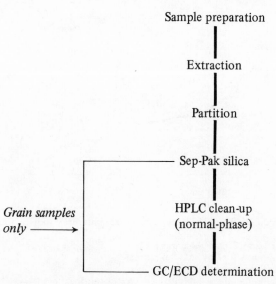

Scheme 4. Scheme for the analysis of wheat grain and straw for BARNON

the use of a Sep-Pak C18 cartridge pre-clean-up step in conjunction with reversed-phase HPLC clean-up followed by normal-phase HPLC determination.

Fig. 3 shows chromatograms obtained for BARNON recovered from wheat grain, and illustrates the adequacy of clean-up using Sep-Pak silica cartridges without the need for further clean-up by normal-phase HPLC although the latter does give further improvement. In contrast further clean-up by HPLC is needed for wheat straw following a pre-clean-up with Sep-Pak cartridges (Fig. 4).

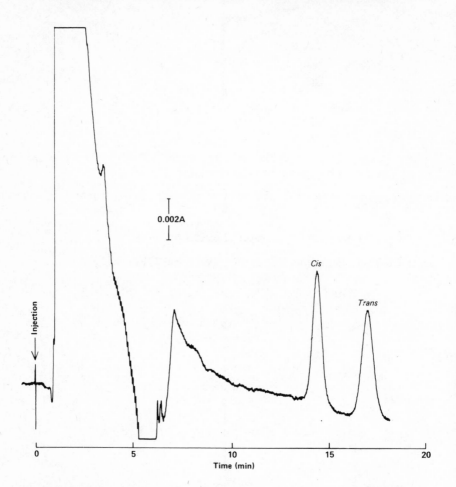

Fig. 2 – HPLC chromatogram of WL 47133 (*cis/trans*-isomer mixture) added to lettuce.
Column: Partisil–10 PAC (10 μm), 20 cm × 4.5 mm i.d.
Mobile phase: methanol + hexane (3:97 by vol.) 2.0 ml/min.
Detection: by Cecil CE 212 UV monitor 210 nm.
Volume injected: 20μl in dichloromethane equivalent to 0.5 g of starting material, viz. lettuce spiked with WL 47133, 0.5 mg/kg.

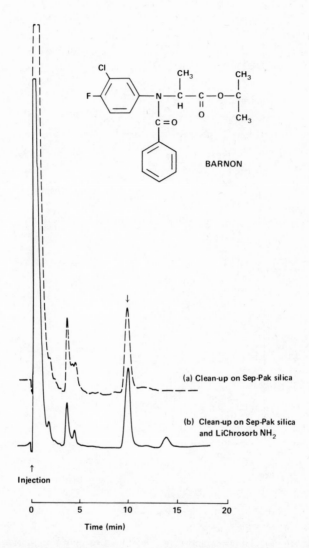

Fig. 3 – GC traces illustrating the clean-up of BARNON in wheat grain extracts.
Column: OV-225, 2% on Gas Chrom Q (100/120 mesh), 0.90 m × 4.0 mm i.d. at
200°.
Carrier gas: nitrogen, 45 ml/min.
Detector: electron capture (^{63} Ni).
Sample: extract equivalent to 2.5 mg of grain fortified with 0.1 mg/kg of BARNON.

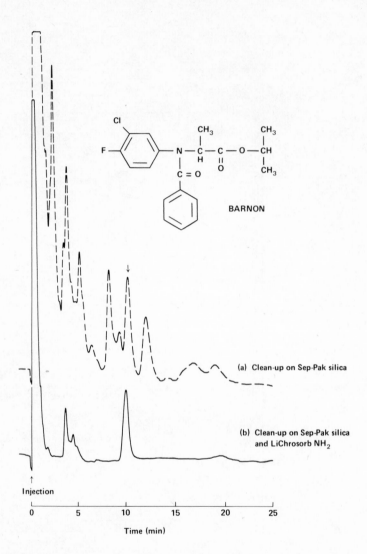

Fig. 4 – GC chromatograms illustrating the clean-up of BARNON in wheat straw extracts
Conditions as in Fig. 3.
Sample: extract equivalent to 2.5 mg of straw fortified with 0.1 mg/kg BARNON.

Vacuum sublimation

An additional clean-up technique that has proved valuable in conjunction with the chromatographic procedures described is that of vacuum sublimation. The technique is based on an apparatus constructed in our own laboratory (Fig. 5). Sample is deposited on a glass sinter, and the pesticide residue is volatilized by heating under vacuum to between 80 and 195° depending on the compound. A nitrogen carrier gas flowing at 15–40 ml per min carries the residue off the sinter, and during 10 min the pesticide is collected quantitatively on three cooler sinter traps situated above the heated one. The pesticide is desorbed from the trapping sinters with petroleum spirit and the extract, which may or may not require additional clean-up by column chromatography, is used for determination of the residue. In the complete apparatus there are 6 sinter tubes concentrically arranged so that up to six samples may be run simultaneously.

A number of pesticides in different crops and soils have been cleaned up by vacuum sublimation; but the technique has proved most useful for the clean-up of oily extracts. Compounds ranging from the volatile organophosphorus insecticide mevinphos-E to the less volatile synthetic pyrethroids such as cypermethrin have been recovered quantitatively from cotton-seed oil and extracts of cotton seed.

REVIEW OF TRENDS

Although the literature on methods for pesticide residue analysis is extensive, many of the papers discuss in detail the development of clean-up, derivatization and determination steps while making only limited reference to the sample handling and extraction stages. This is reflected by the balance of references in two recent reviews of pesticide residues by Thornburg [3,4]. Where extraction procedures are discussed, use of laboratory and field-aged samples extracted with a range of solvents is normally described [5,6], and only a few authors mention the use of radiolabelled compounds [7-9]. It is our view that extraction procedures should be given greater consideration.

Thier [10] in a recent status report on clean-up procedures for pesticides reviewed several residue extraction and clean-up techniques, giving references to original work. He includes a section on HPLC in which he comments that many of the publications deal only with the application of the technique as a determination step. Clean-up using microparticulate materials is described; but only one reference is made to the application of microparticulate bonded-phase packings [11]. In view of the advantages now described for these materials, it is expected that much greater use of such packings for clean-up of pesticide residues will be made in the future.

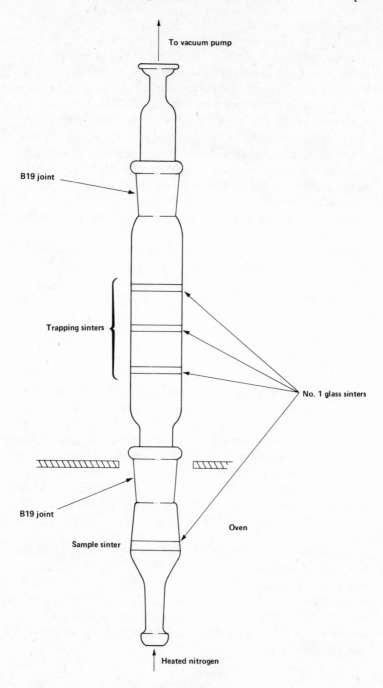

Fig. 5 – Details of a sublimation tube.

References

[1] Roberts, T. R. & Standen, M. E. (1977) *Pestic. Sci.,* **8**, 305–319.
[2] Publication F82, *Sep-Pak Cartridges for Rapid Sample Preparation,* available from Waters Associates.
[3] Thornburg, W. (1977) *Anal. Chem.,* **49**, 98R–109R.
[4] Thornburg, W. (1979) *Anal. Chem.,* **51**, 196R–210R.
[5] Smith, A. E. (1978) *Pestic. Sci.,* **9**, 7–11.
[6] Smith, A. E. (1979) *J. Agric. Food Chem.,* **27**, 428–432.
[7] Elgar, K. E., Wallace, B. G. & Woodbridge, A. P. (1975) *Rep. Prog. Appl. Chem.,* **60**, 387–404 (published 1976).
[8] Lichtenstein, E. P., Katan, J. & Anderegg, B. N. (1977) *J. Agric. Food Chem.,* **25**, 43–47.
[9] Wheeler, W. B., Thompson, N. P., Andrade, P. & Krause, R. T. (1978) *J. Agric. Food Chem.,* **26**, 1333–1337.
[10] Thier, H.-P. (1978) I.U.P.A.C. Commission on Pesticide Residue Analysis, Report to the 12th Meeting, Limburgerhof, July 1978, Project No. 5.
[11] Mittelstaedt, W., Still, G. G., Duerbeck, H. & Fuhr, F. (1977) *J. Agric. Food Chem.,* **25**, 908–912.

#NC(C) Notes and Comments related to Samples for Residue and other Non-pharmacological Studies

#NC(C)-1 *A Note on*
STABLE ISOTOPES AS MS TRACERS IN A STUDY OF THE METABOLITES AND RESIDUES FROM A PESTICIDE

P. HENDLEY and J. LAM, ICI Plant Protection Division, Jealott's Hill Research Station, Bracknell RG12 6EY, U.K.

An experiment was designed using an important metabolite and photoproduct of the insecticide Pirimicarb (I, R = CONMe$_2$); its object was to investigate for ourselves the practical problems involved in stable isotope studies [*see also* 1] and to estimate the costs/benefits of inserting these labels.

1,3–^{15}N$_2$–2–^{13}C–5,6-dimethyl-2-dimethylaminopyrimidin-4-ol (I, R = H) was synthesized from ^{13}C–^{15}N$_2$–cyanamide and admixed with unlabelled and 2–^{14}C-material. The mixture was administered to rats, and their excreta was collected and 'cleaned-up' by a combination of XAD-2 resin, normal-phase column chromatography, TLC, reverse-phase HPLC, and trimethylsilylation techniques. GC/MS on the products gave spectra in which the peaks of interest could easily be distinguished from those due to endogenous compounds by their characteristic P/P+3 doublet patterns. Approximately 10 metabolites were detected in this way and from these, compounds II–VII were tentatively identified by their mass spectra. The time saved compared with earlier studies using ^{14}C-label only was considerable; no isotope effects were observed.

In addition a simple computer search system was set up which automatically selected MS scans containing P/P+3 doublets in the appropriate ratio from GC-MS data. This proved a more sensitive technique for identifying peaks due to a pesticide metabolite than radiogas chromatography.

Limited ^{13}C-NMR studies on raw rat urine have demonstrated some potential for monitoring the numbers of major metabolites; an advantage is the non-destructive nature of this technique.

In conclusion we found that the combined use of a mass spectral label *in combination with* a ^{14}C-label was very effective. In future if there is no natural MS label (e.g. chlorine) in a molecule we shall endeavour to introduce a stable label(s), preferably so that there are at least two extra mass units. We feel that the potential for simplifying metabolism studies by MS and NMR in addition to the possibilities offered by stable labelled internal standards for residue determinations [2,3] justifies the expense of such labelling.

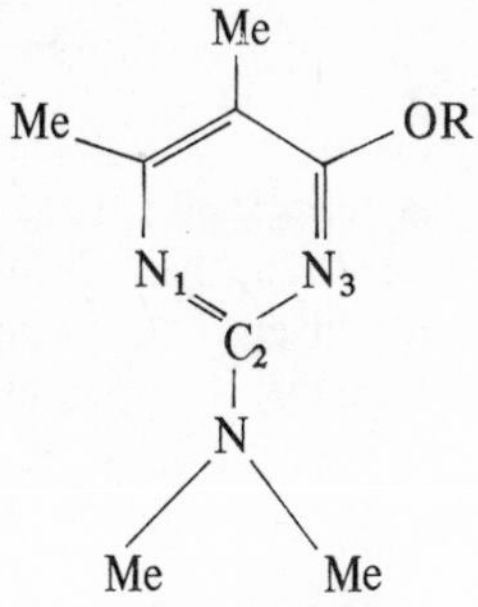

I

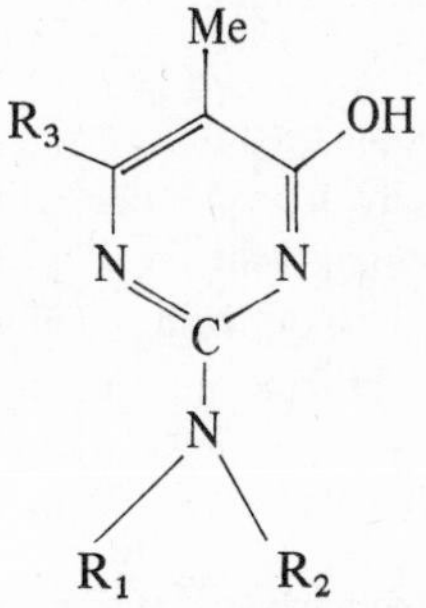

II R_1 = H, R_2 = Me, R_3 = Me
III R_1 = R_2 = H, R_3 = Me
IV R_1 = Me, R_2 = CH_2OH R_3 = Me
V R_1 = Me, R_2 = CH_2OMe, R_3 = Me
VI R_1 = H, R_2 = CH_2OMe, R_3 = Me
VII R_1 = Me, R_2 = H, R_3 = CH_2OH

References

[1] Hawkins, D. R. (1977) *Prog. in Drug Metab.*, 2, 163–218.
[2] Millard, B. J. (1975) in *Assay of Drugs and other Trace Compounds in Biological Fluids* (Vol. 5, *this series*), (Reid, E. ed.), North-Holland, Amsterdam, pp. 1–7.
[3] Baillie, T. A. (1977) in *Blood Drugs and other Analytical Challenges (Vol. 7, this series)*, (Reid, E. ed.), Horwood, Chichester, pp. 141–151.

#NC(C)-2 *A Note on*
ILLUSTRATIVE PROCEDURES FOR
CONTAMINANTS IN FOODSTUFFS

M. J. SAXBY, Leatherhead Food Research Association, Randall's
Road, Leatherhead, T22 7RY, U.K.

The following outline [drafted by the Editor] of some established procedures
serves to reinforce the preceding contributions. It is concerned particularly with
'sample preparation', in a wider sense (solvent extraction, then clean-up) than
used by A. P. Woodbridge [*this vol.*]. Consideration is given firstly to approaches
for organochlorine pesticides, as typifying non-ionizable compounds of low
polarity.

ORGANOCHLORINE PESTICIDES, EXEMPLIFYING PRE-MEASUREMENT PROCEDURES

Solvent extraction

Samples of low water content, e.g. cereals, are ground and, after water addition,
are extracted with 50% (v/v) acetone in petroleum (hexane).

Samples of high water content, e.g. fruit or vegetables, are macerated in this
solvent.

Samples of high fat content are petroleum-extracted, with boiling if necessary—
or [Editor's note] a procedure effective for fruit or vegetables may also be applied
to meat or sausage [1]: maceration is performed with acetone (50 ml copes with
0.7 g fat) to which 4 vol. of hexane is then added.

Milk may, after adding 0.5 g potassium oxalate and 50 ml ethanol, be extracted
with 50% diethyl ether in petroleum.

Clean-up

A suitable first step may be to partition between a polar and a non-polar solvent
[A. P. Woodbridge, *this vol.*], and possibly dilute with water and re-partition
into petroleum. Column clean-up may then be performed with water and re-
partition into petroleum. Column clean-up may then be performed with Florisil,
or else [1] with activated alumina—which is more reproducible and needs less
eluent. Illustrative approaches are now outlined.

'Shell' method: two partitionings are performed from hexane into hexane-satd.
CH_3CN, each with addition of 2% aq. Na_2SO_4 to give a small hexane layer: the
combined hexane layers are run on 2 g of 8% deactivated Florisil with 4% ether
in hexane.

Method of Telling and co-workers [1]: an acetone/hexane extract [*see above*] is
washed with aq. Na_2SO_4, and the hexane layer in an amount found to contain

0.5 g of fat is concentrated in a Kuderna–Danish evaporator (Fig. 1) and then chromatographed on 22 g of alumina (activity '4') with 150 ml of hexane.

Nestle method [2] : after extraction with warm hexane (where the fat content is high) or, for insoluble samples such as cereals or cocoa beans, after adsorption onto Florisil by grinding in the presence of hot water, the material is applied to a column (25 g) of 3% deactivated Florisil and eluted with 500 ml of CH_2Cl_2/hexane (1:4); GC or TLC is then performed.

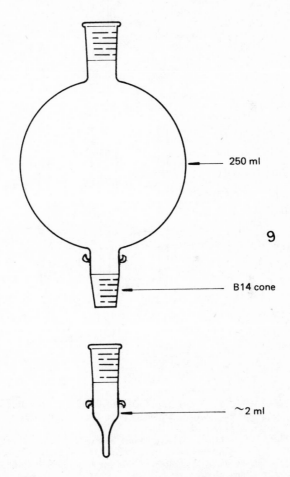

250 ml

9

B14 cone

~2 ml

Fig. 1 – The Kuderna–Danish evaporator [3], for concentrating extracts of volatile organics.
The apparatus, available from the Kontes Glass Co. (Vineland, NJ, U.S.A.), is used in conjunction with a Snyder column, possibly of micro type [*cf. Editor's opening article*, #0]. The extract is placed in the flask, which is heated in a water bath.

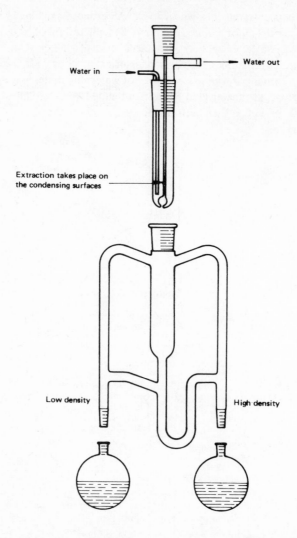

Fig. 2 – The distillation apparatus of Likens & Nickerson [5], entailing phase separation and solute accumulation in the distillation reservoir containing the appropriate phase.

The sample solution is placed in one flask and the extracting solvent in the other flask. Both solutions are heated to boiling and the vapours co-condense on the cold finger. Volatile compounds are thus transferred from the aqueous phase to the organic phase.

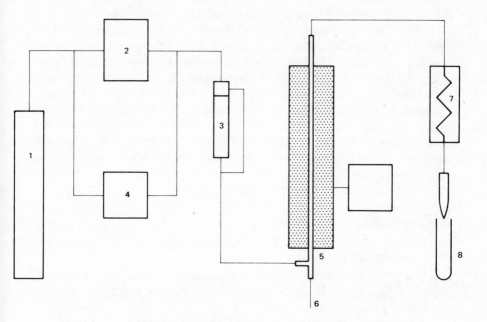

Fig. 3 – A commercially available apparatus ('Tracesep') for sweep co-distillation
e.g. of organochlorine pesticides.
The sample is initially distributed as a film on a packing of 'Anskron ABS' or
glass beads (*see text*) in the separation column. Inert carrier gas (1), controlled by
a needle valve and monitored by a flowmeter (2), reaches the column through the
mixing chamber (3), which provides independent continuous addition of the
sweep solvent (4) to the carrier gas. The volatile substances from the injected
sample (6) are separated in the heated column (5) by the carrier gas/solvent
vapour mixture and reach the condenser tube (8) *via* a cooling system (7).

This description is reproduced with acknowledgement to the manufacturer, viz.
Innovativ Labor AG (Feldblumenstrasse 76, CH–8134 Adliswil, Switzerland).

VARIOUS TYPES OF CONTAMINANT

Established analytical approaches for certain types of pesticide are outlined in Table 1, the extent of clean-up being determined partly by the specificity of the measurement method. The tabulation also deals with some of the diverse types of mycotoxin (fungal metabolites), some of which, e.g. ochratoxin A, have an ionizable group (carboxyl) in contrast with the aflatoxins.

Where the compound to be determined possesses significant volatility, the choice of sample-handling approaches includes headspace analysis by GC, possibly continuously with the Pavelka apparatus; extraction with the Likens-Nickerson apparatus [5]; Soxhlet extraction; and, in the case or acids or bases, steam-distillation into appropriate trapping solutions. Another advantageous approach is 'sweep co-distillation' [6,7]. Here the injected sample is distributed in the form of a thin film on the surface of a suitable packing in a heated form. The volatiles are distilled out from this film by the combined action of a carrier gas with a solvent vapour at elevated temperature, and after condensation are collected in a receptacle.

References (see also foot of Table 1)

[1] Telling, G. M., Sissons, D. J. & Brinkman, H. W. (1977) *J. Chromatog.,* **137**, 405-423.
[2] Stijve, T. & Cardinale, E. (1974) *Mitt. Geb. Lebensm. Hyg.,* **65**, 131-150.
[3] Gunther, F. A., Blinn, R. C., Kolbenzen, M. J. & Borkley, J. H. (1951) *Anal. Chem.,* **23**, 1835-1842.
[4] Lawrence, J. F. & Leduc, R. (1978) *J. Chromatog.,* **152**, 507-513.
[5] Likens, S. T. & Nickerson, G. B. (1964) *Proc. Amer. Soc. Brewing Chem.,* p.5.
[6] Storherr, R. W., Murray, E. J., Klein, I. & Rosenberg, L. A. (1967) *J. Assoc. Offic. Analyt. Chemists,* **50**, 605-615.
[7] Malone, B. & Burke, J. A. (1969) *J. Assoc. Offic. Analyt. Chemists,* **52**, 790-797.

Table 1. Illustrative assay procedures.

Requirement	Propoxur in cocoa beans	Fenitrothion in cocoa	Carbofuran & metabolites in crops	Mycotoxins			
				Aflatoxins	Ochratoxin	Zearalenone	Various
End-step	GC-ECD (3% QF-1 + 2% OV-17, 200°)	GC with thermionic det.	HPLC	TLC (easy, but limited capability), GC-FID (non-specific), GC-MS (excellent but costly), or HPLC with fluorescence det. (good but somewhat costly)			
Extraction	Hexane, Soxhlet	Hexane, macerate	Acetone, macerate	(a) MeOH; (b) aq. CH₃CN	as for (b) on left	as for (a)	H₃PO₄ in CHCl₃
Clean-up	Partition → CH₃CN; → neutral alumina + 2 mm charcoal, & elute with hexane/CH₂Cl₂ (2:1); treat with fluoro-2,4-dinitrobenzene, pH 11	Florisil, & elute with benzene/EtAc	Partition → hexane/CH₂Cl₂; → 2% deactivated Florisil; elute with 15% acetone in hexane; hydrolyze with 0.1 M Na₂CO₃; dansylate	(a) 20% (NH₄)₂SO₄ in aq. MeOH/chlorobenzene; or (b)→CHCl₃; then TLC; or (c)→silica, & elute with MeOH/CHCl₃	as for (b) on left, or as for (c); but AcOH/benzene	(d) Partition from aq. MeOH → benzene; →Sephadex LH-20 & run with benzene then 10% MeOH in benzene; or as for (c) but acetone/benzene	
Ref.			4	(a): $JAOAC$* (1976) 59, 106–109; (b) N. T. Crosby, #C-2, $this\ vol.$; (c) $JAOAC$ (1968) 51, 74–78; (d) $JAOAC$ (1977) 60, 272–278.			$JAOAC$ (1977) 60, 1369–1371

*$J. Assoc. Offic. Anal. Chemists.$

#NC(C)–3 *A Note on*
PREVENTION OF CONTAMINATION IN THE RESIDUE LABORATORY

A. P. WOODBRIDGE, Shell Biosciences Laboratory, Sittingbourne, Kent ME9 8AG, U.K.

Contamination in the extracts used for the determination stage of a residue analysis can give rise to interference which obscures the residue or, more seriously, to false positive results and a totally invalid analysis. Potential sources of contamination have recently been highlighted by Telling in a paper on 'Good analytical practice in pesticide residue analysis' [1]. Contamination may be caused by specific chemicals, most importantly a large amount of the compound being determined, or by a non-specific source such as vacuum grease. It can arise either from the laboratory environment itself or from the procedure being used.

LABORATORY ENVIRONMENT

It is considered essential that the residue laboratory should be physically separated from areas where macro-chemistry such as formulation development or analysis is being carried out even though the two laboratories may be located in the same building. Laboratory design and fittings should minimize the chance of contamination. It is important that a wash-up facility is available just for residue work. Separate location of areas within the laboratory for sample storage, extraction, clean-up, determination and storage of reference materials and certain reagents is desirable. Within the individual areas, appropriate fittings such as non-porous bench surfaces to allow complete clean-up of any spillages and plastic trays for storage of extracts should be used. A long list of potential sources of contamination from the working environment can be compiled, e.g. barrier creams, bench polish, perfumes and cosmetics, natural products from the skin, greases, plasticizers, rubber bungs and tubing, oil from air lines, filter papers and cotton wool. In some of these cases, it is for the individual operator to establish his or her own satisfactory operating procedures.

ANALYTICAL PROCEDURE

Procedures and methods must be developed which avoid sources of contamination. The possibility of contamination of glassware, sample-processing equipment, syringes and columns (including GC) arising from previous samples must be recognized. Few residue analysts who have used GC-ECD will not have experienced the unexpected peaks resulting from a septum or syringe contaminated from a previous injection of a concentrated solution. The analysis of reagent blanks from time to time can be used to highlight problem areas such as these.

All chemical reagents, absorbents and solvents must be of sufficiently high purity for residue work, and the use of distilled solvents is normally essential. Several manufacturers now supply specially purified grades of solvents for residue analysis. Water, even if purified, can sometimes still be suspect, and interference from this source in HPLC residue determinations, especially in gradient work, can be a problem.

Many additional examples could be cited, but it is important in all residue analysis to recognize the potential for contamination as an ever-present problem whilst adopting procedures which offer the best chance of avoiding it.

Reference

[1] Telling, G. M. (1979) *Proc. Analyt. Div. Chem. Soc.,* **16**, 38–42.

#NC(C)–4 *A Note on*
MEASUREMENT OF A PLASTICIZER (DEHA) IN TISSUE

S. FAYZ and R. HERBERT, School of Pharmacy, Sunderland Polytechnic, Sunderland SR1 3SD, U.K.

Requirement	*(1) Assay for plasticizer di-2-ethylhexyl adipate (DEHA) in tissue, especially liver and lung, at levels of 0.1–20 µg/g.*	*(2) Assay for adipic acid in tissue (liver and lung) at levels of 0.1–50 µg/g.*
End-step	*GC-FID on OV-1; inject in acetone.*	*GC-FID of esterified acid on Carbowax 20 M. SCOT capillary column used in difficult cases.*
Sample handling	*Homogenize tissue, denature extract with methanol, centrifuge, decant, dilute the extract with saline and extract into chloroform, dry down, column-chromatograph on Kieselgel, elute with petrol/ether and dry down.*	*Homogenize, denature and extract with several small volumes of 2 M HCl. Centrifuge, decant, raise pH of combined extracts to approx. 3 and evaporate water. Extract residue with methanol and dry down. Esterify residue from methanol with BF₃/methanol of BF₃/butanol; extract ester into chloroform and dry down.*
Comments	*(1) High risk of contamination from reagents and apparatus [1–3].* *(2) Chromatographic clean-up stage rather slow but essential.* *(3) Very rapid metabolism of DEHA by tissues samples even at 0° prevents meaningful measurements.*	*(1) Water solubility and low solvent solubility of adipic acid prevent the use of partition techniques.* *(2) Evaporation of aqueous extract slow but involves minimal manipulations and gave consistently better recoveries than column-chromatographic procedures.*

The need for the assays arose during an investigation into the fate of the plasticizer DEHA when introduced into the blood of animals and also humans during the use of artificial kidney machines [4]. Initial recoveries from homogenized bovine liver spiked with DEHA were very poor despite vigorous extraction techniques. Recoveries of di-2-ethylhexyl phthalate (DEHP) under the same and other conditions were good [1-3]. The cause of the poor recoveries was eventually traced to the rapid tissue-catalyzed hydrolysis of the ester which occurred quite rapidly even at $0°$. Typical half-lives of DEHA in bovine liver homogenate were 2 min and 7.5 min at $20°$ and $0°$ at initial levels of 200 μg/g (Fig. 1). The initial V_{max} of this hydrolysis was found to be 217 μg/min/g of tissue at $20°$. Other adipates (dimethyl and dibutyl) were degraded similarly but the rates were very much greater. It was thus not feasible to attempt to estimate levels of DEHA in actual samples.

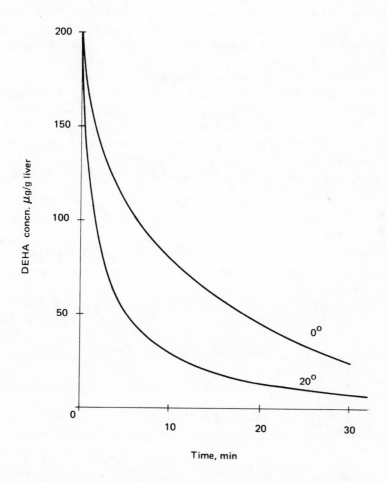

Fig. 1 – Rate of disappearance of DEHA from bovine liver samples.

It was considered that useful information could be obtained by the examination of the tissue for adipic acid, and the development of a tissue assay for this compound was attempted. Difficulties were encountered which were associated with the high aqueous solubility and the unfavourable partition characteristics of this acid [see also 5]. Eventually the method outlined above gave recoveries of 95–105% relative (60–80% absolute) at levels of 10 μg/g. Other less drastic methods of esterification [6–8] invariably gave poor recoveries.

In some tissue samples further difficulties were experienced owing to the co-extraction of other water-soluble acids which formed esters with essentially the same GC characteristics as the adipates. This was true on a variety of columns as either the dimethyl or dibutyl ester. In these cases, the problem was best overcome by the use of a SCOT capillary column. In our experience the choice of internal standards is critical for capillary column work of this nature. The vapour pressure curves of the standard and test substances must be almost identical if reproducible results are to be obtained. The reason for this we believe is associated with the small volumes of solution injected into the chromatograph. This seems to allow significant fractionation to occur in the syringe needle during injection, which causes varying amounts of the less volatile components to be lost when the syringe is withdrawn. Increasing the temperature of the injector port did not significantly improve matters and could be a definite disadvantage with the plunger-in-needle syringes essential for this type of column. A septum purge or splitter can aggravate this problem, which was resolved only by the use of a more appropriate internal standard.

References
[1] Poole, C. F. & Wibberley, D. G. (1977) *J. Chromatog.*, **132**, 511–518.
[2] Mes, J. & Campbell, D. S. (1977) *Bull. Environ. Control Toxicol*, **16**, 53–60.
[3] Takeshita, R. & Takabatake, E. (1977) *J. Chromatog.*, **133**, 303–310.
[4] Fayz, S., Herbert, R. & Martin, A. M. (1977) *J. Pharm. Pharmacol.*, **29**, 407–410.
[5] Chalmers, R. A., Lawson, A. M. & Watts, R. W. E. (1972) *Analyst*, **97**, 958–967.
[6] Thenot, J. P., Horning, E. C., Stafford, M. & Horning, M. G. (1972) *Anal. Lett.*, **5**, 217–223.
[7] Mlejnek, O. (1972) *J. Chromatog.*, **70**, 59–65.
[8] Levitt, M. J. (1973) *Anal. Chem.*, **45**, 618–620.

#NC(C)–5 *A Note on*

ASSAY OF BLOOD FOR TRACES OF ANIONIC AND NON-IONIC SURFACTANTS

P. THACKERAY and D. HOAR, Unilver Research Ltd., Colworth House, Sharnbrook, Bedford MK44 1LQ, U.K.

Requirement	*Safety-evaluation method capable of measuring 50 ng/ml blood (taking account of rapid metabolism), on small samples such as arise in toxicological testing.*
Chemistry & metabolism	Example of anionic: *primary alcohol sulphates*, $CH_3 \cdot (CH_2)_x -SO_3Na$ *(e.g. sodium dodecyl sulphate, SDS); ω-oxidation to CO_2H and then β-oxidations [1] take place rapidly [2], the excretion of the metabolite (sulphated fatty acid) being 40% at 1 h after i.v. injection and 80% after 6 h.* Example of non-ionic: *primary ethoxylates*, $CH_3 \cdot (CH_2)_x \cdot (C_2H_4O)_y$ OH *(where* x *and* y *may each vary); similar metabolic path to the above [3].*
End-step	*GC-ECD, after rendering the compounds amenable to GC and ECD-responsive by forming alkyl iodides and extracting with hexane.*
Sample preparation	CH_3CN *extraction from plasma saturated with NaCl, then TLC and finally scrape off the silica containing each surfactant (Table 1).*
Comments	*Shown with radiolabelled surfactants that blood content entirely in plasma, largely as metabolite in case of SDS. Successful derivatization hinges on use of HI, with hypophosphorus acid present to re-convert any iodine back to HI. GC separation depends merely on alkyl chain length, hence the need for pre-separation of the compounds by TLC. The approach could be useful for various compounds besides the surfactants now studied.*

DERIVATIZATION

The first approach attempted involved the conversion of the surfactants to their corresponding alkyl alcohols which could be subsequently derivatized to form EC-sensitive compounds. However, the formation of heptafluorobutyrate esters [4] from alcohol sulphates following this derivatization route was found to give insufficient sensitivity, while silylation using halogenated silylating reagents [5] was found to be non-reproducible.

An alternative to alcohol derivatization was the direct reaction of the surfactants with a halogen acid. Reaction with HBr had been successfully performed [6], but the alkyl bromides formed were insufficiently ECD-sensitive. However, reaction with HI had also been reported [7], and it was found that as little as 15 pg of alkyl iodide could be detected by GC-ECD.

The reaction was, however, found to be no more than semi-quantitative, and consideration of the chemistry led us to believe that it was being hindered by the presence of iodine formed by the decomposition of the HI. This problem was overcome by the addition of hypophosphorus acid to the reaction mixture, which reduced any iodine formed back to HI and thus stabilized the reaction. Optimum yields were obtained by using 0.5 ml HI and 0.2 ml hypophosphorus acid and heating the mixture with the surfactants at 110° for 1 h.

SEPARATION BY TLC

Separation of the surfactants from each other and from their correspoinding alcohols needed to be carried out prior to derivatization. TLC was used to separate the surfactants according to their active hydrophilic groups (Table 1).

Table 1 R_f values of surfactants in various TLC systems. E denotes an ethoxy moiety.

Compounds	Ethyl methyl ketone/H_2O	$CHCl_3$/MeOH (100:3 by vol.)	$CHCl_3$/MeOH (10:1 by vol.)
$C_{12}SO_4$ (SDS)	0.30	0.00	0.00
$C_{12}E_3$	0.80	0.48	0.66
$C_{12}E_6$	0.75	0.24	0.61
$C_{12}E_{10}$	0.45	0.10	0.48
Dodecanol	0.80	0.64	0.66

The alcohol sulphates were detected by spraying with fluorescein and observing under UV light, while the ethoxylates were visualized using a modified Dragendorff reagent [8]. Sulphate and ethoxylate markers were spotted on to each plate next to the samples and visualized after elution to show the surfactant positions.

The ethyl methyl ketone/water system (Table 1), viz. the top layer obtained by shaking together equal quantities of the components, separated the alkyl sulphates well; but the chloroform/methanol systems were preferred for the ethoxylates as giving a manageable spectrum from a mixture.

STEPS IN THE METHOD

1.   Centrifuge 1 ml of heparinized blood and collect the plasma quantitatively.

2.   Saturate the plasma with NaCl and extract surfactants using 3 X 3 ml acetonitrile. The acetonitrile also deproteinizes the plasma while the NaCl aids the breakdown of any surfactant–protein bonds and gives a 2-phase system, the separation being aided by centrifugation.

3.   Evaporate the combined acetonitrile extracts down to 0.2 ml under a stream of nitrogen, and spot onto a TLC plate.

4.   Wash the test-tube with 1 ml hot methanol, evaporate to 0.2 ml and apply onto the spot.

5.   Wash the test-tube with 0.2 ml hot methanol and apply onto the spot. Unless these washing procedures are carried out, surfactant losses of up to 70% can occur owing to adsorption onto the glass test-tube.

6.   Apply surfactant markers in the edge positions. Elute with the appropriate solvent system to a height of 10 cm.

7.   Dry plate at $100°$ for 15 min.

8.   Visualize markers and thereby locate the sample constituents.

9.   Scrape appropriate areas of silica from the plate and place in a 15 ml test-tube.

10. Add 0.5 ml HI and 0.2 ml hypophosphorus acid, stopper tightly and heat for 1 h at $110°$.

11. Cool and extract the alkyl iodides with 3 X 3 ml hexane.

12. Inject 5 μl of hexane extract into a GC-ECD fitted with a 1.25 m X 4 mm i.d. 3% SP 2250 column at $180°$.

13. Quantitate each iodide by peak-height comparison with an external (directly injected) standard prepared from stock solutions of pure alkyl iodide in hexane.

The recoveries from a series of plasma samples to which surfactant had been added over a range of 50–1000 ng show good reproducibility (Table 2). Yields are the same at the 50 ng level as they are at the 1000 ng level. The ethoxylate yields become lower as the displaced ethoxylate groups become longer, owing to the lowering of the mol. wt. on iodination. However, yields relative to the external iodide standard are quoted, since iodinating a commercial ethoxylate mixture where many ethoxylate chain lengths are present precluded allowing for the mol. wt. change. A blank level of *ca.* 20 ng of alkyl iodide has been allowed for in the recovery data, but precautions must be taken to minimize this level.

Table 2 Alkyl iodide yields from surfactants added to plasma over the range 50–1000 ng. yields are given relative to the starting wt., and thus are inherently < 100% for ethoxylates (*see text*).

Compound	Maximum theoretical iodide yield, %	Iodide yield achieved, % (± S. E.)
Sodium dodecyl sulphate	100	70 ± 4
Sodium tridecyl sulphate	100	54 ± 5
Sodium tetradecyl sulphate	100	48 ± 8
Sodium pentadecyl sulphate	100	40 ± 6
Dodecyl triethoxylate	93	51 ± 4
Dodecyl hexaethoxylate	66	47 ± 3
Dodecyl decaethoxylate	47	28 ± 3

MINIMIZING THE BLANK LEVEL

1. Distil all solvents from glass.

2. Soak glassware overnight in 'RBS' solution and thoroughly rinse with tap-water, distilled water and hot acetone before drying.

3. Develop TLC plates in the solvent system to be used, prior to actual use. Reactivate by heating to 110° for 30 min.

The omission of any of these precautions can lead to blank iodide levels of up to 100 ng.

OBSERVATIONS WITH RADIOLABELLED SDS *IN VIVO*

Comparisons were made between the GC and radioactive assays of blood obtained from a rat injected with [^{14}C] labelled SDS. The metabolite (sodium butyric acid-4-sulphate) was determined by hydrolyzing it to form γ-butyrolacetone which was extracted into chloroform and analyzed by GC. Table 3 shows the balance between the parent surfactant and its metabolite, and a comparison with the total radioactivity found in the blood.

Table 3   The mass balance between sodium dodecyl sulphate (SDS) and its metabolite in plasma. The blood samples were taken from an adult rat, 2 h after i.v. injection of 5 mg of [^{14}C] SDS (labelled at 1-C).

Compound	Amount found/ml plasma, μg	
	by GC	by radioactivity (liquid scintillation)
SDS, first analysis-(a)	1.53	1.55
repeat analysis-(b)	1.38	1.38
Metabolite (see text), as SDS wt. equivalent	9.67	10.56
Total SDS + metabolite, first analysis-(a)	11.20	12.11
repeat analysis-(b)	11.05	11.94
Directly determined total [^{14}C] activity expressed as SDS		11.78, 12.75

SUMMARY OF THE OUTCOME

The various analytical problems posed by this kind of determination have been overcome. The low detection limit has been achieved by derivatization (cf. ECD-responsiveness). Specificity has been achieved by TLC separation of the classes of detergent and then GC separation according to alkyl chain length.

References

[1] Denner, W. H. B., Olavesen, A. H., Powell, G. M. & Dodgson, K. S. (1969) *Biochem. J.*, 111, 43-51.
[2] Burke, B., Olavesen, A. H., Curtis, C. G. & Powell, G. M. (1975) *Xenobiotica*, 5, 573-584.
[3] Drotman, R. B. (1977) in *Cutaneous Toxicity*, (Drill, V. A. & Lazar, P., eds.), Academic Press, New York.
[4] Walle, T. & Ehrsson, H. (1970) *Acta. Pharm. Suec.*, 7, 389-406.
[5] Brooks, J. B., Liddle, J. A. & Alley, C. C. (1975) *Anal. Chem.*, 47, 1960-1965.

[6] Luke, B. G. (1973) *J. Chromatog.*, **84**, 43–49.
[7] Lee, S. & Puttnam, N. A. (1966) *J. Am. Oil Chem. Soc.*, **43**, 690.
[8] Ginn, M. E., Church, C. L. & Harris, J. C. (1961) *Anal. Chem.*, **33**, 143-145.

Comments on #C-1, M. D. Osselton — ENZYMIC LIBERATION OF DRUGS

M. D. Osselton, *responding to questions and to a remark* about smelliness if pancreatin is used to digest tissue: any smell under our conditions, with the efficient enzyme subtilisin used at 55° for 60 min in a fume chamber, has disappeared before we take the digests into the open laboratory for extraction. Substrates which are well digested with subtilisin include cardiac muscle, skin injection sites (if the incubation time or enzyme amount is increased), and even soil proteins; other workers have been successful with lung and kidney, but efficiency is poorer for hair and epidermal proteins. Interferences with blood are obviated by using butyl acetate for drug extraction, and with liver by adding a minute quantity of sodium tungstate in the case of 'acidic drug fractions', which are most prone to show endogenous interferences; with brain the extracts are remarkably clean. Liver extracts to be analyzed by UV spectrophotometry indeed need clean-up, although drug recoveries remain notably good; but with HPLC as in the case of benzodiazepines, it suffices to replace the top of the column packing occasionally. GC-ECD is less tolerant of impurities, removal of which can be achieved by partitioning the drug into aqueous acid if basic, or *vice versa* if acidic.

Answers to W. Dünges and P. Hendley. — Subtilisin does not attack glucuronides, but it would proteolyse glucuronidases; hence use of an 'enzyme cocktail' to hydrolyse conjugates in a digest would have to be preceded by blockage of the action of subtilisin, as would also be necessary in the case of an immunoassay. Drugs in general, if not proteinaceous, have not shown susceptibility to subtilisin action, although alkali-susceptible compounds may show some breakdown during the usual incubation at pH 10.5 for 60 min, circumventable by incubating at, say, pH 8.0. — The enzyme is active over a remarkably wide pH range (pH 6-11).

Comment by A. C. Moffat (subsequent to the Forum). — In view of the risk of chemical degradation of amides and esters at pH 10.5, their stability at pH 7.0 and 10.5 with or without the enzyme is now under study; meanwhile, the preferred incubation conditions have become pH 7.0 or 7.4 for 1 h at 60°.

Comment on #C-2, N. T. Crosby — PROCEDURES FOR FOODS
Reply to B. Stavric, who had found incomplete recovery of radiolabelled amaranth from intestinal sites in feeding trials. — This result, if not due to

metabolism of the compound, may be comparable to our incomplete recovery of ochratoxin unless, in the case of intestine as compared with kidney, lipolytic as well as proteolytic digestion was performed.

Comments on #NC(C)-1, P. Hendley — MS TRACERS,
                    & #NC(C)-5, P. Thackeray — SURFACTANTS
Reply to B. Woollen. — Isotope clusters could indeed help elucidate the MS fragmentation pattern for a compound whose metabolism, unlike that of pirimi-carb, is little known. *Remark by* J. Chamberlain. — At Hoechst we use ^{13}C material, the main objection to the use of deuterium being the risk of misleading results (as also with tritium) due not to chemical lability — which can be allowed for — but to possible biological lability.

Remark by J. S. Fritz [cf. #B-1]. — In a new method developed (with J. J. Richard) for sulphonate surfactants, they are concentrated on a small anion-exchange bed, eluted with HCl in diethyl ether, derivatized with diazomethane, and analyzed as the methyl esters by GC-ECD.

Editor's citation: relevant to #NC(C)-2 (also to #F-3).—In a useful survey [1] of pesticide residue analysis as aided by HPLC, there is mention of lipid removal by gel permeation chromatography, or by passage through a silica-bead pre-column which is finally back-flushed.

[1] Schooley, D. A. & Quistad, G. C. (1979) *Prog. Drug Metabolism,* **3,** 1–113.

#D Approaches for Biological Fluids, Especially Drug-Related

#D-1   ENZYMATIC APPROACHES TO TRACE-ORGANIC ANALYSIS*

G. GUILBAULT, Chemistry Department, University of New Orleans, New Orleans, LA 70122, U.S.A.

The use of immobilized enzymes for the assay of biomedically and environmentally important compounds is discussed. Electrochemical and fluorimetric methods are described for the assay of substrates such as penicillin, amino acids, creatinine and amygdalin, as well as of enzyme inhibitors such as organophosphate pesticides. Sample preparation is minimal.

Excellent chemical analysis can be performed with enzymes. In immobilized form, they have many real advantages in analyses using electrochemical probes or other measurement methods. One advantage of the immobilized enzyme is a pH shift, enabling the pH optimum to be shifted to that region at which one wants to make a measurement, by choosing the right support for immobilization. Thus, for an enzyme with a narrow pH range of, say, 6 to 8, this can be shifted on immobilization down to the acidic side or, conversely, up to the basic side. The enzymes are furthermore much more stable. In some work we did at Edgewood Arsenal, we actually heated our enzymes to 65° and brought them back down to room temperature, with very little loss in activity. No soluble enzyme could be treated in this fashion.

There is an advantage, often overlooked, that better selectivity can be realized with the enzyme when immobilized; this insolubilized reagent becomes much more selective for an inhibitor, and only the most powerful inhibitor can actually attack the enzyme. We demonstrated this several years ago in an immobilized-cholinesterase alarm for the assay of organophosphorus compounds in air and water. No other common interferants disturbed the alarm. It responded only to organophosphorus compounds.

---

* Notwithstanding the section title, this contribution is applicable to various matrices besides biological fluids, and to various analytes besides drugs. —*Ed.*

ENZYME ELECTRODES

Enzyme electrodes represent the most recent advance in analytical chemistry. These devices combine the selectivity and sensitivity of enzymatic methods of analysis with the speed and simplicity of ion-selective electrode measurements. The result is a device that can be used to determine the concentration of a given compound in solution quickly and a method that requires a minimum of sample preparation. In constructing an enzyme electrode one need only (a) pick an enzyme that reacts with the substance to be determined, (b) obtain that enzyme from commercial sources or isolate it oneself, (c) immobilize the enzyme by standard techniques or buy it already immobilized, if possible, and (d) place the immobilized enzyme around the appropriate electrode to monitor the reaction that occurs, this probably being the limiting factor in the construction of an enzyme electrode since steps (a)-(c) are always feasible. 'Physical' immobilization entails entrapment in polyacrylamide gel, 'chemical' immobilization entails attachment to glutaraldehyde with albumin, to polyacrylic acid, or to acrylamide followed by physical entrapment.

Enzyme electrodes can be either potentiometric or amperometric devices with the appropriate sensor: we can measure the response either by a steady-state, i.e. equilibrium, method measuring millivolts or microamperes, or by a rate method which senses the change in millivolts or microamperes per min. Measurements of substrate can be performed by either approach. Sensors may respond, for example, to pH, NH_4^+, NH_3, CO_2, O_2, H_2O_2, quinone, CN^-, or I^-. Appropriate sensors and enzymes, with information on feasible concentration ranges, have been tabulated elsewhere [1] for assay of urea, uric acid, lactic acid, triglycerides and various other substrates (not necessarily at trace level, and including heavy metals) besides those in the examples below. Thus glucose can be assayed with glucose oxidase, the products being hydrogen peroxide and gluconic acid. One can measure the uptake of oxygen with a gas membrane electrode, a technique pioneered by Clark and perfected by Hicks and Updike, or record the peroxide or oxygen polarographically. There are other ways: one can measure the gluconic acid by a pH change, as Mosbach showed very nicely at low buffer capacity, or use an iodide membrane (the latter is much less recommended). The essential point is that there are many ways to measure a particular substrate, and one should choose whichever is best for ones's application. For example, one would not choose to measure urea in biological fluid with an ammonium cation elctrode, because of the interference of potassium and sodium; one would choose, preferentially, an ammonia gas membrane electrode with which there is no such interference.

The stability of the electrode depends on the type of entrapment. Here again, literature on stability after immobilization abounds in ambiguous information. Some individuals store dry for a long period, and then report a fantastically long lifetime. One should realistically define the immobilization characteristics and the stability of the enzyme in terms of dry storage *and* of use storage. The lifetime of most soluble enzymes, except perhaps in the case of some types of glucose oxidase which are quite stable in the crude form, is generally about one week or 25-50 assays. However, one must realize that with soluble enzymes there are potential interferences that are not found with entrapped enzymes.

We demonstrated this in an enzyme alarm for organophosphorus pesticides using immobilized horse serum cholinesterase: although the soluble enzyme is either inhibited or activated by fully 100 substances, the insolubilized enzyme is affected only by the organophosphorus compounds.

A physically entrapped enzyme lasts about 3-4 weeks, or 50-200 assays. For a chemically bound enzyme, 200-1000 assays is a reasonable expectation. In many cases we, and others, have achieved at least this. Stability is in fact very good for many commercially available enzymes that are bound onto nylon tubes, e.g. those to be marketed by Technicon for use on SMAC or the Auto-Analyzer, those Boehringer has been experimenting with, or those Miles is selling under the trade name 'Catalink', which are very stable. These tubes have been shown capable of 10,000 assays.

Interferences can arise from the sensor itself, or from the presence of other substrates for the enzyme. For example, alcohol oxidase can be used for an excellent acetic acid electrode; the native substrate, ethanol, could interfere, as could inhibitors of the enzyme. Here the enzyme immobilization is an asset, in making the enzyme much less susceptible to environmental factors. Rather than present an overall view of what has been done and how broad the field really is, I now touch on illustrative applications from our recent work.

AMINO ACIDS

As a simple and rapid approach, amplified elsewhere [2,3], to assessing the protein quality of foods, we have induced and isolated specific amino acid enzymes that act on only one of the key amino acids, and use a CO_2 or ammonia electrode as a base probe to measure the reaction product. Thus, phenylalanine ammonia lyase isolated from Wisconsin potatoes following irradiation, and then immobilized, enables L-phenylalanine to be assayed specifically. L-arginine and L-lysine can be measured specifically with carefully purified decarboxylases from *E. coli*. For histidine and tyrosine, certain strains furnish specific decarboxy-lases.

To prepare the enzyme electrode, one simply smears a glutaraldehyde solution of the amino acid decarboxylase or lyase directly onto the tip of a CO_2 or ammonia membrane, as a very thin film such that the return to baseline is quite fast (about 10 min); response times are typically 1-3 min. If the enzyme layer is too thick, the return to baseline is slow. We and others have shown that one does not have to wait for the complete return to baseline in the case of high substrate concentrations; 2 min may suffice. Typical calibration curves are linear from 10^{-1} to 5×10^{-5} M for most of the amino acid electrodes; the C. V. is about 2.5%.

For determining methionine in hydrolyzates from grains, a sensor has been developed in our laboratory [3]. Methionine-γ-lyase is covalently coupled directly onto an ammonia electrode, the reaction (at pH 8) giving rise to α-ketobutyrate, methyl mercaptan and ammonia. The rate of the change of the electrode potential (mv/min) was linearly proportional to the concentration of methionine in the range 10^{-4}-10^{-2}M; each analysis required about 20 min. Of the common amino acids, only L-asparagine, L-glutamine, L-threonine, and L-cysteine interfere [3]; the first two will be decomposed during the acid hydrolysis of the grain sample. Work is in progress to improve the performance of the electrode and to purify the enzyme.

CREATININE

The spectrophotometric methods widely used for routine determination of creatinine in serum, based on the Jaffé reaction between creatinine and picric acid in alkaline solution, are rather time-consuming, require deproteinization and incubation to develop colour, and have the serious defect that serum contains other materials which react positively. After Szulmajster's isolation from Pseudomonas of a pure creatininase (EC 3. 5. 4. 2), which hydrolyzes creatinine to N-methyl hydantoin and ammonia, it has been introduced by Beckman for enzymatic determination of creatinine.

Aiming at the higher sensitivity of a fluorimetric method, we investigated coupled enzymatic reactions whereby NADH was converted into NAD^+ through reaction of the ammonia with α-ketoglutarate in the presence of glutamate dehydrogenase. For the sake of simplicity, speed and economy, a specific semi-solid surface fluorimetric pad method was sought [4]. All of the reagents are applied onto the surface of a silicone-rubber pad; only 45 μl of reagent is necessary with a total sample volume of 85 μl. The rate of disappearance of NADH fluorescence at 460 nm (excitation wavelength 340 nm) is proportional to serum creatinine concentration. The whole assay takes less than 5 min and the calibration curve is linear up to 8.2 mg creatinine per 100 ml. The results obtained correlated well with the modified Jaffé method as run on a Technicon SMA 12/60 (correlation coefficient 0.998); the recovery was 97–100%.

An immobilized enzyme electrode for creatinine was prepared by covalent coupling directly onto the NH_3 sensor as in the work of Anfalt and White; during use over 32 days for 100 assays, no significant activity loss has been detected. This electrode can tolerate temperatures up to 40°, where the rate is twice that at room temperature (25.2°). The calibration curve is linear up to 5 mg %, showing a change of 12 mV/2 min at 2.4×10^{-5}M. The free ammonia in serum is the only interference.

Finally, we immobilized creatininase onto alkylamine porous glass beads with glutaraldehyde as cross-linking reagent. A small stirring bar loaded with the beads was used for fluorimetric determination of serum creatinine at 25°.

The bar survived more than 150 assays over 6 months. A larger stirring bar with higher activity was developed for enzymatic measurement of creatinine using an ammonia electrode as sensor. This approach can eliminate the free ammonia interference, this being an advantage over the creatinine-specific enzyme electrode, but is less sensitive to the same concentration of substrate.

PENICILLIN; AMYGDALIN AND LAETRIL

Penicillin electrodes were described by Papariello et al. [15] and Mosbach et al. [6]. The probes were formulated by layering insolubilized penicillinase onto the surface of a pH (glass) electrode. The H^+ liberated (penicillinoic acid) is measured in a weakly buffered solution. A plot of pH vs. log penicillin is linear over the concentration range 10^{-2}-10^{-4}M. The electrodes have been used to monitor penicillin formation in culture baths.

Rechnitz [7] has described an electrode useful for assay of amygdalin (and hence its derivative laetril). A cyanide probe was used, coated with a layer of entrapped β-glucosidase:

$$C_6H_5.CHCN.O.glucoside \xrightarrow{\beta-glucosidase} C_6H_5.CHO + CN^-$$

Rechnitz et al. proposed a pH of 10, justified by the argument that free CN^-(to which the CN^- electrode responds) is produced only at high pH. Great instability and long response times resulted. Mascini and Liberti [8], on the other hand, used a low pH (7) and observed greater enzyme stability and much faster response times. A linear range of 10^{-5} to 10^{-2}M was observed, with response times at pH 7 less than 1 min.

PESTICIDES

Guilbault et al. [9] have proposed an automated electrochemical system for assay of organophosphorus compounds at ppb (10^{-8}M) levels. The system uses an immobilized horse serum cholinesterase (which is totally specific for organophosphate pesticides) held between two platinum electrodes (Fig. 1). A 5×10^{-4} M solution of butyrylthiocholine iodide in tris buffer, pH 7.4, is passed over the electrodes at a flow rate of 1.0 ml/min, together with a flow of air at 1 l/min. As long as the enzyme is active, the substrate, acetylthiocholine iodide, is hydrolyzed to thiol, and a low potential (about 0.2 V; Fig. 2) is obtained. However when the enzyme is inhibited by organophosphate pesticides hydrolysis is prevented, and the potential rises to the level of I^- (Fig. 2). The system is sensitive to extremely low levels of pesticide or nerve agents, and has a very fast response time.

The immobilized enzyme sample could be used for up to 12 h of continuous operation, and was applied to the assay of various pesticides such as Systox, Parathion and Malathion in soils. A linear range of about 10ppb–10 ppm was obtained, with a reproducibility of about 3%. The device responds to many organophosphorus pesticides, so the nature of the substance must first be determined before a quantitative procedure can be developed. An instrument to measure pesticides in air and water is sold by Midwest Research Institute, Kansas City.

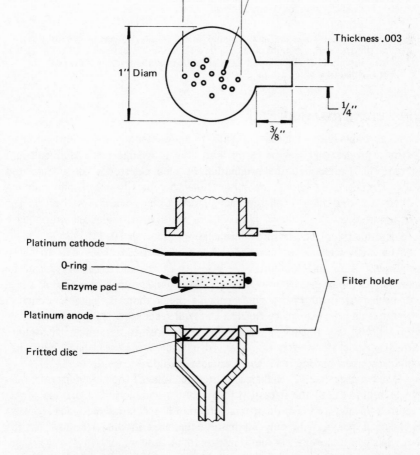

Fig. 1 – Details of enzyme pad, O-ring, and grid electrode assembly. *From ref.* [9], by permission (likewise Fig. 2).

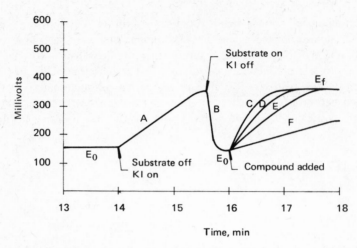

Fig. 2 – Typical operation and response curves of experimental apparatus. A: flow rate, 1ml/min: substrate solution, off: KI solution, on, B: flow rate, 1 ml/min: substrate solution, on: KI solution, off. C–F: various concentrations of the pesticide Systox added to a water stream.

CONCLUDING REMARKS

Several attempts have been made recently to design stable, self-contained enzyme electrode probes that can be easily fabricated in a large scale. Guilbault and Lubrano [10] described the production of such electrodes for glucose and L-amino acids, using various membrane films. A generally useful mild coupling method for enzyme or collagen membranes, using acylazide activation, has been reported by Coulet *et al.* [11]. Stable, very sensitive glucose sensors have been described that can measure concentrations as low as 10^{-8}M.

The analytical use of immobilized enzymes, already applicable to heavy metals also, has a great potential; the quantity of enzyme is conserved, and the same material may be used over and over. With some samples over 10,000 assays have been performed with a single enzyme preparation. By suitable induction procedures an enzyme can be isolated that will react specifically and sensitively (10^{-7}M) with any organic substance one might wish to determine. The immobilized enzyme, moreover, offers an advantage of increased specificity over soluble enzymes, making it a highly desirable analytical reagent.

Disadvantages are (1) the need to prepare some of the newer enzymes oneself, as required; (2) the necessity to prepare the immobilized enzyme, which requires some skill; and (3) the special attention to technique required for precise biochemical assays. With only minimum experience in this area, however, the researcher will discover these disadvantages to be quite minor.

In the next decade we can expect both immobilized enzyme instruments, as well as self-contained enzyme electrode probes, to be available for the assay of most substrates of analytical interest. Moreover, immobilized enzyme tubes (nylon tubes with immobilized enzymes) are already available from two companies [Miles Laboratories, Stoke Poges, England, 'Catalinks'; and Technicon Inc., Tarrytown, U.S.A.] for glucose (hexokinase-glucose phosphate dehydrogenase) and uric acid (uricase)]. Many more can be expected from these companies, as well as the advent of more industrial concerns in this market.

References

[1] Guilbault, G. (1977) *Enzymatic Methods of Analysis,* Dekker, New York.
[2] Guilbault, G. & White, C. (1978) *Anal. Chem.,* **50**, 1481–1486.
[3] Fung, K., Kuan, S., Sung, H., Guilbault, G. (1979) *Anal. Chem.,* **56**, 2319–2323.
[4] Guilbault, G., Kuan, S. & Chen, B., (1980) *Clin. Chim. Acta,* **100**, 21–31.
[5] Papariello, G., Mukerji, A., & Shearer, C. M. (1973) *Anal. Chem.,* **45**, 790–792.
[6] Mosbach, K., (1976) *Meths. Enzymol.,* **44**, 144–168.
[7] Rechnitz, G., & Llenado, R., (1971) *Anal. Chem.,* **43**, 1457–1461.
[8] Mascini, M., & Liberti, A., (1974) *Anal. Chim. Acta,* **68**, 177–184.
[9] Guilbault, G., Kramer, D. N., Bauman, E. V. & L, Goodson., (1965) *Anal. Chem.,* **37**, 1378–1381.
[10] Guilbault, G., & Lubrano, G., (1978) *Anal. Chim. Acta,* **97**, 229–236.
[11] Coulet, P., (1977) *J. Biol. Chem.,* **252**, 7919; (1977) *Biochim. Biophys. Acta,* **391**, 272–281.

#D-2   DRUGS IN SALIVA

ROKUS A. de ZEEUW, Department of Toxicology, Laboratory for Pharmaceutical and Analytical Chemistry, State University, 9713 AW Groningen, The Netherlands

There is growing interest in the secretion of drugs in saliva. For some drugs their concentration in saliva was found to be proportional to that in plasma, but for other substances such correlations were less satisfactory, irreproducible, or apparently non-existent. Some major factors involved in the salivary excretion of drugs are now reviewed, together with the requirements for obtaining reliable analytical data from saliva; if these are properly assessed and taken into account, for some drugs excellent correlations between concentrations in saliva and plasma ['S/P ratio'] can be found. For most drugs, however, additional investigations are required to determine the usefulness of saliva as a substitute for plasma in therapeutic monitoring or in pharmacokinetic studies.

The determination of drugs in plasma or serum in order to monitor their fate in the body or to optimize therapy usually poses a number of problems:

a. The complexity of the biomatrix. Plasma and serum are relatively rich in organic material, notably lipids and proteins, which may easily interfere and/or which may call for extensive clean-up.

b. The taking of samples. Most subjects find this an unpleasant and sometimes painful experience. Moreover, it poses a number of risks and it has distinct limitations with regard to sample size, frequency of sampling, need for skilled personnel, etc. Some of these factors are even more important in geriatric and paediatric investigations and in studies requiring numerous serial samples.

c. In general, the total concentration of drug is determined, including both protein-bound and free drug, whereas the concentration of free drug is considered to be related to the effect.

Thus, there is a need for a substitute body fluid for drug analysis which would give less endogenous interference, would be more easily and freely available, and would give an indication of the free, unbound drug concentration in plasma. An obvious pre-requisite is that there be a fixed and well defined relationship between the drug concentrations in that body fluid and in plasma which is concentration-independent and constant for a given individual. In addition, it should be ascertained that the analytical methodology is amenable to the substitute body fluid.

Lately, as recently reviewed by Danhof & Breimer [1], there has been strong interest in saliva as a potential substitute for plasma, reflected by many publications dealing with the saliva/plasma concentration ratio of a wide variety of drugs. For some drugs, excellent and consistent correlations between saliva and plasma concentrations could be found, with the saliva concentration seeming to reflect the free drug concentration in plasma. Yet, for other drugs correlations were less satisfactory: some showed large inter- and intra-subject variations, and in some instances no correlations at all seemed to exist.

Therefore it seems appropriate to evaluate here some important factors in the salivary excretion of drugs, and to review the basic analytical requirements for saliva assays, taking account also of some recent investigations [2-5].

SALIVARY SECRETION AND COMPOSITION

Saliva is secreted by the salivary glands situated in the buccal cavity and its neighbourhood. These glands are made up of parotid, submandibular, sublingual buccal and retromolar glands, together with glands of the lips and tip of the tongue and small isolated mucous glands of the anterior surface of the soft palate and on the margins and the root of the tongue. In humans, the parotid and the submandibular glands are the major excretory glands. Total saliva secretion in 24 h in adults may range from 500 to 1500 ml [6]. Whole saliva is a mixture of secretions of the major and minor salivary glands plus a small amount of albumin-rich gingival crevicular fluid [3]. Parotid saliva usually is a non-viscous, clear liquid, containing little insoluble matter. Submandibular and mixed saliva are usually more viscous, turbid liquids with a higher amount of insoluble matter, consisting of protein-containing material, bacterial cells and epithelial

Table 1 Concentrations of some organic substances (mg/100 ml) in saliva and in plasma. Mean values of Mandel [7] for normal adults. Salivary samples obtained after stimulation with citric acid.

Constituent	Parotid saliva	Submandibular saliva	Plasma
Total lipid	2.8	2.0	500
Cholesterol	< 1.0		160
Fatty acids	1.0		300
Amino acids	1.5		50
Proteins	250	150	6000

cell debris [3]. Although the composition of saliva depends on a number of factors such as age, sex, diet, time of the day, various diseases and also upon the stimulation of salivary flow, the average concentrations of lipids, fatty acids and proteins are considerably lower than those in plasma [7], as shown in Table 1. So if such endogenous compounds interfere in the analysis of plasma the use of saliva may circumvent this problem.

SALIVA COLLECTION AND STIMULATION OF SALIVARY FLOW

Mixed saliva samples are easily obtained by having the subjects expectorate into glass collection tubes or beakers. Techniques and devices for collecting saliva from individual glands are also available [4,7,8]. Although detailed comparative studies are scarce, it has been suggested [3,8] that parotid saliva should be preferred because of its more constant composition, its homogeneity and lower content of insoluble material and because it can be obtained in a relatively simple way without contamination from the rest of the oral secretions, food constituents or traces of orally administered drugs.

The presence of a higher content of insoluble material in mixed saliva samples is an important factor as it was recently shown that phenytoin could be bound to a considerable extent to the insoluble mucoid sediment in mixed saliva, resulting in a significantly lower phenytoin concentration in mixed saliva supernatant than in homogenized mixed saliva samples. No difference was found between parotid saliva supernatant and homogenized parotid saliva [3]. Nevertheless these findings emphasize the role of homogenized samples.

As the secretion of saliva is a relatively slow process (the production of 5 ml saliva may take of the order of 10 min), many authors prefer to stimulate the salivary flow. This can be done in a number of ways, e.g. by means of citric acid [8,9] or flavoured lozenges [4], by having the subjects chew on insoluble and hopefully inert materials like Parafilm [2, 5, 10-12], Teflon [13-16], paraffin wax [17,18] or glass beads [19], or by tongue-and-cheek movements [2]. Other advantages of stimulation have been reported to be a more constant and higher pH of stimulated saliva (pH 7.0-7.8) as compared to non-stimulated saliva (pH 5.1-6.8) [7, 20, 21] and a smaller inter-subject variation in the saliva/plasma ratio. It should be noted that the pH of stimulated saliva is closer to the pH of plasma (7.4) than that of non-stimulated saliva, and the above advantages may be related to the pH-dependence of the salivary excretion of ionizable substances (*see below*). On the other hand it has been argued that the drug concentration in stimulated saliva may give an underestimate of the plasma concentration when equilibration between plasma and saliva is slow [22,23]. Adsorption of drugs to the so-called inert chewing materials may also occur, as was reported for butaperazine and chlorpromazine [24].

MECHANISMS OF DRUG EXCRETION IN SALIVA

Saliva may be considered as a natural ultrafiltrate of plasma, such that drug concentrations would in some way reflect the concentration of non-protein-bound drug in plasma which is considered to have a more direct bearing on drug effect than does the total concentration of drug in plasma. This assumption may not always hold, e.g. when saliva production is stimulated (*see above*) or in single dose studies during the adsorption phase of the drug [1,25,26].

From the information available at this stage it seems that in the concentration ranges studied most drugs enter saliva by a passive diffusion process. However, indications for active transport mechanisms have been found for lithium [27,28], phenytoin in the rat [29] and in humans [3], and penicillin [30]. Lipophilic substances are thought to reach saliva mainly by diffusion through lipoid membranes, whereas more polar substances may reach saliva through the water-filled pores in these membranes [17,21].

It is now generally accepted that for neutral drugs and for drugs that are largely unionized at plasma pH and saliva pH, paired saliva and plasma samples can be used to calculate the fraction of unbound drug in plasma without applying a correction for differences between saliva and plasma pH [2,31,38]. Yet, for ionizable drugs such as acids and bases it is well known that their distribution across a membrane is pH-dependent, and in those cases pH differences between plasma and saliva must be taken into account in the interpretation of S/P ratios and in the calculation of the unbound drug fraction in plasma. The relationships between saliva and plasma concentrations can be expressed as follows [2].

For acidic drugs:
$$\frac{C_s}{C_p} = \frac{1 + 10^{(pH_s - pK_a)}}{1 + 10^{(pH_p - pK_a)}} \cdot \frac{f_p}{f_s} \tag{1}$$

For basic drugs:
$$\frac{C_s}{C_p} = \frac{1 + 10^{(pK_a - pH_s)}}{1 + 10^{(pK_a - pH_p)}} \cdot \frac{f_p}{f_s} \tag{2}$$

in which C_s and C_p = concentration of drug in saliva and plasma, respectively; pK_a = pK_a of drugs; pH_s and pH_p = saliva pH and plasma pH, respectively; f_s and f_p = fraction of unbound drug in saliva and plasma, respectively.

In these equations, saliva pH is the major variable. There exists a considerable inter- and intra-subject variation in saliva pH, which can be further enhanced by differences in flow rate [2,20,32]. The plasma pH can be considered to be constant at 7.4 although slight inter-individual differences can occur. The binding of drugs to plasma generally shows little intra-individual changes but it should be emphasized that changes may occur due to food constituents or to co-medicated drugs.

Inter-individual differences in plasma-protein binding are, of course, known to exist and should be taken into account. The available data on drug binding to saliva proteins indicate that this is negligibly small ($<$ 4%), so that $f_s \simeq 1$ [2,15, 33,34]. Table 2 exemplifies the impact which alteration of saliva pH can have on the ratios of saliva concentration to unbound plasma concentration. It can be

Table 2   Predicted ratios of saliva concentration to unbound plasma concentration for drugs of different pK_a values in the physiological range of saliva pH. Plasma pH assumed to be constant at 7.4. *After* Mucklow *et al.* [2].

	Ratios for acidic drugs			Ratios for basic drugs		
pK_a	pH 6	pH 7	pH 8	pH 6	pH 7	pH 8
2	0.04	0.40	3.98	1.00	1.00	1.00
4	0.04	0.40	3.98	1.01	1.00	1.00
6	0.08	0.42	3.87	1.92	1.06	0.97
8	0.81	0.88	1.60	20.28	2.21	0.40
10	1.00	1.00	1.00	25.06	2.51	0.25
12	1.00	1.00	1.00	25.06	2.51	0.25

seen that very weak acidic drugs with a $pK_a > 9$ do not need correction for salivary pH, whereas for weak basic drugs this applies to substances with a $pK_a < 5$. For other acidic and basic drugs the salivary pH is an important variable with a large impact on S/P ratios, as is further demonstrated in Fig. 1 which shows predicted S/P ratios over a range of salivary pH for a number of neutral, acidic and basic drugs. The pK_a values and f_p values for these drugs are given in Table 3. It can be seen from these data that, depending on saliva pH, drug concentrations in saliva not only can be higher than the concentrations of unbound drug in plasma but also higher than the total concentration in plasma (Fig. 1; procainamide, pethidine), despite the occurrence of protein binding; it does not have to be assumed that there are active transport mechanisms.

Thus, for those acidic and basic drugs that are largely ionized at normal plasma pH it is imperative that pH measurements in saliva be carried out or that experiments be done under standardized conditions with regard to salivary flow rate and pH. However, it should be noted that even after correction for saliva pH, saliva-to-plasma correlations may remain rather poor for certain

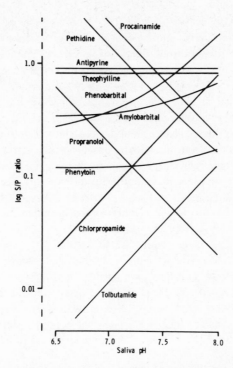

Fig. 1 – Predicted S/P ratios over a range of saliva pH values for a selection of drugs with known pK_a and f_p values, assuming plasma pH = 7.4 and $f_s \cong 1$. Semilogarithmic scale. *From [2], with permission of The Mosby Co.*

Table 3 Summary of pK_a values and f_p values (fraction of unbound drug in plasma) for a series of acidic and basic drugs as depicted in Fig. 1. *From* Mucklow *et al.* [2].

Substance	pK_a	f_p	Substance	pK_a	f_p
Acidic drugs			*Basic drugs*		
Chlorpropamide	4.8	0.2	Theophylline	0.7	0.85
Tolbutamide	5.4	0.03	Antipyrine	1.4	0.92
Phenobarbital	7.2	0.60	Pethidine	8.7	0.60
Amobarbital	7.95	0.42	Procainamide	9.4	0.85
Phenytoin	8.3	0.13	Propranolol	9.5	0.16

drugs as was found for chlorpropamide [2]. One factor responsible for such deviations could be a difference in pH between saliva at the time it is formed and saliva as collected. Adsorption of drugs to buccal membranes or inadequate analytical procedures at lower concentrations may also be involved.

Although for neutral, weakly acidic or weakly basic drugs the predictability of S/P ratios seems much better at this stage, it may be that we still are not aware of all factors involved. Anavekar *et al.* [3], in a detailed study on phenytoin in non-stimulated saliva, found close linear relationships between salivary and serum ultrafiltrate concentrations in various saliva fractions. Yet, surprisingly, all salivary concentrations were significantly and consistently lower than the corresponding serum ultrafiltrate concentration. The smallest underestimation was 16% in parotid saliva, whereas in mixed saliva supernatant it was 25% due to binding of phenytoin to the insoluble mucoid sediment. For theophylline, a very weak base with a pK_a of 0.7, values for the unbound drug fraction in plasma range from 0.41 to 0.85 [2,10], whereas S/P ratios range from 0.52 to 0.85 [10,25]. Finally, it should be stressed that also for weakly acidic and weakly basic drugs we have to be careful in simply assuming that the concentration of these drugs in saliva corresponds to the free, unbound plasma drug concentration. For example, S/P ratios for phenobarbital are usually found to be fairly constant and independent of concentration, ranging from 0.30 to 0.40 [2,18,35-37], whereas the free fraction of drug in plasma was found to be 0.60 [2].

ANALYTICAL METHODOLOGY

The usefulness of saliva determinations clearly depends also on the application of analytical procedures adequate for the assay in saliva. In general, one would be inclined to select a method that has been originally developed for plasma or serum; but the applicability of such methods to saliva must be fully ascertained with regard to sensitivity, accuracy, precision and interferences.

Sensitivity is an important factor in that many organic drugs appear in saliva in lower concentrations than in plasma as a result of plasma protein binding and/or pH-differences between saliva and plasma. The saliva concentrations, which may be lower than in plasma by a factor 100, may therefore require special analytical adaptions. For the more potent drugs which even in plasma are in the low ng/ml or pg/ml range and which are largely bound to proteins, saliva analysis may become impossible because of the lack of sufficiently sensitive detection systems. The sensitivity limitation will also hold for the salivary analysis of drug metabolites that are present at very low levels in plasma.

Accuracy and precision may also represent problems in saliva determinations. An inherent problem when dealing with smaller concentrations is that recoveries tend to become lower and less reproducible. Therefore, these parameters should be critically investigated in saliva over the entire expected concentration range and with a sufficiently high number of replicate analyses to draw statistically valid conclusions. Unfortunately these factors seem to have had little attention so far, and some of the observed discrepancies in saliva concentrations may well have been due to inadequate accuracy and precision of the analytical methodology.

Because saliva contains relatively small amounts of endogenous organic material, compounds such as lipids, fatty acids or proteins are less likely to interfere. Yet, there remains a risk of adsorption to mucoid proteins present, as was shown for phenytoin [3]. Adsorption losses onto the inert material used to stimulate saliva flow have also been reported [24], and adsorption to the oral mucosa—although not yet reported—cannot be excluded, in particular when dealing with low drug concentrations. Obviously, contamination of saliva by food constituents or remaining traces of orally administered drugs should be avoided.

Finally, the type of correlation between plasma concentration and saliva concentration should be established experimentally. This must be done over as wide a concentration range as possible, and include both inter- and intra-subject variability. In addition, the possible influence of type of saliva, saliva flow and pH, collection mode, sampling time, age, sex, diseases and co-medication should be investigated if relevant.

For a few drugs such as phenobarbital [2,18,35–37], carbamazepine [5], ethosuximide [37] and antipyrine [2,37,38] sufficient data are available to conclude that saliva is a reliable substitute for plasma. For carbamazepine, which appears to behave as a model drug, this is exemplified in Figs. 2 & 3 and

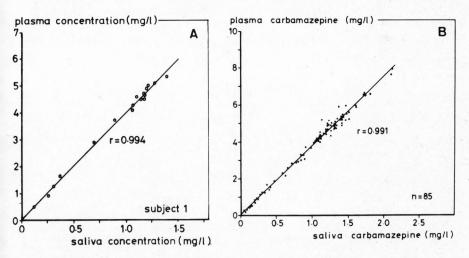

Fig. 2 – Intra-subject correlation (A) and inter-subject correlations in 7 subjects (B) between the saliva and the plasma concentrations of carbamazepine in paired samples.

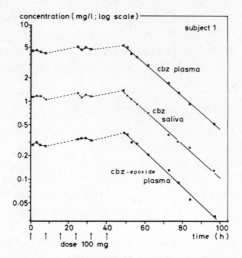

Fig. 3 – Semilogarithmic concentration *vs.* time curves for carbamazepine in plasma and saliva and for carbamazepine epoxide in plasma showing that the half-life of carbamazepine can be determined from saliva determinations. Steady-state concentrations were measured during the first two days when carbamazepine was administered as indicated.

in Tables 4 & 5. The S/P ratio was found to be linear over a wide concentration range with little inter- or intra-subject variation, and was independent of saliva flow and pH, collection mode, sampling time and co-medication. The saliva concentration appeared to be directly correlated with the concentration of unbound drug in plasma, the free fraction being of the order of 27% [39].

Table 4 Concentration ratios of carbamazepine in saliva and plasma in 7 subjects.

Subject	mean S/P ratio ± S.E.M. (& no. of abstractions)
1	0.25 ± 0.02 (17)
2	0.26 ± 0.01 (17)
3	0.28 ± 0.01 (15)
4	0.26 ± 0.02 (16)
5	0.26 ± 0.01 (8)
6	0.27 ± 0.01 (4)
7	0.26 ± 0.01 (8)

Table 5   Half-lives of carbamazepine as determined in plasma and in saliva in 4 subjects.

Subject & sex	Half-life, h	
	Plasma	Saliva
1, F	14.2	14.3
2, M	20.9	19.5
3, F	10.7	10.3
4, F	11.3	11.9

The metabolite carbamazepine epoxide could not be detected in saliva, probably because the HPLC method used has a detection limit of 0.4 $\mu g/ml$.

For some other drugs such as phenytoin [2,3,18,35-37], primidone [35-37]. diazepam [4], procainamide [2,11] and tolbutamide [2,15], there are various indications that a satisfactory relationship between saliva and plasma concentrations can be obtained, although not all variables have yet been investigated, and there are still divergences amongst different reports. However, information on other drugs is rather contradictory or quite non-existent, and thus further and more comprehensive investigations are necessary to fully assess the usefulness of drug analysis on saliva samples.

References

[1] Danhof, M. & Breimer, D. D. (1978) *Clin. Pharmacokinetics,* **3**, 39–57.
[2] Mucklow, J. C., Bending, M. R., Kahn, G. C. & Dollery, C. T. (1978) *Clin. Pharmacol. Therap.,* **24**, 563–570.
[3] Anavekar, S. N., Saunders, R. H., Wardell, W. M., Shoulson, I., Emmings, F. G., Cook, C. E. & Gringeri, A. J. & *Clin. Pharmacol. Therap.,* **24**, 629–637.
[4] DiGregorio, G. J., Piraino, A. J. & Ruch, E. (1978) *Clin. Pharmacol. Therap.,* **24**, 720–725.
[5] Westenberg, H. G. M., de Zeeuw, R. A., van der Kleijn, E. & Oei, T. T. (1977) *Clin. Chim. Acta,* **79**, 155–161.
[6] Scientific Tables (1970), 7th edn (Diem, K. & Lentner, C., eds), Ciba-Geigy, Basle, pp. 643–646.
[7] Mandel, I. W. (1974) *J. Dental Res.,* **53**, 246–266.
[8] Stephen, K. W. & Speirs, C. F. (1970) *Br. J. Clin. Pharmacol.,* **3**, 315–319.

[9] Reynolds, F., Ziroyanis, P. N., Jones, N. F. & Smith, S. E. (1976) *Lancet,* 2, 384-386.

[10] Koysooko, R., Ellis, E. F. & Levy, G. (1974) *Clin. Pharmacol. Therap.,* 15, 454-460.

[11] Koup, J. R., Jusko, W. J. & Goldfarb, A. L. (1975) *J. Pharm. Sci.,* 64, 2008-2010.

[12] Levy, G. & Lampman, T. (1975) *J. Pharm. Sci.,* 64, 890-891.

[13] Graham, G. & Rowland, M. (1972) *J. Pharm. Sci.,* 61, 1219-1222.

[14] Bochner, F., Hooper, W. D., Sutherland, J. M. & Eadie, M. J. (1974) *Arch. Neurol.,* 31, 57-59.

[15] Matin, S. B., Wan, S. H. & Karam, J. H. (1974) *Clin. Pharmacol. Therap.,* 16, 1052-1058.

[16] Boxenbaum, H. G., Berkersky, I., Mattaliano, V. & Kaplan, S. A. (1975) *J. Pharamacokin and Biopharmaceutics,* 3, 443-456.

[17] Hoprich, P. D. & Wahrshhauser, D. M. (1974) *Antimicrobial Agents and Chemotherapy,* 5, 330-336.

[18] Troupin, A. S. & Friel, P. (1975) *Epilepsia,* 16, 223-227.

[19] Inaba, T. & Kalow, W. (1975) *Clin. Pharmacol. Therap.,* 18, 558-562.

[20] Schmidt-Nielsen, B. (1946) *Acta Phys. Scand.,* 11, 104-110.

[21] Feller, K., LePetit, G. & Marx, U. (1976) *Pharmazie,* 31, 745-746.

[22] Borzelleca, J. F. & Putney, J. W. (1970) *J. Pharmacol. Exptl. Therap.,* 174, 527-534.

[23] Gruneisen, A. & Witzgall, H. (1974) *Europ. J. Clin. Pharmacol.,* 13, 357-360.

[24] Chang, K. & Chio, W. L. (1976) *Res. Comm. Chem. Pathol. Pharmacol.,* 13, 357-360.

[25] Knop, H. J., Kalafusz, R., Knolls, A. J. F. & van der Kleijn, E. (1975) *Pharm. Weekblad,* 110, 1297-1299.

[26] De Blaey, C. J. & De Boer, A. G. (1976) *Pharm. Weekblad,* 111, 1216-1221.

[27] Burgen, A. S. V. (1958) *Canad. J. Biochem. Physiol.,* 36, 409-411.

[28] Spring, K. R. & (1958) Spirtes, M. A. (1969) *J. Dental Res.,* 48, 546-549 and 550-554.

[29] Allen, M. A., Wrenn, J. M., Putney, J. W. & Borzelleca, J. F. (1976) *J. Pharmacol. Exptl. Therap.,* 197, 408-413.

[30] Borzelleca, J. F. & Cherrick, H. M. (1965) *J. Oral Therap. Pharmacol.,* 2, 180-187.

[31] Dvorchik, B. H. & Vesell, E. S. (1976) *Clin. Chem.,* 22, 868-878.

[32] Dawes, C. & Jenkins, G. N. (1964) *J. Physiol.,* 170, 86-100.

[33] Rasmussen, F. (1964) *Acta Pharmacol. Toxicol.,* 21, 11-19.

[34] Killmann, S. A. & Thaysen, J. H. (1956) *Scand. J. Clin. Lab. Invest.,* 7, 86-91.

[35] Schmidt, D. & Kupferberg, H. J. (1975) *Epilepsia,* 16, 735-741.

[36] Cook, C. E., Amerson, E., Poole, W. K., Lesser, P. & O'Tuama, L. (1976) *Clin. Pharmacol. Therap.,* **18**, 742-747.

[37] Horning, M. G., Brown, L., Nowlin, J., Lertratanangkoon, K., Kellaway, P. & Zion, T. E. (1977) *Clin. Chem.,* **23**, 157-164.

[38] Vesell, E. S., Passananti, G. T., Glenwright, A. & Dvorchik, B. H. (1975) *Clin. Pharmacol. Therap.,* **18**, 259-272.

[39] Hooper, W. D., Dubetz, D. K., Bochner, F., Cotter, L. M., Smith, A., Eadie, M. J. & Tyrer, J. H. (1975) *Clin. Pharmacol. Therap.,* **17**, 433-440.

#D-3   SOLVENT–PARTITIONING STRATEGY IN DRUG ASSAYS*

KENNETH H. DUDLEY, Department of Pharmacology, University of North Carolina School of Medicine, Chapel Hill, North Carolina, 27514, U.S.A.

Some factors that affect the distribution of a solute are summarized, with allusion to the behaviour of phenobarbital. A weakly acidic or basic drug may be extractable from an aqueous medium in which the drug is largely ionized. Consideration is given to solvent-extraction strategies that may minimize lipid interferences or select out individual analytes.

The knowledge of how a drug distributes between buffered aqueous solutions and organic solvents can be useful not only in the design of an efficient extraction scheme for a drug assay, but also in the challenge to eliminate potential pitfalls in the assay [1–4].

DISTRIBUTION OF A SOLUTE BETWEEN TWO PHASES

At a given temperature, the *partition ratio* may be expressed as $K' = C'_{org}/C'_{aq}$ where the concentration terms include *all species of solute* present in the two phases, respectively. For the *partition coefficient, K,* correction is made in the experimental data for dissociation, ionization, and association of the solute in the two-phase system [5], the concentration terms referring to a single species of the solute in each phase. However, K' values determined within 10% suffice for the analyst who is primarily concerned only with the recovery of total solute and the design of an efficient extraction scheme.

Obviously there is a net movement of solute as the relative volume of the organic phase is increased. Thus, if 3 μg of a hypothetical solute with $K' = 2$ in a given two-phase system were partitioned between equal 1-ml volumes, 2 μg of the solute would be distributed in the organic phase and 1 μg in the aqueous phase. If the phase volumes were 3 ml and 1 ml respectively, the amounts of solute in the two phases would be 2.57 μg and 0.43 μg respectively.

---

*Editor's note: The introductory part of the author's text has been shortened, and distribution values altered from lighter phase *vs.* (as denominator) heavier phase to organic phase *vs.* aqueous phase.

For a solute that ionizes by gain or loss of a proton in an aqueous medium K' will depend on pH. Thus K' for a weak organic acid will decrease as the pH of the aqueous phase is increased from 2 units below to 2 units above pK_a. A useful expression, due to Butler [6], to calculate K' for a weak acid as a function of pH is $\log(K/K' - 1) = pH - pK'_a$ where K refers to the undissociated form of the weak acid, assumed to have a constant distribution over the pH range studied, and K' to the distribution of total weak acid. In practice, K is determined at a pH where the weak acid is undissociated, and the pK'_a is determined by a conventional technique. K' can then be calculated for any desired pH. This point has been illustrated elsewhere [4] for phenobarbital, a weak dibasic acid. For example, if the drug is partitioned between ether and phosphate buffer of pH 8.80, a buffer in which it exists as its monobasic anion to the extent of 97.5%, K' = 1.3. Thus, with an equal volume of ether, about 56% of the total amount of drug originally present in the pH 8.80 buffer can be recovered; with a two-fold volume of ether, 72% of the drug can be extracted; with a three-fold volume, 80%.

EXTRACTION STRATEGIES

Fatty acids are normal constituents of plasma which, unless removed by appropriate manipulations, can be devastating to the reliability of gas chromatographic (GC) drug assays [as is also considered by de Zeeuw, #F-3–*Ed.*] especially where the residue from a dried-down solvent extract is reconstituted in a small volume of solvent or derivatized. A simple clean-up extraction step, which eliminates fatty acids, involves partitioning of the residue between hexane and methanolic HCl [7-9]. Any fatty acids or lipoidal components present in the residue distribute into the hexane phase (which is discarded), and the drug(s) distribute into the methanolic HCl phase, which is processed for quantitation of the drug. A hexane/aqueous methanol system has also been used for the same purpose [10]. Such a clean-up step, in GC assays where fatty acids could interfere, can eliminate serious pitfalls provided that the drug and its internal standard have a low partition ratio in the hexane/methanolic HCl or hexane/methanol systems.

Svensmark and Kristensen [11] studied the distribution of phenytoin (pK'_a 8.3) and phenobarbital (pK'_a 7.2) between equal volumes of chloroform and borate buffers of different pH (Fig. 1). In the chloroform/pH 8.8 buffer system, about 88% of the phenobarbital was present in the aqueous phase, and about 88% of the phenytoin was retained by the chloroform phase. If two such extractions were carried out successively, each drug could be recovered in a yield of 77% and was contaminated by the other only to the extent of 1.4%. The recovery and purity of each drug was sufficient to permit spectrophotometric quantitation of the drugs when both were present in the same plasma sample.

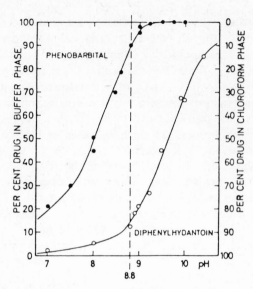

Fig. 1 – Distribution of phenytoin (diphenylhydantoin) and phenobarbital between equal volumes of 0.1M borate buffers of different pH and chloroform. *From [11] by permission.*

The GC method of Schäfer [12] for the simultaneous determination of underivatized antiepileptic drugs is an example of a systematic study of partitioning properties that was made to furnish the basis for the extraction scheme. His multiple-step extraction scheme resulted in excellent recoveries of the drugs, the lowest recovery being about 75%.

On-column methylation is a popular technique for the simultaneous determination of multiple antiepileptic drugs in plasma samples [13]. Very precise results can be obtained with an assay scheme that has been properly designed to eliminate interfering substances and problems with on-column methylation reactions [14]. In the scheme of Dudley *et al.* [15], the overall recoveries are 88% for phenytoin, 98% for phenobarbital, 34% for primidone, and 79% for ethosuximide. The recovery of pyrimidone, although low, is adequate for a precise determination of it. In the design of their assay care was taken to eliminate all interferences in the extraction scheme, and to employ, for quantitation of primidone, an appropriate internal standard, 4-methylprimidone; it has partitioning and chromatographic properties similar to primidone under the conditions adopted.

References

[1] Craig, L. C. & Craig, D. (1956) in *Techniques of Organic Chemistry,* Vol. 3, Part 1 (Weissberger, A., ed.), Interscience, New York, pp. 149–393.

[2] Bush, M. T. (1961) *Microchem. J.*, **5**, 73-90.

[3] Jones, C. R. (1976) in *Assay of Drugs and Other Trace Compounds in Biological Fluids* [this series, Vol. 5] (Reid, E., ed.), North-Holland, Amsterdam, pp. 107-113.

[4] Dudley, K. H. (1978) in *Antiepileptic Drugs: Quantitative Analysis and Interpretation* (Pippenger, C. E., Penry, J. K. & Kutt, H., eds.), Raven Press, New York, pp. 43-54.

[5] Leo, A., Hansch, C. & Elkins, D. (1971) *Chem. Rev.*, **71**, 525-616.

[6] Butler, T. C. (1956) *J. Pharmacol. Exp. Ther.*, **108**, 11-17.

[7] Pippenger, C. E. & Kutt, H. (1978) in *Antiepileptic Drugs: Quantitative Analysis and Interpretation* (Pippenger, C. E., Penry, J. K. & Knutt, H., eds.), Raven Press, New York, pp. 199-208.

[8] Kupferberg, H. J. (1970) *Clin. Chim. Acta*, **29**, 283-288.

[9] Sichler, D. W. & Pippenger, C. E. (1978) as for 7, pp. 335-338.

[10] Pynnönen, S., Sillanpää, M., Frey, H. & Iisalo, E. (1976) *Epilepsia*, **17**, 167-172.

[11] Svensmark, O. & Kristensen, P. (1963) *J. Lab. Clin. Med.*, **61**, 501-507.

[12] Schäfer, H. R. (1975) in *Clinical Pharmacology of Antiepileptic Drugs*, (Schneider, H., Janz, D., Gardner-Thorpe, C., Meinardi, H. & Sherwin, A. L., eds.), Springer-Verlag, New York, pp. 124-130.

[13] Pippenger, C. E., Penry, J. K. & Kutt, H. (eds.) (1978) as for 7, 367 pp.

[14] Dudley, K. H. (1978) as for 7, pp. 19-34.

[15] Dudley, K. H., Bius, D. L., Kraus, B. L. & Boyles, L. W. (1977) *Epilepsia*, **18**, 259-276.

#D-4   SAMPLE PREPARATION FOR TRACE DRUG ANALYSIS

J. A. F. de SILVA, Department of Pharmacokinetics and Bio-pharmaceutics, Hoffman-La Roche Inc., Nutley, N. J. 07110, U.S.A.

Consideration is given to the design of procedures for the selective extraction and 'clean-up', in relation to the choice of measurement method, of drugs in a biological matrix. The chemical structure and the pharmacokinetics of a compound govern not only the sensitivity and specificity requirements of the assay, but also the most suitable biological specimen for its quantitation. For acidic or amphoteric drugs, serum or plasma is preferable to blood. Sample preparation should aim to optimize various factors for the sake of a reliable and validated method for the compound, suitable for use in clinical therapeutic monitoring.

The context is a drug in a complex biological matrix from which it must be selectively extracted and cleaned up prior to quantitation [1]. Its chemical structure influences choice of measurement method, while its ionizable groups determine its acid/base character (pK_a) and its extractability [2] which is also influenced by protein binding. The dose administered, the bioavailability of the dosage form, and the pharmacokinetic profile of the drug govern the absolute concentration of drug and/or metabolites to be quantitated [3]. These criteria influence the ultimate sensitivity and specificity or any compromise thereof of the analytical method required [4].

Basic drugs generally exhibit a high volume of distribution (Vd) which results, especially at low doses (< 1 mg/kg), in low blood concentrations that may be further reduced owing to extensive 'first pass' biotransformation and elimination. Acidic and amphoteric drugs are generally protein-bound and tend to remain in the central compartment, i.e. the plasma, and are also extensively eliminated in the urine (e.g. salicylates, antibiotics) [3,5]. Thus, while whole blood suits for basic drugs, plasma, serum and/or urine is preferred for the analysis of acidic or amphoteric compounds.

The solubility, pH of the medium, and the partition coefficient of a drug into a particular solvent influence its extractability [K.H. Dudley, #D-3, *this volume*]. Solvents lighter than water facilitate phase transfer and minimize contamination of the organic extract with the aqueous phase during the transfer step. Aqueous molar buffers cause protein denaturation due to their salting-out effect. Basic drugs are generally extracted at pH > 7.0, while acidic drugs require protein precipitation and are best extracted below pH 5.0, preferably from plasma or serum (which already has the bulk of the haemprotein removed).

The degree of clean-up required depends on the analytical method used (commonly GC or HPLC) and on the tolerance of the detection system to contamination. The options available for processing a biological specimen must be tailored not only to the method itself, but also to the sensitivity and specificity required of it. Factors responsible for compound losses during sample preparation (adsorption, stability) are critical at low concentrations; hence maximizing the overall recovery is essential to good precision and reproducibility.

Enhancing the sensitivity and specificity of detection through derivatization, where feasible [6,7], has certain inherent advantages even if the intrinsic sensitivity is adequate and/or blood concentration is not a limiting factor. With minimization of endogenous interferences, the sample volume extracted can be reduced from ml to μl amounts, and/or the aliquot of the final residue analyzed can be reduced. All the above factors should be optimized in the overall development of a reliable and validated method for eventual clinical evaluation.

The range of analytical methods [4] includes chromatographic techniques with a variety of selective detectors, and non-chromatographic techniques, all of which are capable of sensitive quantitation at ng to pg levels. The type of assay selected will also govern the amount of sample preparation and clean-up required, and the biological fluid best suited to it, i.e. whole blood, plasma, or urine. Faeces and tissues generally have to be analyzed also in drug-metabolism or toxicological studies.

SAMPLE PREPARATION FOR WHOLE BLOOD AND TISSUES

As indicated in Scheme 1, acidic drugs are in general extractable at pH < 6.0 while basic drugs are extractable at pH > 6.0. The propensity to and the extent of plasma-protein binding (albumin) of a drug determines the actual free or pharamacologically effective fraction of the total amount present in whole blood. The free concentration can be determined by equilibrium dialysis [cf. C. J. Coulson & V. J. Smith, this volume] or by the analysis of a body fluid such as saliva which is in equilibrium with blood [cf. R. A. de Zeeuw, this section]. Adjustment of the natural pH of whole blood by the use of a buffer solution may denature the protein and release bound drug, disrupting the equilibrium of free circulating drug. Thus, chemical manipulation of a biological specimen yields total drug concentration = free (pharmacologically active) + protein-bound (inactive) drug.

Tissue specimens must first be homogenized in water, saline or a buffer solution to achieve a homogenous suspension, followed by denaturation of the protein prior to extraction at a suitable pH (Scheme 1). Whole blood suits for the direct extraction of a basic drug with use of appropriate molar buffers ranging in

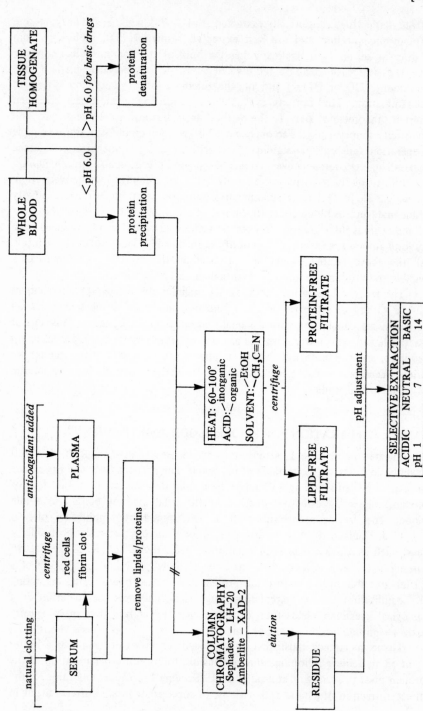

Scheme 1 Sample preparation approaches for drugs in whole blood or a tissue homogenate.

pH from 7.0 to 14.0. The ionic strength (molarity) of the buffer effects protein denaturation to form a loose floc, minimizing physical entrapment or co-precipitation losses of the drug, while the pH selected (usually 2 pH units above the pK_a) renders the drug > 99% unionized and hence lipophilic and extractable into a suitable solvent (monophasic or multiphasic) into which it has a high partition coefficient, e.g. diethyl ether, benzene (toxic!), benzene-dichloromethane (9:1 by vol.), n-heptane + 1.5% isoamyl alcohol. Solvents lighter than water (s.g. < 1) facilitate phase transfer following centrifugation, and minimize contamination of the organic extract with the aqueous phase during the transfer step. Centrifugation at 0–5° results in excellent phase separation, since the denatured haemprotein compacts down to a dense mat at the interface, enabling nearly quantitative phase transfer, and tends to minimize the solubility of water in the organic phase and consequent contamination.

The extraction of acidic drugs at pH < 6.0 results in extensive protein precipitation in whole blood with attendant co-precipitation losses; hence the use of a fraction separated from freshly drawn whole blood (i.e. plasma or serum) is preferred. The separation of serum requires natural clotting of fibrinogen, which precipitates out the bulk of the haemprotein and lipoproteins; this can result in co-precipitation losses in the fibrin clot. The separation of plasma requires the addition of an anticoagulant (oxalate, citrate, heparin or NaF) to prevent clotting, followed by centrifugation to separate the red cells from the lipids and lipoprotein fraction; thus, plasma is a lipid/lipoprotein-rich medium. The separation of either fraction from whole blood can result in co-precipitation losses; moreover, some drugs permeate the red cells [O. Borgå, *this volume*] and can be irreversibly bound to the protein within, as in the case of coumarin analogues.

In processing plasma or serum it may help to remove the lipids/lipoprotein prior to extraction of acidic drugs [*cf.* a later contribution by R. A. de Zeeuw, *this volume*]. This may be accomplished by passing the specimen through a column of Sephadex-LH-20 or XAD-2 resin which retains the drug compound, while the lipids pass through and are further washed out with distilled water. The adsorbed drug is stripped off the column with methanol and is ready for analysis (Scheme 1). Alternatively, denaturation of the lipids/lipoprotein in serum or plasma or protein precipitation in whole blood may be accomplished either by direct heating in a 60–100° water-bath, or by the use of an acid [inorganic acids such as HCl, $HClO_4$, H_3PO_4 or tungstic acid, or organic acids such as trichloroacetic acid (TCA)], or by the use of organic solvents such as absolute ethanol or acetonitrile, the latter being preferred due, again, to the formation of a loose floc. The pH of the lipid/protein-free filtrate (PFF) can be adjusted to optimize the extraction of the drug in question.

Thus, whole blood or a tissue homogenate may be processed to ultimately yield a protein-free filtrate (or supernatant) whose pH may be adjusted appropriately to permit exhaustive analysis of the sample (generally done in forensic toxicological analysis) for acidic, neutral and basic drugs, while the non-extractable amphoteric compounds may be determined directly in the PFF. Tissue homogenates can also be subjected to enzymatic incubation (digestion) at 55° for 60 min with a proteolytic preparation (M. D. Osselton, *this volume*) without damage to the compounds of interest, a very useful tool for the forensic toxicologist.

SAMPLE PREPARATION FOR URINE AND FAECES

A flow diagram for the extraction of urine and faeces is shown in Scheme 2.

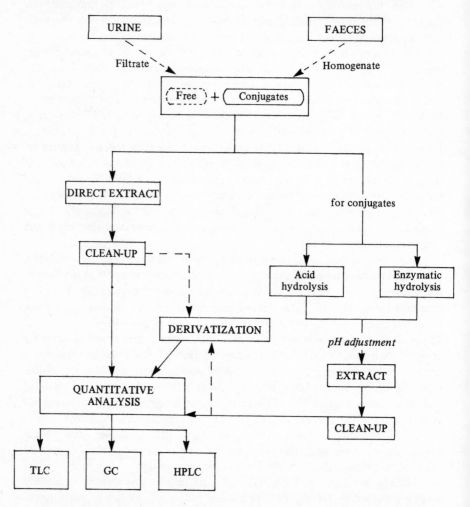

Scheme 2 Sample-preparation approaches for drugs in urine and faeces.

The analysis of these two media requires an aliquot of a representative sample, i.e. an aliquot of urine from a total voidance volume collected over a known excretion period (e.g. 24, 48 h), and for faeces an aliquot of a homogenate of a total voidance collected similarly. The sample is filtered to remove particulates and analyzed for the free (directly extractable) and bound (conjugated) fractions of drug and metabolites present.

Basic extracts of urine or faeces are not as heavily contaminated with endogenous compounds as are acidic extracts, which have an abundance of phenolic and indolic acids, and require additional clean-up, e.g. column chromatography. The conjugated or bound fraction has to be hydrolyzed suitably with acid, e.g. to cleave hippurates and other amino acid conjugates, or with β-glucuronidase/sulphatase at 37° for 2–12 h, suitably in a Dubnoff incubation shaker. The products can be extracted after appropriate pH adjustment, and the extract cleaned up by chromatography (column or TLC) or liquid–liquid partition (acid/base).

Depending on the final determinate step, the residue of the final extract may have to be derivatized (silylation of hydroxyl groups, esterification of carboxylic acids) for GC analysis, or analyzed *per se* (Scheme 2). HPLC is generally the method of choice, since the components can usually be analyzed without derivatization and concentration is not limiting. Resolution of endogenous impurities not removed by previous clean-up may be a limiting factor.

SAMPLE PROCESSING *vs.* ANALYTICAL DETERMINATE STEP

The degree of sample preparation and clean-up required is usually a function of the analytical method to be used and the tolerance of the specific detection system to contamination. The options available are outlined in Scheme 3. The biological specimen (whole blood/plasma/tissue homogenate or proteolytic digest) undergoes a protein precipitation step, followed by pH adjustment and selective extraction into a suitable solvent. It can then be processed in a way depending on the analytical method to be used [4].

Radioimmunoassay (RIA) can be performed either directly on the biological sample (plasma or serum) or on the residue of the solvent extract which can also be used for HPLC analysis. The residue is reconstituted in 50–100 μl of the mobile phase to be used and passed through a 40 μm millipore filter to remove colloidal lipids/lipoproteins which precipitate out if a reverse phase (RP) system is used.

HPLC analysis is advantageous: amongst the options (Scheme 3) is ion exchange as used for highly polar zwitterionic biochemical molecules.

GC analysis may require more extensive sample preparation depending on the drug to be analyzed. The drug in the solvent extract can be derivatized directly by extractive alkylation (e.g. for *N*-1-desalkyl-1, 4-benzodiazepin-2-ones) followed by a clean-up step, or the intact moiety can be further purified by

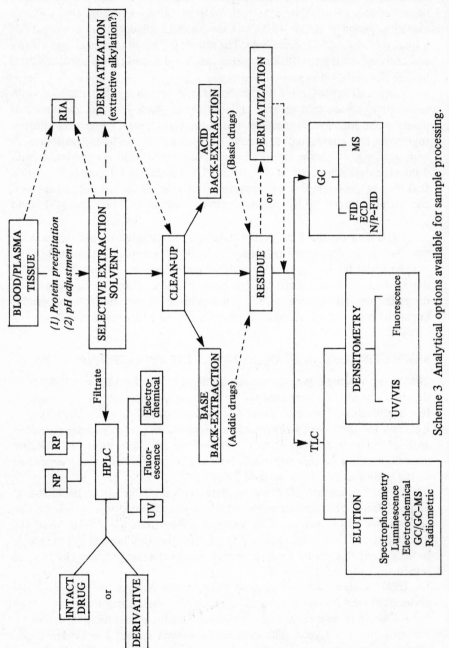

Scheme 3 Analytical options available for sample processing.

back-extracting and then re-extracting into a solvent (Scheme 3), the residue from which may be derivatized (silylation, esterification, alkylation) prior to GC analysis suitably with a highly selective detector.

TLC analysis may be performed directly on the residue from the initial solvent extract, unless pre-chromatographic clean-up is required due to a derivatization step. TLC separation can be used, *per se,* as an effective clean-up step since multiple development may be performed, first in a non-polar lipophilic solvent system to move the endogenous contaminants from the origin to the solvent front, and then in a more polar solvent to resolve the compounds of interest from one another. Scheme 3 indicates options for quantitation.

FACTORS CAUSING COMPOUND LOSSES

Compound losses during sample preparation may be due to several factors (*see panel*) which can usually be corrected to optimize overall recovery and so ensure the accuracy of the assay. Adsorption on glass or metal surfaces or on the rubber stoppers of collection vessels ('Vacutainers') can be a problem and may be corrected by siliconization (Siliclad^R), which confers more resistance to hydrolysis than silanization. Adsorption losses are usually severe with primary aliphatic or aromatic amines and thio compounds. Whereas there may be *coprecipitation* with red cells or fibrin clots (*see above*), extraction of buffered whole blood yields recovery of cell-bound drug since the cells are ruptured and the drug is desorbed.

ADSORPTION	— Glass/metal surfaces
CO-PRECIPITATION	— Red cells/fibrin clot
CHEMICAL BREAKDOWN	— Instability, chemical or biological — Photochemical — Thermal
DERIVATIZATION	— Incomplete reaction — By-products — Removal of excess reagents
CHELATION	— Heavy metals — Intercalation
EVAPORATION	— Insoluble residue — Volatility of drug

Factors that may cause compound losses during sample preparation.

Chemical breakdown losses can be attributed to chemical or biological instability of the drug (e.g. enzymatic hydrolysis of esters), to photochemical instability, e.g. isomerization of carotenoids, or to thermal instability especially during neutralization of strong acid or base during clean-up steps, when the heat of neutralization can cause breakdown. Compounds that undergo spontaneous ring opening in acid or base could also be lost due to partial hydrolysis. *Derivatization* losses are usually due to only partial conversion into the desired product, to the formation of unwanted by-products, or to evaporative removal of excess reagents causing the derivative to azeotrope with the solvent. *Chelation* losses are rare and may be due to complex formation with heavy metals and or to intercalation of the drug molecule with endogenous macromolecules, e.g. psoralens intercalating with DNA. *Evaporation* may lead to unforeseen losses due to volatility of the drug (e.g. amphetamines) or a derivative, as has been documented [8], or to insolubility of the residue formed. In the day-to-day use of a method one has to be constantly aware of the interplay of these factors, of which co-precipitation and chelation are the least serious.

PHARMACOKINETIC FACTORS WHICH INFLUENCE SAMPLE PREPARATION

A. Biotransformation
The relevance of elucidation, at an early stage in drug development, of the *in vivo* metabolite profile with radioisotopically labelled drug is evident from a study [cited in ref. 4] with clonazepam in man. Comparison of the plasma ^{14}C-profile with that for ether-extractable ^{14}C-clonazepam indicated the presence of polar non-extractable metabolites. For whole blood the clonazepam concentrations showed good accord between TLC-radiometry and GC-ECD, confirming the accuracy and specificity of the latter technique, but were somewhat lower than that of ^{14}C-clonazepam in plasma, perhaps warranting the use of plasma for enhancing the sensitivity of the GC-ECD assay due to an artificial concentration effect (1 ml of plasma corresponding to about 2 ml of whole blood).

Chromatography is necessary to reveal the relative amounts of the parent drug and any major metabolite(s) present, any 'first pass' biotransformation being reflected in the ratio; indeed the metabolite may be the only measurable component present and dictate the choice of the biological sample to be used, as when there is rapid hydrolysis of an ester to an acid. Chromatography also indicates the specificity of the extraction procedure (pH and solvent used), and radioisotopic data indicate the feasibility of developing a chemical assay ultimately adequate in sensitivity and/or specificity.

b. ELIMINATION
The rate and extent of elimination of a drug and/or its metabolites in urine and faeces dictate the utility of analyzing these media. Drugs that are extensively

metabolized are eliminated as the glucuronide/sulphate/hippurate conjugates. Their concentrations are usually sufficiently high to warrant their analysis in either urine or faeces as in bioavailability/bioequivalence studies for dosage-form evaluation.

ANALYTICAL CRITERIA WHICH INFLUENCE SAMPLE PREPARATION

A. Analysis by GC

Although analysis of the intact molecular moiety (underivatized) is preferred to specificity, derivatization is often an analytical need. The idiosyncrasies of selective detectors have to be considered during sample preparation so as not to introduce detrimental contaminants. Thus, the N/P detector is susceptible to severe interference by residues of silylating reagents and by phosphate plasticizers contained in plastic syringes and blood collection tubes which leach into the biological sample and are extracted. Hence selection of appropriate syringes and collection tubes becomes vital for assay development and for a satisfactory outcome of the clinical studies.

Where the concentration of a major metabolite is sufficiently high, urine analysis can exploit the excellent sensitivity of the ECD as enhanced by derivatization and the use of small sample size, to minimize sample clean-up. We have shown this for midazolam, an imidazo-1,4-benzodiazepine (pK_as 1.7, 6.15) which is being developed as a short-acting hypnotic-anxiolytic, infused intravenously at doses of the order of 10mg/70 kg. The compound yields the 1-hydroxymethyl analogue as the major urinary metabolite; this, after conjugate hydrolysis, is analyzed as its silyl derivative, whose ECD response is so good that its quantitation is, through sample dilution, unimpaired by endogenous interfering peaks and minor metabolites.

B. Analysis by HPLC

HPLC is a most effective means of determining solvent-extractable drugs with little or no clean-up, usually, the residue from the sample extract is dissolved in the mobile phase itself, and an aliquot injected directly. Thermally unstable compounds and amphoteric zwitterionic compounds (notably antibiotics) which are difficult to extract can be anlyzed by the direct injection of the biological sample. Plasma or serum can be analyzed directly following protein precipitation with acetonitrile and partitioning of the filtrate with n-hexane to remove colloidal lipids. Urine is filtered to remove salts and colloidal materials, extracted with diethyl ether or ethyl acetate if clean-up is needed, and an aliquot diluted in the mobile phase and analyzed directly by RP-HPLC.

Amoxicillin [a β-lactam antibiotic (pK_a = 2.4, 9.6) structurally related to ampilillin, *see panel*, I], and its benzyl-penicilloic acid (inactive metabolite) [II] were analyzed directly in urine following post-column derivatization with

I   II

fluorescamine [9]. (Fig. 1). The fluorescence mode (excitation 385; emission 490 nm) was so selective as to circumvent the need for extensive clean-up. The clinical utility of the assay was evident from the urinary excretion profile of the two compounds in a human following the oral administration of a single 250 mg dose; comparison was made with the microbiological assay which is specific for the parent drug (active antibacterial agent). Virtually quantitative recovery of the dose was achieved in a 12-h pooled urine collection, making the method a valid means of assessing the bioavailability of oral dosage forms [9].

THERAPEUTIC MONITORING
Two recent books [10. 11] are relevant to the tailoring of method development, especially sample preparation, to the needs of clinical therapeutic monitoring.

Acknowledgements
The author is indebted to Messrs T. Daniels and R. McGlynn for drawing figures, to *Journal of Pharmaceutical Sciences,* Washington, D. C., for permission to reproduce Fig. 1; and especially to Ms V. Waddell for her untiring assistance in the preparation of this manuscript.

References
[1] Schill, G. (1978) *Separation Methods for Drugs and Related Organic Compounds,* Swedish Academy of Pharmaceutical Sciences, Stockholm 182 pp.
[2] Schwart, M. A. & de Silva, J. A. F. in *Principles and Perspectives in Drug Bioavailability,* (Blanchard, J. Sawchuk, R. J. & Brodie, B. B., eds.), Karger, Basle, pp. 90-119.
[3] Tozer, T. N. (1979) as for 2., pp. 120-155.
[4] de Silva, J. A. F. (1978) in *Blood Drugs and Other Analytical Challenges, [Methodological Surveys (A),* Vol. 7] (Reid, E., ed.), pp. 7-28.
[5] Ritshel, W. A. (1976) *Handbook of Basic Pharmacokinetics,* Drug Intelligence Publications, Hamilton, Illinois, 370 pp.
[6] Blau, K. & King, G. (eds.) (1977) *Handbook of Derivatives for Chromatography,* Heyden, London, 576 pp.

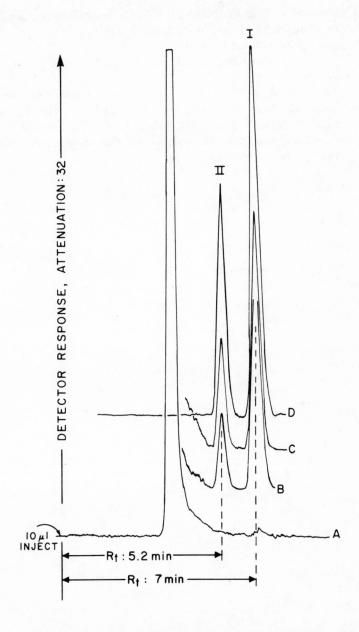

Fig. 1 – HPLC analysis of amoxicillin [I] and its benzylpenicilloic acid [II] in urine by fluormetric detection following post-column derivatization with fluorescamine. (For formulae, *see text*).

[7] Lawrence, J. F. & Frei, R. W. (1976) *Chemical Derivatization in Liquid Chromatography,* Elsevier, Amsterdam, 213 pp.

[8] Reid, E. (1978) as for 4., pp. 61–76.

[9] Lee, T. L., D'Arconte, L. & Brooks, M. A. (1979) *J. Pharm. Sci.,* **68**, 454–458.

[10] Kalman, S. M. & Clark, D. R. (1979) *Drug Assay – The Strategy of Therapeutic Drug Monitoring,* Masson, New York, *(see* pp. 97-120).

[11] Sadee, W. & Beelen, G. C. M. (eds.) (1980) *Drug Level Monitoring – Analytical Techniques, Metabolism and Pharmacokinetics,* Wiley, New York, 495 pp.

#NC(D) Notes and Comments related to Approaches for Biological Fluids, especially Drug-Related

#NC(D)-1 *A Note on*
MASS SPECTROMETRY AND TRACE-ORGANIC ANALYSIS IN BIOLOGICAL FLUIDS

L. E. MARTIN and R. J. N. TANNER, Glaxo Group Research Ltd., Ware, Herts., SG12 0DJ, U.K.

A well-established technique for the determining drugs and hormones in biological fluids is solvent extraction and analysis by GC coupled to a mass spectrometer (MS) which is operated at a low resolving power (R.P.) in the selected ion monitoring (SIM) mode [1,2]. The internal standard used in the assay may be a stable-isotope-labelled analogue or a homologue of the compound to be determined. Endogenous substances or reagent impurities which in a biological extract can be in a concentration 10^8 times that of the analyte can interfere in an assay if they co-elute with it and give ions of the same nominal mass as the ion selected for the determination of the analyte. Extensive chemical clean-up of such extracts will be required before quantitative low-resolution GC-MS can be undertaken. Such procedures take time and usually entail loss of sample. An alternative procedure is to develop MS instrumental methods to improve sensitivity and specificity.

CHEMICAL IONIZATION

One procedure is to use alternative modes of ionization such as chemical ionization, (CI), possibly negative [3], with selected reagent gases to produce high-mass ions suitable for SIM. For example, phencylidine [4] and phenformin [5] have been quantified in biological fluids at a concentration of 1 ng/ml by positive CI using methane as the reagent gas. If the sample has good electron-capturing properties then the analysis may be carried out by negative CI. The limits of detection of some polychlorinated dibenzo-*p*-dioxins by this procedure are in the range 10 to 100 pg [6].

HIGH RESOLVING POWER SELECTED ION MONITORING

Since biological extracts can be complex mixtures, the analyte may be a minor component of the GC peak eluting into the MS. Other components may interfere in the quantification by giving rise to ions of the same nominal mass but different atomic composition to the sample ion being monitored. Thus, at R.P. = 1000, monitoring at mass 500, all ions with a mass within ±0.25 mass units of the selected ion will be detected to some degree by the MS. At R.P. = 1000 the limit of detection of salbutamol as its TBDMS ether is equivalent to 1 ng salbutamol/ml plasma. An interfering compound in the control plasma sample has a similar GC rentention time to, and yields an ion of the same minimal mass as TBDMS-salbutamol. This compound gives a signal equivalent to 0.3 ng salbutamol/ml plasma and prevents the sensitivity of the assay being improved (Fig. 1). The specificity of the SIM technique can be improved by operating the instrument at a higher R.P. Commercial quadrupole MS instruments have a resolution of up to 1200; hence for analyses at higher resolution it is necessary to use a suitable magnetic sector MS and to increase the electronic gain of the instrument, because the number of the selected ions reaching the detector is reduced as the resolution is increased.

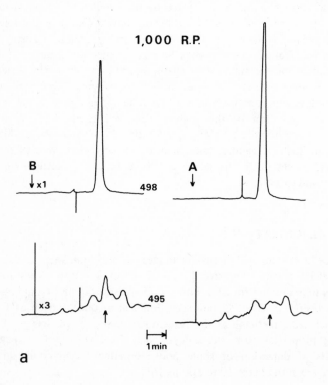

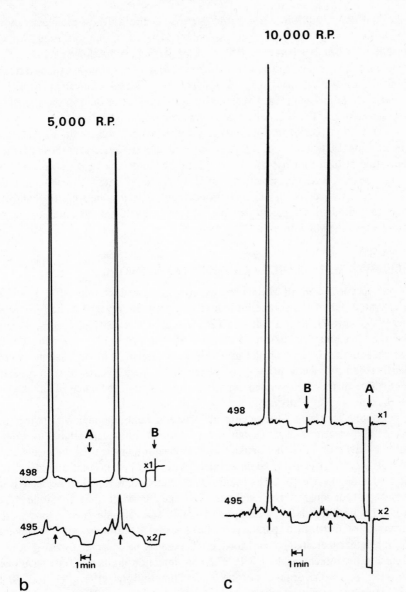

Fig. 1 – Selected ion monitoring traces for the determination of salbutamol as the tris-TBDMS ether in plasma extracts at different resolving powers: (1a) 1000 R.P.; (1b) 5000 R.P.; (1c) 10,000 R.P. Ions monitored: m/z 498 from TBDMS-tris-deutero salbutamol (internal standard) and m/z 495 from TBDMS-salbutamol. [*Note by Ed.:* z = e = charge.] Sample A, control from plasma containing 32.5 ng tris-deutero salbutamol/ml. Sample B, from plasma containing 32.5 ng tris-deutero salbutamol and 1 ng salbutamol/ml.

At R.P. = 5000 there is a 5-fold decrease in the width of the mass window monitored, and therefore a corresponding increase in the specificity of the technique when compared to R.P. = 1000. At R.P. = 5000, ions within ±0.05 mass units of the selected ion will be detected. At this resolution the interfering peak in the salbutamol assay is reduced to a signal equivalent to 0.1 ng salbutamol/ml plasma (Fig. 1b). Thus by increasing the resolution the specificity and sensitivity of the salbutamol assay has been improved.

At R.P. = 10,000 ions with a mass within ±0.025 mass units of a selected ion of mass 500 will be detected. At this R.P. the electronic noise of the instrument operating at high gain has become the limiting factor, the signal-to-noise ratio is not increased, and the sensitivity of the salbutamol assay is much improved (Fig. 1c). However, with MS operation at high R.P. it is possible to determine trace amounts of sample in complex mixtures by direct introduction of the sample into the MS without a GC [7].

SELECTED METASTABLE TRANSITION MONITORING

If an ion formed in an MS source has a half-life greater than 10^{-6} it will be accelerated out of the source into the flight tube. However, if its half-life is less than 10^{-5} this parent ion will undergo a metastable transition, decomposing to form a daughter ion before detection. If the metastable transition occurs immediately before the magnet in a sector instrument, a broad peak due to the daughter ion is observed in the mass spectrum. This peak is centered at an apparent mass value of m^*. For the decomposition, $m_1^+ \rightarrow m_2^+$, the value of m^* is given by the equation $m^* = m_2^2/m_1$.

Double focussing instruments also have a field-free region between the source and electric sector. Daughter ions formed during metastable decomposition in this region may be focussed at the collector by increasing the accelerating voltage [8], or by using an instrument with a linked magnetic field/electric field scanning device [9]. The focussed metastable ions can then be selectively monitored for quantitative analyses. The specificity of the technique arises from the need to define both parent (m_1^+) and daughter (m_2^+) ions in the selection of instrument parameters. Commercial quadrupole MS instruments do not detect metastable transitions; but these may be observed by using double focussing instrument of low R.P. Thereby detection limits of 20 pg have been reported for 5-dihydrotestosterone t-butyldimethylsilyl ether [9] and 30 pg for testosterone methyloxime t-butyldimethylsilyl ether [8]. The sensitivities are similar to those quoted for high-resolution selected-ion monitoring.

A disadvantage of the selected metastable monitoring technique is that the particular molecule under study may not give rise to an ion with a half-life that will undergo a suitable metastable transition. However, the number of metastable transitions may be increased by using a collision gas to induce fragmentation in the first field-free region.

CONCLUSIONS

High-resolution MS instruments are more expensive than low-resolution instruments; but the extra cost can be justified for the analyses of biomedical and environmental samples because of the increase in specificity and sensitivity. Those compounds that generate ions which undergo suitable metastable transitions may be determined using a double-focussing MS of low R.P.

The authors thank Dr. D. Millington, V. G. Organic Ltd., who carried out the salbutamol analyses at 5000 and 10,000 R.P.

References

[1] Millard, B. J. (1976) in *Assay of Drugs and other Trace Compounds in Biological Fluids* [this series, Vol. 5] (Reid, E., ed.), North-Holland, Amsterdam, pp. 1-7.

[2] Millard, B. J. (1978) *Quantitative Mass Spectrometry,* Heyden, London.

[3] Brandenberger, H. (1978) in *Blood Drugs and Other Analytical Challenges,* [this series, Vol. 7] (Reid, E., ed.), Horwood, Chichester, pp. 173-184.

[4] Lin, D. C. K., Fantiman, A. F., Foltz, R. L., Forney, R. D. & Sunshine, I. *Biomed. Mass Spectrum.,* 2, 206-214.

[5] Matin, S. B., Karam, J. H., Forsham, P. H. & Knight, J. B. (1974) *Biomed. Mass Spectrom.,* 1, 320-322.

[6] Hass, J. R., Friesen, M. D., Harvan, D. J. & Parker, C. E. (1978) *Anal. Chem.,* 50, 1474-1479.

[7] Boulton, A. A. and Majer, J. R. (1970) *Nature (London)* 225, 548.

[8] Gaskell, S. J., Finney, R. W. & Harper, M. E. (1979) *Biomed. Mass Spectrom.,* 6, 113-116.

[9] Gaskell, S. J. & Millington, D. S. (1978) in *Quantitative Mass Spectrometry in Life Sciences* (Ronucci, R. & van Petergrem, C., eds.) Vol. 2, Elsevier, Amsterdam, pp. 135-142.

#NC(D)-2 *A Note on*
DRUG BINDING TO PLASMA PROTEINS

C. J. COULSON and V. J. SMITH, May & Baker Ltd., Pharmaceutical Research Laboratories, Dagenham, RM10 7RS, U.K.

Plasma protein binding may have a considerable influence on excretion and biological activity [1,2]. Usually only the free drug is active; so highly bound drugs may have narrow therapeutic dose ranges. Interference in binding between two ligands may lead to undesirable side-effects by increasing the circulating level of free drug [3]. In addition, one might expect that a drug which was tightly bound to albumin might be more extensively hydroxylated by the liver microsomal system than one that was only weakly bound.

    Equilibrium dialysis is reckoned to be the best way to measure protein binding [1, 4]. The procedures currently available are either tedious, if a large number of dialysis bags have to be set up, or expensive, if an instrument has to be purchased. We have developed some equilibrium dialysis units which are simple, easy to use and give reproducible results. We have used these units to measure the binding of (a) low concentrations of doxantrazole, indomethacin and dicoumarol to human serum albumin (HSA), and (b) radioactive acebutolol and diacetolol to human plasma proteins.

    Dicoumarol is very tightly bound in human serum, e.g. 99.8% at 10 μg/ml [5]. The concentration of free drug is therefore extremely low when dicoumarol is equilibrated with plasma at therapeutic concentrations. If radioactive material is available, the level can be determined with ease. Alternatively low concentrations of drug and HSA can be used, and the results obtained extrapolated to give % free drug in equilibrium with whole plasma.

    HSA (0.25-1.0 mg/ml in 0.05 M phosphate buffer, pH 7.4) is placed on one side and drug (15-100 μM in the same buffer) on the other side of a Spectrapor no. 2 membrane in a unit. After equilibration at 37° for 8 h, the units are allowed to cool to 25° for a further 8 h. This period of time is to ensure full equilibration although in most cases this is complete in a shorter time. Bound and free drug are measured by UV spectroscopy using a wavelength, previously determined by difference spectrum, at which there is no net change in absorbance when the drug-binds to HSA. A plot of bound *vs.* free drug gives values for number of binding sites and binding constant by an iterative procedure of non-linear regression analysis, based on the equation:

$$Y = \frac{N_1.K_1.\,T}{1 + K_1.T} + \frac{N_2.\,K_2.T}{1 + K_2.T}$$

where Y is bound drug and T free drug, N_1 and K_1 are the number of binding sites and binding constant for the primary binding sites, and N_2 and K_2 refer to secondary sites respectively. These figures are entered into a second program obtained from R. F. Mais [6] which calculates % bound and free drug in equilibrium with 4.2% albumin (the average concentration in whole plasma [7]. The results obtained for binding constants, number of binding sites etc., for the three drugs are listed in Table 1, together with the % free drug in equilibrium with whole plasma.

Table 1   Binding data for tightly bound drugs. λ_N is the wavelength at which the absorbance change on binding to HSA is zero. The % free is calculated for 10^{-5} M drug and 4.2 % HSA.

Drug	λ_N (nm)	N_1	N_2	K_1	K_2 (Litre/mole $\times 10^{-5}$)	M.S.D.	% free
Dicoumarol	335	3.30	9.65	637.6	0.09	2.39	0.05
Doxantrazole	328	0.95	5.48	10.2	0.50	3.10	0.17
Indomethacin	265	1.34	2.63	1.78	0.37	4.42	0.50

The result for dicoumarol agrees closely with that already noted [5]. The indomethacin result is consistent with published figures, viz. >99% bound to 4.1% bovine serum albumin (BSA) at 37° and pH 7.4 [8] calculated from the binding parameters, and 96.7% bound to human plasma at 10 μg/ml [5]. The latter result was obtained by ultrafiltration, the former by equilibrium dialysis. There is no figure published for doxantrazole [9]. As a drug molecule with a relatively high molecular weight, a large aromatic centre and a negative charge located on the tetrazole ring, doxantrazole resembles the other drugs. It is therefore not surprising that it is also tightly bound to HSA.

The units may also be used to measure the binding of radioactive drugs, e.g. acebutolol and diacetolol, to whole plasma [10] as previously described [11]. Briefly, plasma is obtained from fresh human blood without the use of anti-coagulants, and is placed on one side of the membrane. The drug is dissolved in horse serum ultrafiltrate to simulate the physiological environment *in vivo* [12], in the presence of 0.5% polyethylene glycol 6000 to maintain an osmotic pressure equivalent to serum. The drug solution is placed on the other side of the membrane. The solutions are allowed to equilibrate at 25° for 8 h. Aliquots are taken from both sides of the membrane for liquid scintillation counting and estimation of free and bound drug.

The results obtained for quadruplicate determinations at 4 concentrations between 0.01 and 1 μg/ml show that acebutolol was 11-19% bound, and diacetolol 3–8% bound. The result for acebutolol is in agreement with that of Meffin *et al.* [13], who equilibrated the drug with human plasma proteins. They found a mean value of 25.6% bound at drug concentrations from 0.22 to 9 µg/ml.

Using the equilibrium dialysis units we have obtained results consistent with the published figures for novobiocin, coumermycin A1 and clorobiocin [14] and for salicylate [11]. The units thus give reliable and reproducible results, and are simple to use. They are now available in a modified form (Fig. 1) from Universal Scientific Ltd. (231, Plashet Road, London E13 OYU).

References

[1] Lindup, W. E. (1975) *Biochem. Soc. Trans.,* **3**, 635–640.

[2] Vallner, J. J. (1977) *J. Pharm. Sci.,* **66**, 447–465.

[3] Odell, G. B. (1973) *Ann. N. Y. Acad. Sci.,* **226**, 225–237.

[4] Kurz, H., Trunk, H. & Weitz, B. (1977) *Arzneim. Forsch,* **27**, 1373–1380.

[5] Hucker, H. B., Stauffer, S. C. & White, S. E. (1972) *J. Pharm. Sci.,* **61**, 1490–1492.

[6] Mais, R. F., Keresztes-Nagy, S., Zaroslinski, J. F. & Oester, Y. T. (1974) *J. Pharm. Sci.,* **63**, 1423–1427.

[7] In *Biochemists' Handbook,* Long, C(ed.) Spon, London (1961), p. 1081.

[8] Kaneo, Y., Kai, A., Kiryu, S. & Iguchi, S. (1976) *J. Pharm. Soc. (Japan),* **96**, 1412–1416.

[9] Batchelor, J. F., Garland, L. G., Green, A. F., Hughes, D. T. D., Follenfant, M. J., Gorvin, J. H., Hodson, H. F. & Tateson, J. E. (1975) *Lancet,* **i**, 1169–1170.

[10] Gradman, A. H., Winkle, R. A., Fitzgerald, J. W., Meffin, P. J., Stoner, J., Bell, P. A. & Harrison, D. C. (1977) *Circulation,* **55**, 785–791.

[11] Coombs, T. J. & Coulson, C. J., (1978) *Biochem. Soc. Trans.,* **6**, 1048–1050.

[12] Fox, C. H. & Sanford, K. K. (1975) *Tissue Cult. Assoc. Manual,* **1**, 233–237.

[13] Meffin, P. J., Winkle, R. A., Peters, F. A. & Harrison, D. C. (1977) *Clin. Pharmacol. Ther.,* **22**, 557–567.

[14] Coulson, C. J. & Smith, V. J. (1981) *Biochem. Pharmacol.,* in press.

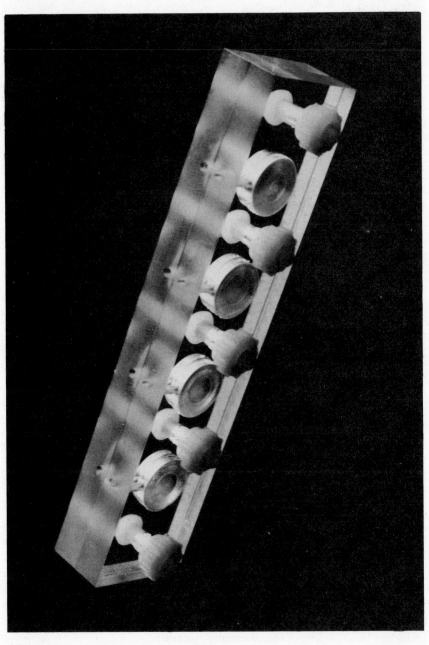

Fig. 1

#NC(D)-3 *A Note on*
MINIMIZATION OF PRE-HPLC PLASMA PROCESSING THROUGH COLUMN-SWITCHING AND SELECTIVE DETECTION*

JAN E. PAANAKKER, JIMMY M. S. L. THIO and **HENK J. M. van HAL**, Organon International B.V., Drug Metabolism R & D Laboratorie, P.O. Box 20, 5340 BH Oss, The Netherlands.

The context of the following schemes is that in developing an HPLC assay for trace levels of drugs in biological matrices, entailing extraction, HPLC separation and detection, problems may arise, especially from the complex nature of plasma in relation to selectivity and sensitivity requirements.

Requirement and chemistry	*Assay of two drugs, Org 4122 and Org 3509 (see formulae below), at levels down to a few ng/ml of plasma.*
End-step	*RP-HPLC with fluorimetric or electrochemical detection and, in the case of Org 4122, with column-switching.*
Sample preparation	*Plasma (citrated; normally 1 ml) extracted with 5 ml of ethyl acetate (Org 4122) or hexane (Org 3509); residue from dried-down extract dissolved in 50 µl of the appropriate eluant for application to the column.*
Comments	*For Org 4122 and Org 3509 respectively the recoveries were typically 60% and 70%, and the detection limits per ml of plasma were about 10 ng and 2 ng or, if 4 ml of plasma were taken (the Org 3509 extracts being clean), 0.5 ng. Consideration is given to how assay simplicity was achieved.*

Choosing and applying a solvent extraction method can be tedious, and all manipulations involved can affect overall accuracy and precision. As there is now a wide choice of highly efficient column materials, selectivity can be attained largely by choosing the most appropriate phase system, a compromise being sought between resolution, speed of analysis and sample capacity. Moreover, a suitable detection made (e.g. fluorimetric or electrochemical) can help give selectivity and sensitivity. A single extraction step at the outset may then suffice.

---

* This contribution has been editorially shortened; it complements those such as #D-4.—*Ed.*

Org 4122: 3-(3',4'-dimethoxyphenyl)-propiono-hydroxamic acid.
Internal std.: 3-(3',4'-dimethoxyphenyl)-butano-hydroxamic acid.

Org 3509: cis-1,2,3,4,10,14b-hexahydro-2-methyl-
          dibenzo[c,f] pyrazino[1,2-a] azepin-10-ol.
Internal std.: the 8-chloro derivative.

APPROACH

The present approach comprises equilibration of the internal standard in the plasma medium, one solvent-extraction step, drying down, residue re-dissolution in the HPLC eluent, reversed-phase chromatography and finally, quantitation by selective detection.

Column-switching and selective detection in HPLC*

With only a single extraction, late-eluting peaks severely impede fast and reliable routine quantitation. The total analysis time can be reduced by applying column switching, governed by a time controller. Liquid leaving the pre-column (5 cm; filled with the same material as the 30 cm analytical column) comes to a pneumatically activated directional valve. A fraction of the injected sample containing analyte and internal standard is allowed to enter the main column; the rest of the sample is directed to waste, after which the valve is switched back to its original position. Thereafter, the compounds of interest are eluted. An example of a column-switching procedure using simultaneous fluorimetric and electrochemical detection is shown in Fig. 1.

Temporary flow interruption in the analytical column, when the flow by-passes it, will affect the responses from the electrochemical and fluorimetric detectors. Upon restoring the flow, the detector signals immediately return to

---

* Relevant articles in Vol. 7 of this series include #D-3 (P. T. Kissinger) and, for detection-mode comparisons, #D-4 (L. E. Martin *et al.*).

their original levels. For each compound and assay involved, the switching time
has to be determined by short-circuiting the pre-column outflow into the detector.

Retention and selectivity by the pre- and analytical column may change
during routine application of the assay method.

A run with a column-switching time different from that in Fig. 1, after
careful reactivation and re-equilibration of the columns, gave better resolution
between the analyte Org 4122 and its internal standard on the one hand, and

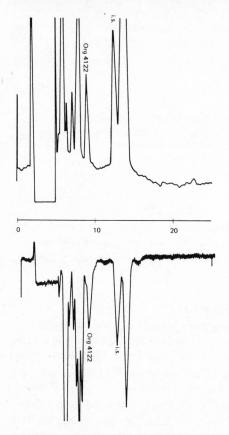

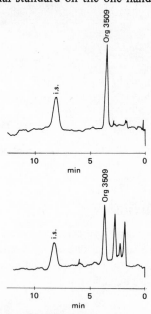

Fig. 2 (above) – Chromatograms of
Org 3509 (20 ng) and its internal standard, i.s. (20 ng) as a standard mixture (*upper trace*) and as recovered by
hexane from spiked human plasma,
with electrochemical detection (+1.2 V).
Mobile phase: H_2O/methanol (3:2),
0.01 M NaH_2PO_4, 0.1 M $NaClO_4$,
adjusted to pH 2.5 with phosphoric
acid; 2 ml/min.

Fig. 1 (above) – Chromatograms of material extracted by ethyl acetate from dog
plasma spiked with Org 4122 and its internal standard, i.s. (resp. 50 and 100
ng/ml) with electrochemical and fluorimetric monitoring in series.

Upper trace: electrochemical detection; oxidation at +1.2 V with a glassy
carbon electrode (Metrohm VA E611 with EA1096/2 flow cell).

Lower trace; fluorimetric detection; excitation: 200 nm, emission: 300 nm
(Schoeffel detector FS 970).

Column: μBondapak C_{18}, 30 cm x 4.6 mm i.d.

Mobile phase: H_2O/methanol (4:1), 0.02 M NaH_2PO_4, 0.2 M $NaClO_4$, adjusted
to pH 3.0 with phosphoric acid; 2 ml/min.

Column-switching (*see text*) was performed at 2 min.

possible endogenous components on the other hand. Furthermore, electro-chemical rather than fluorimetric detection is preferred in the assay for Org 4122 because of the better signal-to-noise ratio, lower detection limit and less inter-ference by possible biotransformation products.

In the development of the assay for Org 3509, column-switching was omitted since hexane extraction resulted in relatively clean blank chromatograms. Fig. 2 shows typical chromatograms, with electrochemical detection.

CHECKING ASSAY QUALITY, AND CONCLUDING COMMENTS

When routinely applying an assay, the requisite quality control is, in our opinion, most conveniently and correctly dealt with by having in the batch the customary spiked plasma samples (relevant to calibration and to accuracy), standard mixtures (a check on detector response), and blank plasma samples for reference purposes; is blank still blank?.

For the Org 4122 assay, with column-switching, the accuracy (difference between true and observed value) at the 100 ng/ml plasma level was 3.6%, and the precision ±7.8% (C.V.).

Circumvention of laborious sample pre-treatment is beneficial to routine operation as well as to accuracy and precision samples. Column-switching (the pre-column serves as a guard for the main column), efficient HPLC phase systems, and selective detection facilitate the reliable quantitation of drugs in plasma.

#NC(D)–4 *A Note (Analytical case history) on*
DETERMINATION OF RANITIDINE AND METABOLITES IN PLASMA AND URINE

P. F. CAREY and **L. E. MARTIN**, Glaxo Group Research Ltd., Ware, Herts., SG12 ODJ, U.K.

Requirement	*Assay sensitive to 20 ng ranitidine and metabolites/ml of plasma or urine.*
Chemistry and metabolism	*Ranitidine: N-[2-[[[5-(Dimethylaminomethyl)-2-furanyl] methyl] thio] ethyl] -N′-methyl-2-nitroethanimidamide (tertiary amine). Metabolites: N-oxide, S-oxide, desmethyl homologue.*
End-step	*RP-HPLC, paired ion; UV detector (320 nm).*
Sample preparation	*Plasma-protein precipitation by NaOH/ZnSO₄. Urine by direct injection.*
Comments	*Part-automated for plasma, fully for urine analysis, using a WISP (Waters) injector, a selective UV detector, an automatic drift-correction circuit and a Spectra Physics 4000 system.*

Ranitidine is a new H_2 receptor antagonist which is about 5 times as potent as cimetidine and is used for the treatment of peptic ulcers. Metabolic studies in animals using ^{14}C-ranitidine suggested that after a therapeutic dose of the drug its plasma concentration in man would be in the 0–400 ng/ml range, and that it was metabolized by oxidation at the tertiary N and the S atoms and by N-desmethylation. The extent and pattern of metabolism varied with the species studied. A method was required for determining ranitidine and metabolites in patients being treated with the drug. There was the problem of determining in plasma and urine a series of compounds with physiochemical properties ranging from that of ranitidine N-oxide. The physiochemical properties of the metabolites precluded the use initially of simple solvent-extraction procedures.

Ranitidine and its desmethyl homologue were not separable by RP-HPLC without ion-pairing. The eluant was a 2:3 mixture (by vol.) of 0.5 mM KH_2PO_4 and methanol containing 5×10^{-3} M sodium lauryl sulphate. The column packing was 5 μm Spherisorb-ODS, and the k' values were: N-oxide, 1.9; S-oxide 3.2; ranitidine 4.0; desmethyl compound, 5.7.

#NC(D)–5 *A Note on*

HPLC DETERMINATION OF HYDROLYSIS RATES OF CORTICOSTEROID ESTERS *IN VITRO* AND *IN VIVO*

M. E. TENNESON and **T. COWEN**, Squibb International Development Laboratory, Moreton, Merseyside, L46 1QW, U.K.

Requirement   *Specific assays for triamcinolone acetonide phosphate, prednisolone hemisuccinate, methyl prednisolone hemisuccinate and their corresponding free steroids in a range of biological fluids, sensitive to 1 µg/ml.*

Chemistry and metabolism   *1,4-diene-3-one steroids with ester side-chain. Hydrolyzed to the corresponding free steroids and then after reduction derivatized to the glucuronide.*

End-step   *RP-HPLC on a C_{18} column with detection at 254 nm.*

Sample preparation   *Immediate solvent extraction with diethyl ether (serum and liver homogenate) or dichloromethane (whole blood) separates the steroid ester (aqueous layer) from the free steroid (solvent layer). The residual ester is then converted into the free steroid with alkaline phosphatase and extracted into ether. The solvent evaporate is dissolved in the HPLC solvent and chromatographed. Aqueous enzyme solutions are chromatographed directly.*

Comments &   *Few or no interferences; each steroid ester is well separated from*
alternatives   *its corresponding free steroid. HPLC assay was chosen because of*
              previous knowledge of the HPLC characteristics of triamcinolone
              acetonide.

Intravenous corticosteroid esters are administered to patients with allergic
shock. When present in the ester form such compounds have no biological
activity and must first be hydrolyzed to their corresponding free steroid before
they show pharmacological activity. Since 50% of allergic shock patients die
within the first five minutes, the rate of hydrolysis of a steroid ester to its free
steroid is of prime importance.

   An HPLC system was first developed which enabled the determination of
the relative concentrations of three corticosteroid esters, triamcinolone acetonide
phosphate, prednisolone hemisuccinate, methyl prednisolone hemisuccinate and
their corresponding free steroids, triamcinolone acetonide, prednisolone and
methyl prednisolone. Extraction procedures were then developed which extracted
the free steroids from serum, liver homogenate and whole blood whilst inactivating
any enzymatic activity in the sample. The esters remaining in the aqueous layer
could then be hydrolyzed before extracting as the free steroid. Experiments
with aqueous enzyme solutions did not require any extraction procedure;
simultaneous measurement of steroid ester and free steroid could be carried
out by direct injection of the sample.

   Studies have been made of the enzymic hydrolysis rates of all three steroids
esters in human serum, mouse liver homogenate and aqueous enzyme solutions,
and in whole blood samples obtained from allergic shock patients given triamcinolone acetonide phosphate i.v.

Comments on #D-1, G. Guilbault — ENZYMIC APPROACHES
In reply to J. C. Swann. — In a biological fluid, it is indeed to the concentration
of the *free* drug or other substrate that enzyme electrodes respond, irrespective
of any that is protein-bound. K. H. Dudley's *comment on a question by* S. H.
Curry: formaldehyde such as arises from *N*-demethylation of drugs might be
measurable with immobilized alcohol dehydrogenase.

Comment on #D-2, R. A. de Zeeuw — DRUGS IN SALIVA
Queries by N. Sistovaris. — In comparing saliva levels with plasma levels, unless
salivary flow is constant should one correct for any stimulation of flow? Also,
salivary excretion of drug being quite high, one wonders whether there might
be reabsorption and hence a need to look for an additional elimination phase.

Comments on #D-3, K. H. Dudley — SOLVENT PARTITIONING [*see also* (iv) in #0 & 2.5 in #NC(F)–4]

Answer to B. Scales. — We have not encountered examples of altered partitioning or faster extraction due to the presence of the biological sample in the aqueous phase. *Remark by* J. Ramsey: use of a commercial 'cartridge' where the aqueous phase is immobilized on a support might overcome the problem of any slowness in coming to equilibrium and hence a long shaking time. D. Westerlund's *comment on a question* by R. Whelpton, concerning a relationship between distribution (partition) coefficient and extraction kinetics. — Studies on the kinetics of ion-pair extraction (T. Nordgren *et al.*) have shown that ion-pairs with high extraction constants are extracted relatively fast, and that there is the relationship K_{ex} (extraction constant) = k_1/k_2 where k_1 and k_2 are the rate constants for the extraction to and from the organic phase respectively; by analogy, similar relationships may well apply also to the extraction of uncharged compounds.

Remarks by C. R. Jones. — Shaking time should not be so long as to ensure equilibrium, insofar as the time taken for the last few percent of the drug to extract may be sufficient to raise the blank values to an unacceptably high level because of the slow transfer of endogenous substances. I do not agree that the partition ratio is independent of volume ratio for the two liquid phases: often this is not the case, especially when back-extracting from organic phase into mineral acid; the volume needed to ensure efficient transfer can be very small. R. A. de Zeeuw (*prompted by a question from* B. Scales). — In considering deviations between theoretical recovery and recovery from biological samples, account has to be taken of the actual amount of drug present in the sample. We find that deviations from 'theory' increase with decreasing amounts of drug, e.g. a drug showing 95% recovery at μg level may show only 10% recovery at ng level, with a large C. V. (Other speakers agreed, although Dudley had no such problem in his work at μg levels.) Such concentration-dependent recovery unfortunately cannot be corrected by using an internal standard, because this assumes a fixed recovery over a wide concentration range. *Remarks by* A. P. Woodbridge. — Non-quantitative extraction of a pesticide is circumvented by performing 3 or 4 extractions. Partitions are checked in the presence of co-extractives; lipid materials have a notable influence.

Editor's citation: METHODS AVAILABLE, ESPECIALLY FOR METABOLITES A forthcoming review [1] deals with solvent extraction and other approaches to isolating metabolites, these including conjugates and typically being more polar than the parent drug. The survey of methodology includes characterization and, for a range of compounds, analytical determination (often entailing HPLC) as also included in a recent book [2].

[1] Martin, L. E. & Reid, E. (1981) *Progr. Drug Metabolism*, **6**, in press.
[2] Sadee, W. & Beelen, G. C. M. (1980) *Drug level Monitoring*, Wiley, New York.

#E   Automated Sample Handling

#E-1   AUTOMATION OF SMALL COLUMN CHROMATOGRAPHY

G. B. BARLOW, Department of Paediatric Surgery, Institute of Child Health, 30 Guilford Street, London, WC1N 1EH, U.K.

Where sample preparation can be accomplished using column chromatography, e.g. for isolation, desalting or removing impurities, a system has been devised for automatic operation. Each of 6 small columns operates sequentially through a 6-step cycle (including regeneration), each column being at a different part of the cycle. The system can be coupled directly into the normal continuous flow-analysis system running at up to 60 samples/h. Alternatively, the prepared samples can be collected separately and analyzed by a discrete analysis system, again at a rate of 60/h. Applications to urine already include analysis for creatinine (without dialyzer module), xanthurenic acid and kynurenic acid, de-salting prior to amino acid analysis and the removal of acidic and neutral amino acids prior to estimating polyamines.

A pre-analytical sample-preparation step typically involves the separation of the required components from interfering substances by methods such as precipitation, extraction or chromatography, which can often be so tedious as to seriously limit their use in routine analysis. To simplify these problems some manufacturers have introduced small ready-buffered ion-exchange columns of disposable plastic, e.g. for estimating creatine phosphokinase isoenzymes, glycosylated haemoglobin and thyroxine. Column chromatography can usually be automated. When a complex mixture has to be analyzed for several components such as amino acids, long columns are used and the eluate monitored continuously, the whole apparatus being very costly. Even when short columns are used for a very few components, it is usually impossible to isolate and estimate more than a limited number of samples per day.

Column chromatography involves a series of steps which can be repeated as a cycle. Thus, one of the simplest cycles for the operation of a column would require six steps (buffer, wash, sample application, wash, elution of required component, regeneration). The component so isolated would then be analyzed separately. These steps can be sequenced automatically. Thus, for the estimation of creatinine [1], one ion-exchange column was used and the liquids involved in the cycle were successively switched on to the column by means of valves; the output of the column was monitored continuously.

DESIGN OF APPARATUS

Clearly there is a need for an automatic short column system so that the columns can be buffered, loaded, eluted and regenerated at such a rate that the eluate can be analyzed either by a flow system or by discrete sample procedures. Ideally, the rate of sample preparation should be comparable to the requirements of the automatic analytical systems normally available in routine laboratories. If the cycle consists of 6 steps, then 6 columns, one in each stage of the cycle, would be needed and the flow of liquids would have to be switched so that there will always be one column being loaded and one column being eluted of the desired component. All that is required is two flat discs (rotor and stator) able to move

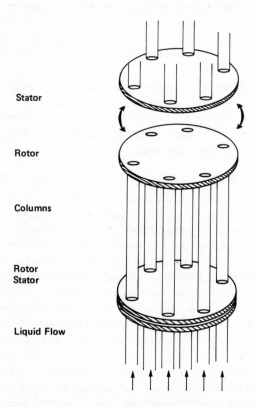

Stator

Rotor

Columns

Rotor
Stator

Liquid Flow

Fig. 1 – Arrangement of the rotor and the two stator discs.

in close contact and having six holes drilled symmetrically around the centre (Fig. 1). Small outlet pipes from the holes would be used for connecting tubes. Two such arrangements are at each end of the 6 column, and with the columns connected to the rotors would allow the flow of liquids to be sequentially altered by rotation of the switching device. There are, of course, other methods of switching the liquid flow. A suitable device was constructed using two glass/ PTFE stopcocks, rotation of the keys being brought about by means of a rotary solenoid. An integral timer enabled the keys to be rotated every 60 sec [2]. Thus, one sample could be loaded and one column eluate examined every min. The eluate from the columns could be fed directly into an automatic flow system, or the eluate could be collected for discrete sample analysis [3]. By this means, sampling rates, separations, regenerations and analyses could be performed at a rate of 60/h.

PERFORMANCE

In the estimation of creatinine by the Jaffé reaction, it is usual to isolate the compound by dialysis to remove interfering material. However, there is still doubt as to the specificity of the method. Creatinine can be absorbed on to a strong cation-exchange resin (e.g. Zerolit 225) and can be eluted with 0.1 M citrate at pH 6.0. The 6 columns (1 cm X 1 cm) containing Zerolit 225 (SRC6; 0.4 g) were pumped in sequence using a multichannel peristaltic pump as follows: (1) 0.3 M NaOH (2.9 ml/min), (2) 0.3 M HCl (2.0 ml/min), (3) water (2.9 ml/min), (4) sample (2.0 ml/min), (5) water (2.0 ml/min) and (6) 0.1 M citrate pH 6.0 (3.4 ml/min). In position (4) the sample was pumped onto the column for 30 sec and washed with water for 30 sec using an automatic sample, e.g. Chemlab 40CS. The eluate from column (6) was mixed with 0.5 M NaOH (1.2 ml/min) and saturated picric acid solution (0.23 ml/min), and a bubble pattern was introduced (air at 0.8 ml/min). The reaction was allowed to proceed using the normal delay coils and the product of the reaction measured at 505 nm. Response was linear at least between 25 mg% and 100 mg%, and reproducibility was ± 6.2% (C.V.). A range of compounds which can be present in urine was tested for possible interferences: albumin (5 mg%), ascorbic acid (10 mg%), glycerol (12 mg%), lithium lactate (50 mg%), Macrodex (60 mg%), sodium pyrurate (10 mg%), urea (200 mg%), glucose (100 mg%) and Vamin/fructose (100 mg%). All were separated clearly from the creatinine, and overall recoveries were 99.9% ± 4.3 (S.D.) (6 estimations per additive).

In a comparison of this method with the Technicon AutoAnalyzer system in a series of 99 pathological urines, 83 (83%) lay within the 95% confidence limits, 9 results being higher with the column arrangement and 7 results lower (Fig. 2). The reasons for these differences have not been investigated although, on balance, one would expect the columns to be the more specific.

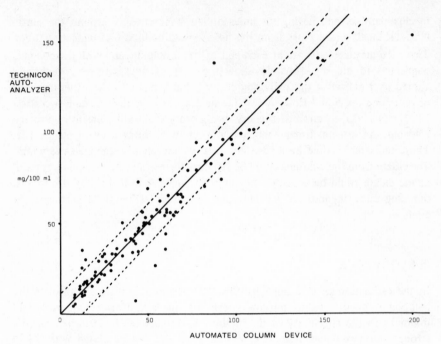

Fig. 2 – Comparison of methods for urinary creatinine: Technicon AutoAnalyser *vs.* automated column device.

COLUMN VARIABLES
The performance of the columns was dependent on several factors including length, volume, bead size and cross-linking in the resin. The size of the column indicated above (1 cm $\times$ 1 cm) appeared to be the most useful although for some uses a longer but narrower column could be desirable. A resin bead size of 52–100 mesh has been satisfactory, but smaller beads might be advantageous in reducing carry-over from one cycle to the next. Cross-linking of the resin, which governs the amount of swelling, noticeably affected the carry-over. The system worked well at 2% and 4% cross-linking but as the degree of cross-linking increased. so the carry-over from one cycle to the next also increased. With 12% cross-linking, carry-over was still appreciable even after 3 cycles.

USES OF THE METHOD
With the same resin columns as used in the creatinine separations (Zerolit 225) a number of other analyses have been done. Xanthurenic acid in urine has been estimated in this fashion, using different buffers, at a rate of 60/h by subjecting the eluate from the column to a coupling reaction with 2,6-dichloroquinone chlorimide [2] (Fig. 3). When the column eluate was collected in an automatic fraction collector synchronized with the rotating columns, it was possible to separate kynurenic acid from urine also at the rate of 60/h [4] (Fig. 4). Both

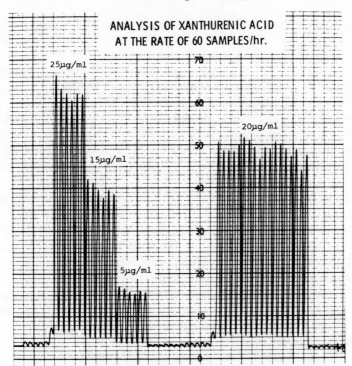

Fig. 3 – Analysis of xanthurenic acid at the rate of 60 samples/h.

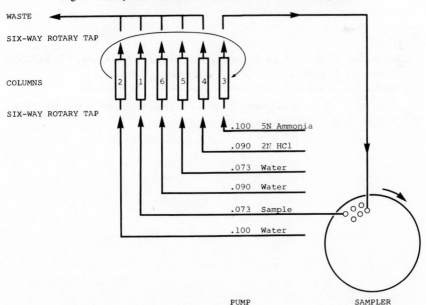

Fig. 4 – Use of the device with fraction collection for analysis in the discrete mode.

of those estimations are tedious by the normal methods. Desalting of urine was also accomplished in this fashion, the resulting solutions being ideal for TLC and automatic amino acid analysis [3].

The analysis of the polyamines (putrescine, spermidine and spermine) in urine can be done on an amino acid analyser; but it is desirable to remove as much of the amino acid content as possible. We were able to remove all but some of the very basic amino acids (His, Lys, Arg) using columns of Zerolit 226 (a carboxylic acid resin), thus speeding up the analysis rate by about 50% [5]. Both the de-salting of the urine for amino acid analysis and the removal of the main bulk of the amino acids prior to polyamine analysis were readily achieved at this high rate (60 samples/h).

The rotating column apparatus would have increased versatility if the taps could hold 8 columns. The increased number of steps would allow greater flexibility in the chromatographic cycle, which would be particularly useful in the separation of isoenzymes. The most interesting immediate application from the clinical point of view could well be in the estimation of the isoenzymes of creatine phosphokinase (EC 2.7.3.2) (of particular concern in cardiac disease) and of glycosylated haemoglobin (for diabetes). Both of these separations and estimations have been achieved normally using short columns, and methods based on this approach are now used clinically. It would seem to be a very short step to automate these analyses.

Potentially, therefore, the system can be used for any small column chromatographic work where the column packing can be regenerated.

References

[1] Corisano, A., Riva, M., & Bonscchi, A. (1970) *J. Chromatog.,* **53**, 517-523.
[2] Barlow, G. B. & Miles, R. (1973) *J. Chromatog.,* **85**, 140-143.
[3] Barlow, G. B. & Ersser, R. S. (1975) *Lab. Pract.,* **24**, 163-164.
[4] Barlow, G. B. & Wilkinson, A. W. (1975) *Clin. Chim. Acta,* **64**, 79-82.
[5] Barlow, G. B. (1980) *Burns* (in the press).

#E-2   AUTOMATED SAMPLE PRE-TREATMENT

P. B. STOCKWELL*, Laboratory of the Government Chemist, Stamford Street, London SE 1 9NQ, U.K.

Our approach to automation has been to cover the analytical problem completely, from sampling, pre-treatment, chemistry, measurement and data reduction, through to the final reporting stage. Of necessity the main thrust of our activity has centred on sample preparation since this requires the most significant use of manpower and offers the greatest potential savings; yet it is a rather neglected topic. Consideration is now given to solvent extraction, especially the phase-separation aspect, and to two applications: non-ionic detergents in water samples, and quinizarin in gas oil. Flash distillation is also considered, with volatiles in beverages as illustrative applications. Automatic approaches centred on GC are discussed. One possible trend is towards measurement methods that need little sample pre-treatment.

Automation in analytical chemistry is a subject which suffers from the lack of an acceptable definition. Instrument manufacturers often describe their products as fully automated when one facet is truly automatic, for example when data handling facilities are provided. The International Union of Pure and Applied Chemistry (IUPAC) has sought through its Commission on Analytical Nomenclature to offer rigorous definitions and provide a common international terminology. They distinguish between mechanization and automation, reserving the term *automation* for those aspects which contain a 'feedback loop'. At first sight this is logically sound but for laboratory automation it is not practical, since the presence or lack of a feedback loop has little or no significance. The acid test for automation is whether staff effort can be saved by the introduction of some mechanical, electronic or computer facilities. The term *automation* has been commonly applied to the advances made in applying such techniques to analytical chemistry.

Detailed consideration is given here to a single aspect of automatic analysis, albeit an important one, that of sample preparation. However, it is important not to consider the subject in isolation. The techniques discussed here must be considered in the context of the complete analytical process. The principal

---

* Present address: Plasmatherm Ltd., 6 Station Road, Penge, London SE20 7BQ.

operations comprising typical chemical analysis are: sampling, pre-treatment, sample measurement including standardization and instrument calibration, control of instrument operating parameters, calculation of results, initially for the analytical measurement and then related to the sample under test, and finally, report generation, distribution and archiving.

In considering the potential benefits of automating an analysis, each item above should be considered, although in many instances sampling will be an external operation not amenable to automation. The majority of publications in the field of automatic analysis have dwelt on particular facets of the total subject. Recently, attention has been focussed on the post-analysis stages, and there has been extensive study of data-processing aspects, using computers, microprocessors or simple calculators. With the introduction of microprocessor technology considerable attention has been given to data analysis and instrument control, whereas sample pre-treatment, which in practice is often the most labour-intensive and onerous task, has been neglected.

When one transfers from a manual regime to an automatic system it is important to define a proper specification of the analytical needs. A simple approach is to mechanize a manual procedure directly. This is not always the best approach, however, because new technology and the use of automated systems can allow alternative, more easily automated procedures, to be adopted. Defining the analytical specification is not a simple task; it must take full recognition of the analyst's needs, his manager's requirements and those of the customer for the results of the analysis. Whilst the designing or purchasing of instruments to meet the analytical specification will require a multidisciplinary approach, the role of the analyst must be the directing factor; he alone can provide the precise information on the chemistry of the system which must often dictate the automation strategy.

Automation has been applied to many sample pre-treatment techniques; but the focus will now be on some of the more relevant developments at the author's Laboratory with examples relating to liquid/liquid solvent extraction, flash distillation techniques and automatic chromatographic applications. Preliminary experience in automatic methods was obtained using conventional Technicon AutoAnalyzer techniques. Initial studies entailed application of the well-known continuous flow technology to the analysis of beer and water samples. In the former area, a range of methodologies were developed for the analysis of alcohol, total sugar and acid content of beer, and a screening procedure for original gravity resulted [1-4].

Automation of water analysis was precluded at that time because sample types were numerous and the analyses required varied widely from sample to sample. Many modifications to the accepted AutoAnalyzer technology were made which greatly extended the application of continuous flow techniques. This required a good background knowledge of the chemistry involved in the methods and samples. The limitations of the technology were also fully exposed.

LIQUID/LIQUID SOLVENT EXTRACTION

Application of solvent extraction techniques are numerous in the scientific literature; but there are few, if any, good commercially available automatic systems. To ensure good reliable results in an automatic system it is important that the analyst has information on the relative solubility of the compounds of interest in the respective phases. In a manual regime this is catered for by multiple extractions, backwashing and the like, carried out by the analyst as the extraction proceeds. However, in an automatic system these parameters must be built into the instrument specification. In two particular areas of work in the author's Laboratory; the analysis of water samples for non-ionic detergent concentration and the extraction of quinizarin from gas oil samples, it was recognized that the automation of the solvent extraction techniques could produce significant staff savings. Various approaches to solvent extraction were considered and preliminary experiments carried out prior to the defining of instrument specifications.

Solvent extraction can be automated in the context of continuous flow analysis. For both conventional AutoAnalyzer and flow-injection techniques analytical methods have been devised incorporating a solvent extraction step. In these methods a peristaltic pump delivers the liquid streams, and these are mixed in a mixing coil, often filled with glass ballotini; the phases are subsequently separated in a simple separator which allows the aqueous and organic phases to stratify. One or both of these phases can then be re-sampled into the analyzer manifold for further reaction and/or measurement. Whilst the sample-to-extractant ratio can be varied within the limits normally applying for such operations, the maximum concentration factor consistent with good operation is normally about 3 : 1.

It is important to design the phase separator correctly and to adjust the rate of removal of the respective phases from it. In the case of aqueous determinations, globules of an organic phase entering the flow-through cell will give false reading. A far greater limitation is presented by the nature of the pump tubes; but recent developments in the plastics industry and the use of displacement techniques can overcome some of the problems [5]. The 'Evaporation to Dryness Module' recently introduced by Technicon can also extend the concentration factors to more acceptable levels. In this, samples extracted into a volatile solvent are placed on a moving inert belt over which air or vacuum is applied. The solvent evaporates and the sample is then re-dissolved into another solvent as the belt moves into a new section of the manifold. This technique is particularly suitable where there is a need to change the solvent matrix to ensure compatibility with the measurement stage, as in liquid chromatography. The application of solvent extraction in flow-injection applications has been described by Karlberg & Thelander [6].

There are various ways of controlling the phase separation stage in discrete analytical systems; both static and dynamic systems have been described. A dynamic approach using centrifugal force was developed by Vallis [7]. The device comprises a cup-shaped vessel mounted on a rotor complete with a porous lid attached to the lip of the cup. In use the device is placed inside a collecting vessel, and if the porous lip is made from a hydrophilic material such as sintered glass, water will pass into the collecting cup at low rotation speeds, leaving the organic phase retained in the cup. An increased rotation speed is then used to reject the organic phase. A hydrophobic interface such as sintered PTFE will allow the organic phase to be rejected. A major problem encountered by such systems is that the interface is unstable and this requires frequent replacement or regeneration, such defects negating many of the advantages inherent in automatic operation. The use of centrifugation to aid an operation has found considerable uses in the General Medical Science/Atomic Energy Commission approach to automation pioneered by Anderson [8], and has also been used recently by Arndt et al. in the design of a solid-liquid extraction system [9].

In static solvent-extraction devices, a wide variety of phase-boundary sensors controlling simple on/off valves to govern the flow of the liquid through them are used to effect the solvent separation. An early system using a pair of conductivity-sensing electrodes, and a pinch-clip valve was described by Trowell [10]. It is not advisable to use such a device when volatile flammable solvents such as ether are required. In such cases there is a potential explosion hazard if a spark should occur between the electrodes, use may be made of alternative phase-boundary detectors that have been developed, with sensors based on conductivity, capacitance and refractive index. These detectors have been reviewed by Stockwell [11]. For preference the detector should be external to the organic/ aqueous phases. A compromise choice must inevitably be made between retaining an unwanted small portion of one of the phases and the loss of a small fraction of the phase of interest. In practice, the speed of stirring is of considerable importance; too rapid a speed could generate an emulsion which either fails to settle or settles after only a prolonged time. A knowledge of the chemistry involved in the analytical procedure and the aims of the analysis are required to optimize the instrument design. No single type of phase-boundary sensor is universally applicable. The potential of many of these devices has been restricted because, whilst there may be considerable differences in (say) the capacity or conductivity of the pure organic and aqueous solutions, thus giving ample scope for controlling the value, in practice the actual experimental values of conductivity or capacitance are greatly different from those of the pure solutions because of impurities and saturating water in the organic phase. Recognizing the change from one phase to another in the real-world situation may be far from an easy task, and phase-boundary detection becomes erratic and unreliable.

EXTRACTION OF NON-IONIC DETERGENTS FROM WATER SAMPLES

Large quantities of water and aqueous effluents are extracted with chloroform to separate the non-ionic detergent. An automated batch-extraction system has been described by Sawyer *et al* [12]. This also incorporates an evaporation stage prior to further analysis by TLC. Effluents from a sample manifold are fed into an extraction vessel (Fig. 1) through a solenoid valve, and 50 ml aliquots of chloroform are added from a simple automatic pipette. The mixture is stirred slowly with a Teflon paddle. For rapid and efficient separation the dimensions and shape of the extraction vessel and the paddle must be determined by experiment, using a knowledge of the chemistry involved. The phases are mixed, allowed to settle and the organic phase drained through a solenoid valve controlled by a pair of phase-sensing electrodes. A gravity feed system with a series of solenoid valves, also controlled by electrode pairs, allows the chloroform to pass through acid and alkali wash vessels prior to collection in the appropriate sample receiver. At a pre-set time in the analytical cycle an electromantle is switched on and most of the chloroform is evaporated.

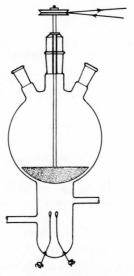

Fig. 1 — Diagram of batch extraction vessel used within LGC for extraction of non-ionic detergents from effluents. Shape of vessel and paddle are particularly important.

An alternative approach, circulating the liquids using a piston pump, a mixing coil and a phase separator, has also be described by Sawyer *et al* [13]. Samples held in a manifold are directed in sequence to the lower half of a two-stage sample vessel; the sample is then pumped by one head of a Hughes piston pump and mixed with a stream of sodium hydroxide pumped *via* a second head of the piston pump from a reservoir. The two streams converge at a T-piece at the base of a mixing coil partially filled with glass beads. Forcing the streams through the beads produces an intimate mixture; the separation of the phases which begins in the unfilled portion of the coils is completed in a double Wier separator. The aqueous stream is then diverted to the upper portion of the sample vessel for re-cycling, or to waste if fully extracted. A series of phase electrodes mounted into the sample vessel (Fig. 2) is used to control the sequence of the extraction operation. The organic phase from the phase separator then proceeds through a series of wash cycles as in the previous instrument, and is again evaporated prior to further analysis.

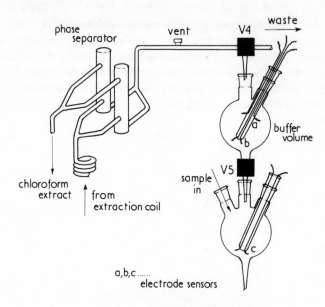

Fig. 2 – Diagram of solvent extraction system employing continuous extraction techniques. Electrodes used to sequence programme, V4 and V5 solenoid control valves.

Both systems are capable of an extraction efficiency exceeding 95% as judged by the manual analysis procedure. An important design consideration for both systems is the choice of valves and other components such as the pump pistons and piston seals. Some components in 'Viton' valve seats can be extracted by chloroform, and interfere with the analytical measurement stage. However, the use of synthetic ruby seats avoids this problem. Similarly, ceramic pistons and inert seals must be used in the piston pumps to avoid the extraction of extraneous materials from the pump-heads.

EXTRACTION OF QUINIZARIN FROM GAS OIL

The two approaches described above have been effectively combined to provide a double extraction system to identify and estimate the quantity of quinizarin in gas oil. A prototype instrument has been described by Tucker *et al.* [14] and prior to routine use this was engineered and a 'stand alone' instrument designed and constructed. Attention to this aspect and correct choice of materials of construction has enabled this instrument to remain in use since its installation in 1971. A second-generation machine is now being designed, as the spectrometer used in the original system is becoming outmoded.

The quinizarin is obtained from the gas oil by continuous extraction followed by phase separation. Prior to this, acid and an organic phase are dispensed into the final extraction chamber. Quinizarin complexed in the aqueous phase carried over into the final extraction chamber is immediately re-liberated by the acid and re-extracted into the organic phase. On completion of the extraction the coil separator is flushed with sodium hydroxide solution. The mixture in the final extraction vessel is then stirred vigorously and allowed to settle. As the aqueous solution is drained to waste, the organic phase is retained and re-sampled into a flow-through cell of a spectrophotometer. Quinizarin, which has a characteristic spectral pattern, is identified by scanning the spectral range between 400–600 nm and it is also quantified by reference to a suitable calibration graph.

Apart from removing the tedium of this operation and easing the work-load upon staff, a further advantage accrues from automating this analytical procedure. This arises from the fact that the sodium salt of quinizarin is unstable and so the manual method must rely for success upon considerable dexterity, protection from sunlight and rapid operation. No such restrictions apply to the automated method. Almost instantaneous liberation of quinizarin by continual feeding into acid and organic solutions in the automated system eliminates problems associated with the manual procedure and introduces a further degree of control. The only modification to the manual procedure needed to transfer it to the automatic approach is to substitute decalin as the final extraction solvent. Cyclohexane has a low flash-point which would provide a possible risk since electrode sensors are used in this instrument.

FLASH VAPORIZATION IN CONTINUOUS FLOW ANALYSERS

In Technicon 'AA1' technology, the option of performing flash distillation has found many applications. It is a simple clean-up procedure which removes a whole range of interfering compounds from the reagent stream of interest. However, since the introduction of 'AA2' technology its value has very much been forgotten. To some degree this can be explained by the lack of an acceptable commercially designed unit. Initial design considerations applied in the author's Laboratory followed early work by Mandl et al. [15] and Keay & Menage [16] and, most closely, the design of a more flexible unit by Shaw & Duncombe [17]. This latter unit was used for the determination of aldehydes and ketones in experimental cultures of microorganisms.

A device described by Sawyer & Dixon [2] was used for the determination of alcohol and acid in beer and stout in the initial development work. Attempts to improve the reliability of this method and to improve the signal-to-noise characteristics of the measurements prompted a critical evaluation of the distillation unit. This, in turn, resulted in an improved design described by Lidzey *et al.* [18]. This unit overcomes many of the fluctuations in results observed with use of the first unit: in this a number of possible sources of surging were indicated and these were not controlled owing to the varying conditions in the coil. In addition, the separation of the waste involatile material from the volatile phase took place outside the heated flask distillation unit. Air bubbles present in the segmented stream were also responsible for considerable surging.

Fig. 3 shows diagrammatically the unit designed by Lidzey *et al.* [18]. The air-segmented sample stream enters the coil which unlike that in the original design, contains no glass beads and hence has limited resistance to flow. At the point of entry the air bubble is released immediately and the liquid meets a carrier gas stream at (B) and flows in a thin film down a coil with air continuously flowing over it; alcohol and some water vapour are vaporized into the gas stream. At the bottom of the coil, residual liquid is pumped to waste *via* the vertical tube (E), and the vapour flows through the capillary junction (D) where it meets a stream of water introduced from the manifold. The gas-and-liquid stream then passes through a vertical condenser and the condensate collects in a liquid trap with a wide exit tube. The liquid is then re-sampled through the manifold and analyzed using a potassium dichromate reaction as described previously [2].

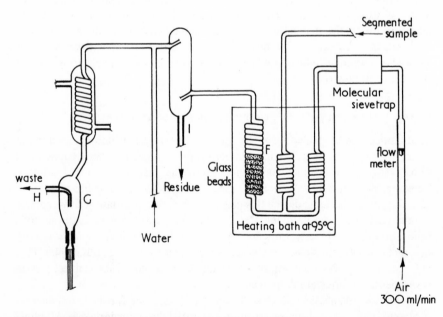

Fig. 3 – Flask distillation unit designed in LGC to minimize resistance to flow inherent in previous designs.

Use of this device for routine analysis of real-world samples again illustrates the need to understand the chemistry involved. Whilst the device performed reliably with alcohol-water solutions, the results originally produced with beer samples were low and erratic. This was because the sample contained certain proteinaceous materials which affected the alcohol distillation rates and hence gave variable results. However, the addition of detergent (0.1% Nonex) to the wash-water stream overcomes this fluctuation, and reliable and consistent results can then be obtained. Modification of the method for the analysis of samples of wine has also highlighted a similar problem. Wines can contain up to 30% w/v of sugar, and such a wide variation has deleterious effects on the distillation rate. However, the addition of a solution of 2% sugar and 2% ammonia to the wash water serves to improve the distillation character-istics, swamp out the variation caused by the sugar and neutralize any acid present.

While a reliable flask distillation unit as described above has many applications for 'AA1' methodology, it can also be used coupled to 'AA2' technology. For the determination of sulphur dioxide in wine and soft drinks it has many advan-tages over the commercial methods based on a gas membrane [19].

GAS CHROMATOGRAPHY APPLICATIONS

Examples from the author's Laboratory using GC serve to illustrate two leading aspects of automation: (a) the need to define a specification of the analytical requirements, and (b) the design of suitable interfaces which link the sample pre-treatment modules to conventional flowing systems.

In addition to quinizarin as mentioned above, furfuraldehyde is also added to hydrocarbon oils as a marker, and its presence and amount have to be ascertained for statutory purposes. The definition of the requirements of this analysis simply dictates that the unambiguous presence of furfuraldehyde should be established. It must be quantified in the range 0-6 μg/g with a precision of $\pm$ 5-10%, the rate of analysis should be 4 or 5 samples per hour and, where possible, the instrument should be constructed from readily available components. In the manual regime two analytical methods were in use, a colorimetric estimation and a TLC separation using a specific spray reagent to establish the presence of furfuraldehyde.

In the automatic instrument developed by Lidzey & Stockwell [20], a preliminary separation is performed on a GC column coupled to a specific colorimetric reagent in a continuously flowing liquid stream. A back-flushing unit is incorporated into the instrument, which extends the life of the column and removes heavy hydrocarbons from the gas stream. The method provides a single-peak chromatogram for samples containing furfuraldehyde which is identified by (a) a response signal and (b) the corresponding retention time. The most important feature of the instrument is the design of the interface

between the GC outlet and the flowing liquid stream (Fig. 4). Design considerations should avoid condensation of the components in the gas stream, provide
adequate mixing to maximize the colour formation, and transfer the solution
quickly into the measurement cell to avoid any undue colour loss. The instrument
uses a conventional GC with an automated syringe injection and with the conventional detector replaced by the interface to the colorimeter. The instrument
provides specific detection which can be readily extended to other analytes:
some possible further applications, including the measurement of aldehydes in
tobacco-smoke condensates, have been described by Stockwell [21].

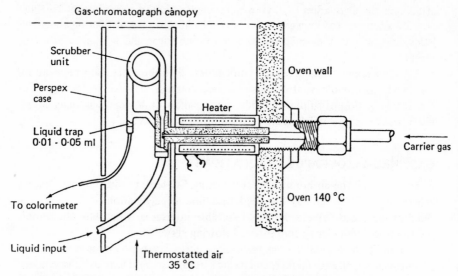

Fig. 4 – Detail of scrubbing unit used in analysis of furfuraldehyde by a hybrid GC-colori-
metric approach.

   The majority of commercial developments which relate to the automation
of GC and HPLC pay little attention to the problem of sample preparation. In
the author's Laboratory there are few examples where some pre-treatment is
not carried out prior to injection onto a column. A fully automated system was
developed for the analysis of the ethanol content of tinctures and essences to
estimate the duty payable. An instrument was designed and patented which
coupled the sample pre-treatment modules, based on conventional AutoAnalyzer
modules, to a GC incorporating data-processing facilities. This has been described
by Stockwell & Sawyer [22]. A unique sample-injection interface shown in
Fig. 5 is used to transfer samples from the manifold onto the GC column. The
pre-treatment performed on the samples as selected from one of those applicable
to the continuous-flow technology, i.e. solvent extraction, dialysis, derivatization
or simple dilution and addition of internal standard. The pre-treated samples are

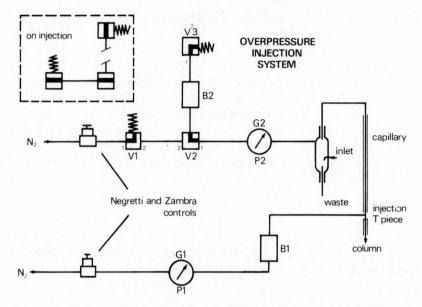

Fig. 5 – Overpressure system designed within LGC to inject 1 μl samples from flowing. AutoAnalyser system on to GC column.

directed to the interface vessel using a simple bidirectional valve. An aliquot (of the order of 1 μl) can then be injected on to the GC column through the capillary tube using a time-over pressure system. To fit this interface, the conventional injection port is removed and replaced by a low dead volume T-piece. This approach therefore is not restricted to any one make of GC. Good plug-profile injection characteristics are readily obtained.

Recently, Burns [23] has described a similar approach for HPLC analysis. A sample valve provides the injection interface in this application.

CONCLUSIONS

Some sample pre-treatment techniques which have been automated in the author's Laboratory have been described above. The coverage is by no means complete, and many other techniques available have been automated and published. Some of the difficulties encountered have been highlighted because few, if any, of the companies offer good automated sample pre-treatment systems along the lines discussed, and the analyst is usually obliged to develop his own devices. Clearly it is important to define the needs for an analysis and to avoid the use of those procedures which are difficult to automate. It is important to keep the approach simple. Complex separation systems are difficult to maintain, and in routine use problems that arise can cause a loss of confidence by the operators. The acid test of a successful automation project is not whether a systems designer can design an automated instrument but whether or not it can be routinely used to give good results.

Alternative analytical approaches, as illustrated by the furfuraldehyde instrument, or the use of alternative techniques which require little or no sample pre-treatment, have much to commend them. The use of reflectance infra-red as developed by Norris [24] could have many applications and reduce the problem of sample pre-treatment. The technique has potential even for trace levels. The instrument was originally developed to analyze grain for its oil, moisture and protein content but is now being applied in the dairy, animal-feed and tobacco industries to remarkable effect.

Acknowledgements

Work on the sample pre-treatment systems described above has been carried out over a number of years. A number of colleagues have contributed greatly to the success of this work and their help is acknowledged.

References

[1] Sawyer, R. & Dixon, E. J. (1968) *Analyst,* 93, 669–679.
[2] Sawyer, R. & Dixon, E. J. (1968) *Analyst,* 93, 680–687.
[3] Sawyer, R., Dixon, E. J. & Johnson, E. (1969) *Analyst,* 94, 1010–1020.
[4] Sawyer, R., Dixon, E. J., Lidzey, R. G. & Stockwell, F. B. (1970) *Analyst,* 95, 957–963.
[5] Carter, J. M. and Nickless, G. (1970) *Analyst,* 95, 148–152.
[6] Karlberg, Bo & Thelander, S. (1978) *Anal. Chim. Acta,* 98, 1–7.
[7] Vallis, G. G. (1967) UK Patent application 14964/67.
[8] Anderson, N. G. (1970) *Am. J. Clin. Pathol.,* 53, 778–785.
[9] Arndt, R. W., Schurmann, W., Bartels, H. & Werder, H. D. (1978) *J. Automatic Chemistry,* 1, 28–32.
[10] Trowell, F. (1969) *Lab. Pract.,* 18, 144–149.
[11] Stockwell, R. B. (1975) *Proc. Anal. Div. Chem. Soc.,* 12, 273–275.
[12] Sawyer, R., Stockwell, P. B. & Tucker, K. B. E. (1970) *Analyst,* 95, 284–290.
[13] Sawyer, R., Stockwell, P. B. & Tucker, K. B. E. (1970) *Analyst,* 95, 879–884.
[14] Tucker, K. B. E., Sawyer, R. & Stockwell, P. B. (1970) *Analyst,* 95, 730–737.
[15] Mandl, R. H., Weinstein, L. H., Jacobson, J. S., McCune, D. C. & Hitchcode, A. E. (1966) *Automation in Analytical Chemistry, (1965) Proc. Technicon Symposium.* Technicon Inc./Mediad Inc., New York, pp. 270–273.
[16] Keay, J. & Menage, P. M. A. (1970) *Analyst,* 95, 379–382.

[17] Duncombe, R. E. & Shaw, W. H. C. (1967) *Automation in Analytical Chemistry, 1966 (Proc. Technicon Symposium)*, Vol. 2, Mediad Inc., New York, pp. 15-18.

[18] Lidzey, R. G., Sawyer, R. & Stockwell, P. B. (1971) *Lab. Pract.*, **20**, 213-216 & 219.

[19] Jennings, N., Bunton, N. G., Crosby, N. T. & Alliston, T. G. (1978) *J. Assoc. Public Analysts*, **16**, 59-70.

[20] Lidzey, R. G. & Stockwell, P. B. (1974) *Analyst*, **99**, 749-754.

[21] Stockwell, P. B. (1978) *Lab. Pract.*, **27**, 715-719.

[22] Stockwell, P. B. & Sawyer, R. (1970) *Anal. Chem.*, **42**, 1136-1141.

[23] Burns, D. A. (1977) *Advances in Automated Analysis, 1976 Proc. 7th Technicon Int. Congress*, Technicon Inc., New York, pp. 332-339.

[24] Norris, K. H. & Hart, J. R. (1965) *Proc. Internat. Symp. (1963) on Humidity and Aloistrine*, Reinhold, New York, pp. 14 & 19-25.

#E-3   **AUTOMATED ANALYSIS IN PHARMACOKINETIC STUDIES**

W. PACHA and **H. ECKERT**, Sandoz Ltd., Basle, Switzerland

In pharmacokinetic studies the large number of samples to be analyzed calls for some degree of automation. A fully automated system has been designed for methods such as fluorimetry, HPLC or GC preceded by liquid/liquid extraction as being a widely applicable approach to sample clean-up. A system where only solid/liquid isolation procedures can be performed seems to us less efficient and adaptable; but with the system now described, batch adsorption followed by elution can also be carried out. Examples are given for automated fluorimetric analysis of thioridazine and an HPLC analysis of an acidic compound (32-692). The apparatus could be rendered simpler and cheaper for routinely performing the pre-chromatographic sample preparation step—the most tedious stage of analysis.

In comparing the sensitivities of different analytical methods [cf. de Silva's Fig. 1 in Vol. 7, this series—*Ed.*], it has to be appreciated that with some modern drugs the amount to be determined maybe below 1 ng/ml. Another problem for the bioanalyst is that bioavailability studies may generate thousands of samples to be analyzed. This forces one to automate the analytical methods. Other advantages of automation include an increase of precision and higher motivation of the individuals involved.

The analytical work entails three stages:

(a) *Sample pre-treatment:*
— isolation of the drug from the biological medium;
— preparation and conservation of the extract for analysis.

(b) *The analytical measurement :*
— introduction into the detection system;
— data collection and storage.

(c) *The final calculations :*
— numeric transformations
— data reduction
— reporting.

In our laboratories, we have reached different stages of automation depending on the method used. Full automation (a)-(c) is available for optical spectroscopy, and stages (b) & (c) for HPLC, GC and GC-MS. We advocate the widely held view that there is a real need for automation of the sample pre-treatment in HPLC and GC. Such an apparatus should be universally applicable and easily adaptable to various types of compound.

AUTOMATIC FLUORIMETRIC ANALYSIS

We have been using for some years a fully automated system for the fluorimetric analysis of biological samples (Fig. 1). The analyzer has been developed according to our specifications (ASA-System; Ismatec S.A., Zurich). It performs an extraction by liquid–liquid distribution, if necessary a reaction to obtain a fluorescent derivative, and the measurement of fluorescence. The system incorporates a central unit which regulates the sequence of all operations (Fig. 2). The following modules are integrated into the system: transport units, dispensers, mixers, centrifuges, transfer units, tube dispensers, heater and sampler. The centrifuge is a 6-place device with swing-out buckets, working in a stop-go mode. At the beginning of each cycle, position 1 is set to the right place automatically. The detection and data sampling devices consist of a Perkin–Elmer Spectrofluorimeter, Model 203, a digital voltmeter, Solartron A200 with recorder drive unit A290, and a paper tape punch, Facit 4070. The calculations and print-out are performed on a Hewlett Packard 9830 desk-top computer in the batch-mode, using the data punched on paper tape. Recently a Basic program has been developed for on-line collection and processing of the data, using a Hewlett-Packard 3353 Laboratory Data System.

The transport units hold the analytical samples in small reagent tubes inserted into open teflon or stainless steel cylinders which are pushed through stainless steel channels. The first part of the transport line (TRAP 1) can be refrigerated and the samples protected from light. The last transport unit (TRAP 3) contains a heatable groove where reaction temperatures of 75° can be attained. From the *dispensers* solvents and reagents can be added into the sample tubes. The *extraction* from aqueous into an organic phase and *vice versa* is effected by stirring the solvent mixture by a fast rotating teflon paddle. The *separation* of aqueous and organic phase is accelerated by automatic *centrifugation*. Transfer of solution between lines is accomplished by the *transfer units* (TRAF).

Compounds lacking intrinsic fluorescence are made fluorescent by *derivatization*, possibly by reacting at elevated temperature in the *heating zone*. Finally, the samples are introduced into the detector by the *sampler*, and the detector signal together with an identification is punched on *paper tape*.

Routinely, the apparatus has been applied to the automatic fluorescence analysis of various drugs using liquid–liquid distribution as the purification step. In pilot studies it was also shown that the residue of an organic extract, after evaporation of the solvent, can be analyzed by GC. There will also be given an

Fig. 1 – Automated analytical system with final fluorimetric measurement.

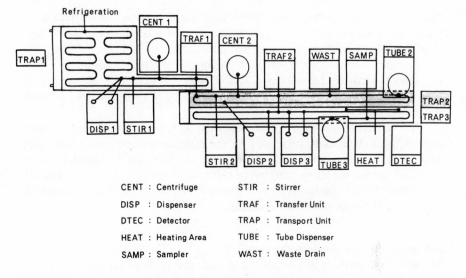

CENT : Centrifuge   STIR : Stirrer

DISP : Dispenser   TRAF : Transfer Unit

DTEC : Detector   TRAP : Transport Unit

HEAT : Heating Area   TUBE : Tube Dispenser

SAMP : Sampler   WAST : Waste Drain

Fig. 2 – Diagram of the automatic system.

example of the purification of samples by adsorption and elution on an Amberlite-resin, XAD-2, and analysis by HPLC.

The following necessities, normally required for automation, are met by this analytical system:

– the ability to carry out all operations of the manual method without intervention by a technician;

– easy and fast adaption to different methods;

– faster operation with at least the same precision as with a manual procedure.

Application to the fluorimetric analysis of thioridazine

Thioridazine was analyzed in spiked pig plasma by a method where the intrinsic fluorescence of the phenothiazines is increased by $KMnO_4$ [1]. The precision of the values obtained with the automat was comparable to the manual results, the C.V's at a concentration of 1 μg thioridazine hydrochloride per ml being 3.3 and 3.5% respectively. The values for a 50–500 ng dilution series fall on a straight regression line ($r = 0.9999$). The sensitivity was also comparable to the manual results, the fluorescence reading for 0.05 μg/ml differing significantly ($p < 0.01$) from the blank value. A throughput of 20–35 samples/h is attainable, whereas with the manual procedure only about 40 samples a day can be analyzed by one person.

SAMPLE CLEAN-UP BY SOLID–LIQUID ADSORPTION

In the previous example a distribution of drug between two liquid phases was used as the purification and isolation step. In our experience, such a preliminary clean-up is always necessary with samples containing very low concentrations of a drug. In agreement with others [2] we consider extraction into organic solvents to be the method of choice for lipophilic substances.

Another way to purify drugs in biological material is adsorption onto adsorbents, such as charcoal [3], XAD-2 resin [4], Extrelut Merck [5] (a diatomaceous earth), ion-exchange resins [6], or the recently described purified cellulose gauze [7], the so-called JET-technique.

An automated ion-exchange resin system with 6 small columns has been applied by G. B. Barlow [*this vol.*] to the estimation of creatinine and other endogenous urinary acids and amines. To obtain a good separation of a lipophilic compound from interfering substances, conditions for quantitative adsorption and for selective elution first have to be elucidated. This optimization can be very tiresome and time-consuming.

Adsorption of pindolol onto charcoal

For the determination of pindolol in plasma and urine a specific fluorimetric method using liquid/liquid extraction has been described [8]. The drug was also used as a model for isolation by activated charcoal (Scheme 1). The linearity of a dilution curve of pindolol depends very much on the amount of charcoal used [cf. survey by A. A. A. Aziz *et al.* in vol. 5, this series—*Ed.*]. For 25–100 ng

1 ml PLASMA containing
pindolol, 10–300 ng

*Take to pH 11–12 and add
deactivated charcoal, 1 mg in 1 ml
of NH$_4$OH, 1 M; centrifuge*

CHARCOAL PELLET

*Add 1 m NH$_4$OH saturated with
solid Na$_2$SO$_4$ (0.5 g), 0.5 ml; then
toluene/ether (8:2 by vol.), 10 ml;
desorb by shaking for 30 min, then
centrifuge; re-extract into aqueous
phase, viz. 0.1 M HCl, 2 ml*

AQUEOUS EXTRACT OF ELUATE

Scheme 1. Isolation of pindolol by use of charcoal. The final steps are reaction with *o*-phthaldialdehyde, then measurement.

of pindolol in 1 ml of *urine* about 1 mg of charcoal is sufficient for extraction. From 1 ml of *plasma* up to 300 ng of pindolol can be adsorbed onto 1 mg of charcoal.

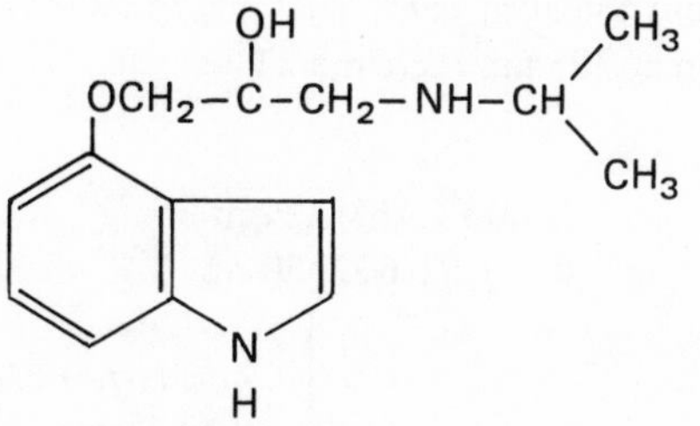

Pindolol

To obtain adequate desorption of the drug the charcoal has to be deactivated at the outset. This can be done by merely storing it in 1 M ammonium hydroxide. The extraction yields are 65-80% from plasma and 80-100% from urine.

This isolation method does not surpass the liquid-liquid extraction method in simplicity, blank reading or sensitivity. The search for optimal conditions was rather cumbersome compared with the development of an extraction method.

Adsorption of the acid 32-692 onto various adsorbents

The compound 32-692 has in a basic milieu an intrinsic fluorescence (excitation at 310/400 nm, emission at 415/510 nm). The detection in plasma is limited, however, to about 100ng by very high blank readings. No improvement in sensitivity was achieved by changing the conditions of extraction, or by using ion-pair extraction with various counter-ions, such as tetraalkylammonium derivatives or fluorescent bases such as aminoanthraquinone, aminonaphthol and quinine.

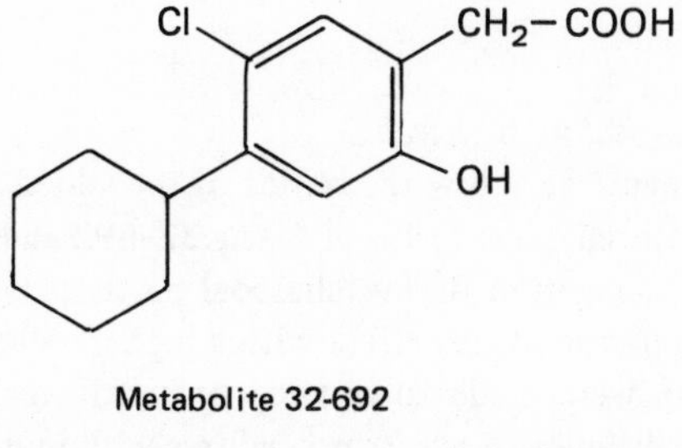

Metabolite 32-692

Trial was made of an approach entailing liquid extraction into benzene/ *iso*amyl alcohol, evaporation of the organic solvent and determination by HPLC with detection at 205 nm. Speedy analysis was precluded since interfering substances from plasma appeared with retention times of 13 min and later, whereas the compound 32-692 elutes early, at 3.8 min. Accordingly, purification of 32-692 was attempted by selective adsorption onto activated charcoal, Extrelut and XAD-2. Thereby it was hoped to eliminate the interfering substances by washing them from the adsorbent. The analysis was performed by HPLC on RP-18 with detection at 205 nm (Scheme 2).

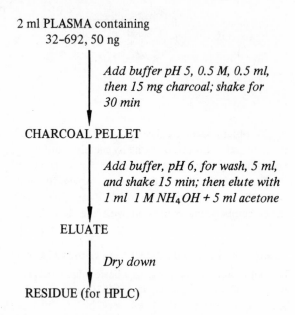

2 ml PLASMA containing
32-692, 50 ng

*Add buffer pH 5, 0.5 M, 0.5 ml,
then 15 mg charcoal; shake for
30 min*

CHARCOAL PELLET

*Add buffer, pH 6, for wash, 5 ml,
and shake 15 min; then elute with
1 ml 1 M NH_4OH + 5 ml acetone*

ELUATE

Dry down

RESIDUE (for HPLC)

Scheme 2. Charcoal absorption of metabolite 32-692. Finally the residue is dissolved in the mobile phase (A + B = 45% + 55%, where A = 1% aqueous H_3PO_4 and B = acetonitrile), and chromatographed on RP-18, 5 μm, 25 cm × 4.6 mm i.d.; flow rate 2 ml/min.

Pre-treatment of 32-692 with charcoal
In the first experiments very low recoveries of the added 32-692 were obtained after adsorption onto charcoal (14% of 50 ng 32-692 added to 2 ml of plasma). In some cases an aggregation of the charcoal particles was observed and there seemed to be a correlation of this effect with a higher yield, of the order of 60%.

Many attempts were made to obtain constantly high recoveries by using various solvents or different washing procedures with inorganic bases and buffer solutions, by applying dextran-coated charcoal, and by varying the time and temperature for adsorption of 32-692. The best results were obtained by the procedure of Scheme 2, but still the reproducibility was not satisfactory (recovery 30-80%).

Extraction of 32-692 with Extrelut

Commercially filled Extrelut columns are tailored to a volume of 20 ml of aqueous test solution, applied onto a dry column and distributed as the stationary phase on the inert porous matrix. Elution is performed by water-insoluble organic solvents, whereby lipophilic compounds are extracted into the organic phase [5].

For the small volumes of plasma available in pharmacokinetic studies the commercially supplied quantity of absorbent has to be reduced. About 1 ml of aqueous phase is held back by 1 g of Extrelut. Diluting 2 ml of plasma containing 2.5 μg of compound 32-692 with water did not give any elution of drug by chloroform from the commercial amount of absorbent.

The optimal conditions we found for extraction of compound 32-692 are given in Scheme 3. To obtain a low detection limit various solvents were tried for washing the columns and for eluting the compound. The best results were achieved after washing the column with 35 ml of chloroform.

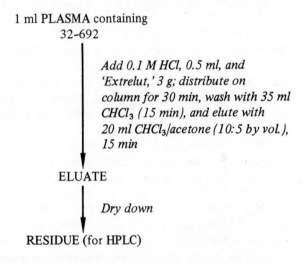

1 ml PLASMA containing
32-692

*Add 0.1 M HCl, 0.5 ml, and
'Extrelut,' 3 g; distribute on
column for 30 min, wash with 35 ml
CHCl₃ (15 min), and elute with
20 ml CHCl₃/acetone (10:5 by vol.),
15 min*

ELUATE

Dry down

RESIDUE (for HPLC)

Scheme 3. 'Extrelut' absorption of metabolite 32-692. Finally the residue is dissolved in mobile phase, and chromatographed on RP-18, as in Scheme 2.

The determination in plasma shows relatively high variability (Table 1). Using the same procedure for the determination in urine very high blank values were obtained. We feel it also to be a disadvantage of the method that such large amounts of solvents are needed for washing the column and for elution of the drug. Automation of this procedure on our system poses real problems.

Table 1. Extrelut extraction of metabolite 32–692 from plasma followed by HPLC analysis (Scheme 3).

32–692, ng spike	Mean peak area	S. D., % (n = 3)	Recovery, %	Comparison with blank; Student's t-test p <
500*	2798	1.9	(100)	
500	2241	8.6	88.4	
250	1113	12.0	87.5	
100	458	19.3	89.3	
50	213	12.1	81.8	0.001
25	87	29.5	64.3	0.01
10	28	36.3	43.5	0.02
0	6	60.1	–	–

*Non-extracted standard, subjected to HPLC direct.

Extraction of 32–692 with XAD–2

Before use this resin had to be pre-washed to eliminate interfering peaks in the analysis. This was done by washing the XAD–2 in a Soxhlet-extraction apparatus for 8 h with acetone containing 0.2% glacial acetic acid. As before, various conditions were tested for optimal results, e.g. pre-washing of the organic extract, adsorption onto the XAD–2 at different acidities, and the use of various solvents for elution of the compound.

There are two ways of performing this analysis (Scheme 4): either by adsorbing the drug from the plasma directly onto the XAD–2 (Method B), or by extracting the compound first into benzene/isoamyl alcohol and then adsorbing it onto the XAD–2 after re-extraction into a basic solution (Method A).

With either method a detection limit of about 10 ng of compound 32–692 in 2 ml of plasma and about 25 ng in 2 ml of urine was obtained.

The XAD–2 adsorption Method A, where the second liquid extraction is replaced by an adsorption/elution step, produces results comparable to or even better than a double extraction procedure with two liquid/liquid distributions.

With the direct adsorption of the drug from plasma (Method B) a greater variability of the results was found (Table 2).

Automation of the XAD–2 extraction method

The combined extraction/XAD–2–adsorption Method A was also adapted to our automated system (Fig. 3). The automatically gained results are comparable with the values obtained manually (Table 3). It should be noted that by improving the conditions for the HPLC-analysis even less variability could have been expected.

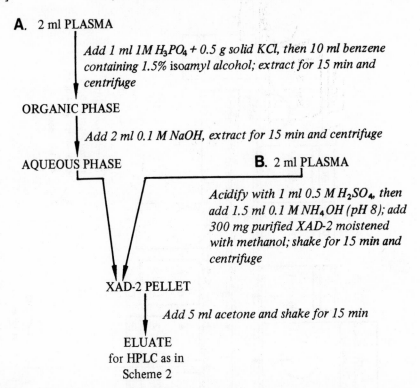

A. 2 ml PLASMA

Add 1 ml 1M H_3PO_4 + 0.5 g solid KCl, then 10 ml benzene containing 1.5% isoamyl alcohol; extract for 15 min and centrifuge

ORGANIC PHASE

Add 2 ml 0.1 M NaOH, extract for 15 min and centrifuge

AQUEOUS PHASE   **B.** 2 ml PLASMA

Acidify with 1 ml 0.5 M H_2SO_4, then add 1.5 ml 0.1 M NH_4OH (pH 8); add 300 mg purified XAD-2 moistened with methanol; shake for 15 min and centrifuge

XAD-2 PELLET

Add 5 ml acetone and shake for 15 min

ELUATE
for HPLC as in
Scheme 2

Scheme 4. Extraction of metabolite 32–692 from plasma: batch adsorption onto XAD-2.

Table 2. Comparison of different methods for extracting 32–692 from plasma

Methods:
    A (as in Scheme 4): Liquid extraction, then adsorption onto XAD–2.
    B (as in Scheme 4): Direct adsorption onto XAD–2.
    C: Two liquid extraction steps.

32-692,	S.D., % (C.V.)			Recovery, %		
ng in 2 ml	A	B	C	A	B	C
500	12.9	2.7	8.6	65.4	66.7	80.0
250	7.3	–	12.0	63.0	–	79.1
100	9.0	16.6	19.6	66.3	68.4	80.6
50	5.2	13.8	7.8	52.8	60.0	74.0
25	7.3	55.1	29.0	84.4	36.0	58.0
10	19.3	–	37.0	84.4	–	40.0
n	3	4	3			

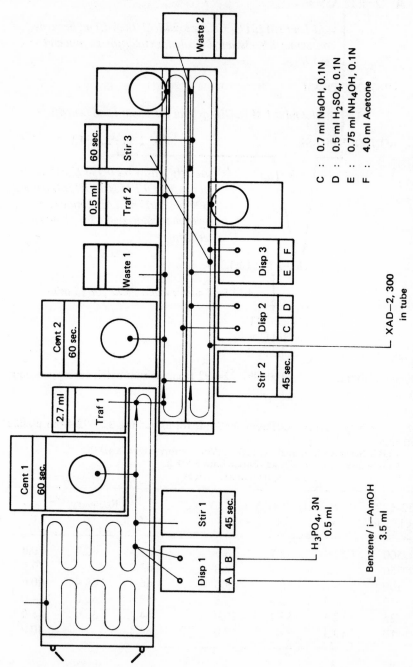

C : 0.7 ml NaOH, 0.1N
D : 0.5 ml H_2SO_4, 0.1N
E : 0.75 ml NH_4OH, 0.1N
F : 4.0 ml Acetone

XAD–2, 300
in tube

H_3PO_4, 3N
0.5 ml

Benzene/ i–AmOH
3.5 ml

Fig. 3 — Arrangement of the automated system for 'Method A', entailing extraction and XAD–2 adsorption.

Table 3. Automatic extraction of 32–692 from plasma (n=4).

Method A (as in Scheme 4): Liquid extraction, then adsorption onto XAD–2; finally analysis by HPLC on RP–18.

32-692, ng in 2 ml	Peak area	S.D., % (C.V.)	Recovery %
500	7261	7.09	89.2
250	3681	6.35	87.5
125	1938	7.08	86.6
50	809	13.68	73.4
25	666	10.88	110.5
0	115	35.55	–

CONCLUSIONS

Since sample pre-treatment is the most time-consuming stage of GC or HPLC analysis, we would hope that an automatic system for this task becomes available in the near future. In our experience sample pre-treatment by liquid/liquid extraction is more easily adaptable and more generally applicable than solid-liquid extraction procedures. The main difficulties with the latter approach proved to be attaining quantitative recovery of minute amounts of drugs from biological media.

Based on the experience with our system we would change several features if we had to develop such a sample preparation system for HPLC/GC. We would replace the sample transport system with its open cylinders. We would use an alternative mixing system for extraction to completely avoid any contamination. We would integrate into the system a device for solvent evaporation.

In addition, we feel that by using a microprocessor as the control unit, instead of the integrated circuitry, some mechanical components could be omitted by using some functions repeatedly. With the cost thereby reduced, one should arrive at a device which can be afforded even by smaller laboratories working in this exciting field of drug analysis.

REFERENCES

[1] Pacha, W. L. (1969), *Experientia,* **25**, 103–104.
[2] Arndts, D. & Rominger, K. L. (1978), *Drug Research,* **28**, 1951–1960.
[3] Meola, J. & Vanko, M. (1974), *Clin. Chem.,* **20**, 184–187.
[4] Mulé, S. J., Bastos, M. L., Inkofsky, D. & Saffer, E. (1971), *J. Chromatog.,* **63**, 289–301.

[5] Breiter, J., Helger, R. & Lang, H. (1976), *Forensic Science,* 7, 131–140.
[6] Dole, V. P., Kim, W. K. & Eglitis, I. (1966), *J. Am. Med. Assoc.,* 198, 349–352.
[7] Lantz, R. K. & Eisenberg, R. B. (1978), *Clin. Chem.,* 24, 821–824.
[8] Pacha, W. L. (1969), *Experientia,* 25, 802–803.

#E–4   AUTOMATED ANALYSIS OF CHLORTHALIDONE BASED ON THE INHIBITION OF CARBONIC ANHYDRASE

M. M. JOHNSTON, M. ROSENBERG, T. E. DORSEY and
R. F. DOYLE, USV Laboratories, Tuckahoe, NY 10707, U.S.A.

Carbonic anhydrase activity in the presence of chlorthalidone or certain other drugs shows a fall which can be the basis of an assay method. With p-nitrophenyl acetate as substrate to enable the enzyme's esterolytic activity to be measured, a continuous-flow automatic system has been developed for assaying chlorthalidone in urinary samples.

Chlorthalidone inhibits carbonic anhydrase (EC 4.2.1.1) by decreasing the rate at which the product is formed from the substrate by the catalytic action of the enzyme [1]. Since the decrease in the catalytic rate is proportional to the chlorthalidone concentration, measurement of this decrease can be used to determine the chlorthalidone level in unknown samples. This principle was utilized previously for the assay of acetazolamide [2, 3] and furosemide [4].

The procedure for the analysis of chlorthalidone in urine samples, unlike those for acetazolamide and furosemide, has been completely automated using commercially available continuous-flow equipment. For convenience of measuring the rate of product formation, the substrate was changed from CO_2 to p-nitrophenyl acetate. Physiologically, carbonic anhydrase catalyzes the hydration of CO_2, and the activity of the enzyme is usually measured by determining the rate of increase in pH during bubbling of CO_2 through a slightly alkaline buffer in the presence of the enzyme [2]. However, the enzyme also has esterolytic activity toward non-physiological substrates [5]. One substrate, p-nitrophenyl acetate, is hydrolyzed to p-nitrophenol; the hydrolysis rate is proportional to the change in absorbance at 400 nm. Since this absorbance change can be readily measured by a spectrophotometer with a flow cell, the esterolytic activity was used.

METHODS

Autoanalyzer II^R equipment was used including a Sampler IV, Proportioning Pump III and UV spectrophotometer [6] (Fig. 1). Absorbance chart paper was

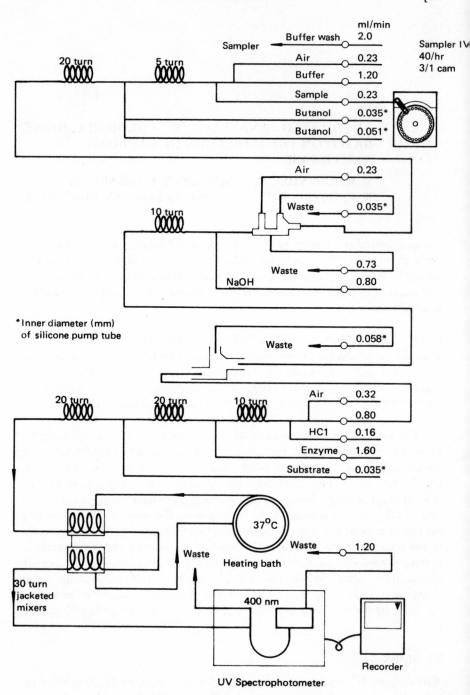

Fig. 1 — The automated system.

used for the recorder. The spectrophotometer was used in the inverse mode in which decreases in absorbance are recorded as positive differences on the chart paper. In this mode, the baseline was set to zero with all reagents pumped through their respective lines and the buffer pumped through the sample line. Substrate solution was kept in an ice-bath. The standard calibration of the spectrophotometer was adjusted so that a 5.0 μg/ml standard chlorthalidone sample gave a reading of 50 absorbance units.

The following reagents were used. The buffer was 0.1 M tris, pH 8.0. The enzyme solution contained 40 mg of bovine erythrocyte carbonic anhydrase/litre of buffer plus 0.1% Triton X-405. The substrate solution was 2.0 mM *p*-nitrophenyl acetate in polyethylene glycol 200. The butanol was saturated by shaking it with 0.4 vol. of glass-distilled, deionized water. Solutions of 0.01 M NaOH and 0.1 M HCl were used.

Samples were assayed using a 40/h cam with a 3:1 sample-wash ratio. After each 15 samples, the baseline was checked using a buffer sample, and the calibration point was checked using a 5.0 μg/ml standard. For unknown samples, the absorbance units were read directly off the chart and divided by 10 to obtain the chlorthalidone concentration. Samples beyond the linear range, >60 absorbance units, were diluted and re-assayed. Figure 2 shows a typical chart recording of a series of standards including the initial 5.0 μg/ml standard used for the spectrophotometer calibration.

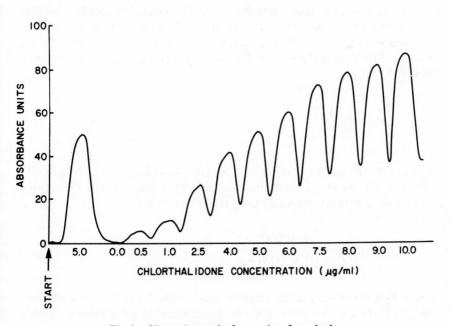

Fig. 2 – Illustrative results for a series of standards.

RESULTS AND DISCUSSION

The assay was linear up to 6.0-7.0 μg/ml. Table 1 shows the reproducibility of the assay when the single 5.0 μg/ml calibration point was used. The sensitivity of the assay was sufficient to measure the chlorthalidone concentrations in urine for approximately ten days after a normal subject had received a single 100 mg dose of the drug.

Table 1   Analysis of urine samples with known chlorthalidone levels.

Chlorthalidone added, μg/ml	Mean ± S.D. of measured values, μg/ml ($n=6$)
0.40	0.41 ± 0.04
1.00	1.00 ± 0.02
2.00	1.99 ± 0.05
3.00	3.03 ± 0.05
4.00	4.03 ± 0.04
5.00	5.02 ± 0.04
6.00	6.05 ± 0.04
7.00	6.98 ± 0.09

The sensitivity limit of this type of assay which uses enzyme inhibition does not depend on the detector sensitivity, as would be the case in an HPLC or colorimetric assay; but rather, it depends on the inhibition constant for the binding of the drug to the enzyme. For an enzyme that is described by Michaelis-Menten kinetics:

$$V_0 = \frac{k_0 [E] \ [S]}{[S] + K_m} \tag{1}$$

where V_0 is the unhibited enzymatic velocity, k_0 and K_m are constants, and $[E]$ and $[S]$ are the concentrations of the enzyme and substrate respectively. For the velocity of a non-competitively inhibited reaction:

$$V_I = \frac{k_0 [E] \ [S]}{(1 + [I]/K_I) \ ([S] + K_m)} \tag{2}$$

where V_I is the velocity of the inhibited reaction, K_I is the inhibition constant, and (I) is the concentration of the inhibitor. Since the parameter measured in this assay is the difference in enzymatic rate in the presence and absence of inhibitor, the equation for the percent of inhibition, $(V_0 - V_I)/V_0$, was derived from equations (1) and (2).

$$\frac{V_0 - V_I}{V_0} = \frac{[I]}{K_I + [I]} \tag{3}$$

Similar equations for the % inhibition can be derived from the equations for competitive and uncompetitive inhibition. All depend only on the concentration of the inhibitor and K_I.

Equation (3) has several implications for the sensitivity limit of this assay for chlorthalidone or other assays involving enzyme inhibition. First, the maximum sensitivity can be calculated from K_I and an arbitrary value for $(V_0 - V_I)/V_0$ such as 0.05 or 0.10 if it is assumed that a 5% or 10% change in enzymatic velocity can be detected. For chlorthalidone, the maximum sensitivity can be calculated using $(V_0 - V_I)/V_0 = 0.10$, $K_I = 3 \times 10^{-7}$ M [from ref. 1] and mol. wt. = 339.

$$[I] = \frac{0.10 \times 3 \times 10^{-7} \, \text{M} \times 339 \, \text{g/mole}}{1 - 0.10} = 11 \, \text{ng/ml} \tag{4}$$

Obviously, 11 ng/ml is much lower than the lowest standard in Table 1. This is because the sample is diluted during the assay and the extraction efficiency is less than 100%. Thus, if maximum sensitivity is desired (it was not necessary for urinary samples), the dilution of the sample would have to be minimized and the extraction efficiency maximized — as in all assay procedures, not merely enzymatic assays.

The second implication of equation (3) is that the substrate and enzyme concentrations have no effect on the sensitivity of the assay. They should be chosen for convenience of measurement. The time of the incubation for the enzymatic reaction should be chosen so that 10% or less of the substrate is converted to product in order to maximize linearity. Factors such as pH and temperature can be chosen for convenience unless there is some indication that they affect K_I, in which case the effect of a changed K_I on equation (3) needs to be considered. The above generalizations should be valid for any assay which uses enzyme inhibition as a means of measuring the concentration of a compound.

References

[1] Beyer, K. H. & Baer, J. E. (1961), *Pharmacol. Rev.*, **13**, 517–562.
[2] Maren, T. H., Ash, V. I. & Bailey, E. M. (1954), *Bull. Johns Hopkins Hosp.*, **95**, 244-255.

[3] Yakatan, G. J., Martin, C. A. & Smith, R. V. (1976), *Anal. Chim. Acta*, **84**, 173.

[4] Navon, G., Shani, J., Panigel, R. & Schoenberg, S. (1975), *J. Med. Chem.*, **18**, 1152–1154.

[5] Pocker, Y. & Stone, J. T. (1967), *Biochemistry*, **6**, 668–678.

[6] Johnston, M. M., Rosenberg, M. & Kamath, B. (1979), *J. Pharm. Sci.*, **68**, 967–970.

#NC(E)   Notes and Comments related to Automated Sample Handling

#NC(E)-1   *A Note on*

AN AUTOMATED FLUORESCENCE ASSAY FOR INDAPAMIDE

P. E. GREBOW, J. TREITMAN and A. YEUNG, USV Laboratories, Tuckahoe, NY 10707, U.S.A.

Indapamide (IPE) is a new agent for the treatment of mild to moderate hypertension [1]. A fluorescence method for the determination of IPE in aqueous solutions and urine has been described previously [2]. This assay has now been automated so that 20 aqueous or 10 urinary samples/h can be assayed.

Urine samples are air-segmented, mixed on-line with 0.5 M tris (pH 8.7), and extracted with dichloroethane. After phase separation, the organic stream is washed with 0.05 M tris (pH 8.5) and then mixed with 0.1 M NaOH/1.0 M NaCl for back-extraction. The aqueous phase is re-sampled and heated at 93° in an oil bath. The emerging stream is cooled and reacted with formaldehyde which enhances the fluorescence intensity. The fluorescence of the product is measured in a flow cell at 356 nm (excitation at 284 nm). Aqueous samples of indapamide are mixed on-line with NaOH, reacted, and the fluorescence measured as described above.

A linear response for 25–200 ng/ml and 2.5–400 ng/ml is observed for the urine and aqueous standard solutions, respectively. This method has been shown to be specific for the unchanged drug in urinary samples, and there was no interference in the assay from endogenous urinary compounds.

References
[1] Onesti, G., Lowenthal, D., Affrime, M., Swartz, C., Shirk, J., Mann, R. & Schultz, E. (1977), *Clin. Pharmacol. Ther.*, **21**, 113.
[2] Grebow, P. E., Treitman, J. A. & Yeung, A. K. (1978), *J. Pharm. Sci.*, **67**, 1117–1120.

#NC(E)-2 *A Note on*
A MULTIPLE-SAMPLE TLC APPLICATOR

J. P. LEPPARD and E. REID*, Wolfson Bioanalytical Unit, Institute of Industrial & Environmental Health & Safety, University of Surrey, Guildford GU2 5XH, U.K.

Manual loading of TLC plates is laborious if the solutions are dilute and need repeated applications for the sake of an adequate solute load. Automatic applicators that incorporate a bank of syringes entail initial filling and final cleaning steps that can be tedious. An applicator developed in our laboratory, of the general appearance shown in Fig. 1, has been described [1] in its original form where the dispensing of successive 1 μl aliquots is achieved by stainless steel

Fig. 1 – General view of the applicator with a TLC plate in place. The capillaries (mounted in bar at left) are barely discernible.

loops with their shanks mounted flexibly in sockets drilled through a support bar. This bar moves to and fro in an arc, with pick-up from sample cups at one extreme and deposition onto the TLC plate at the other, the shanks remaining vertical; at the moment of gentle impact the loops rise up slightly – a patented feature which ensures that the TLC surface remains undamaged. The movement back to the pick-up position is interrupted for a pre-set time to allow the spots

*To whom enquiries about the availability of the applicator should be addressed.

to be dried by an air stream at a chosen temperature. After completion of the chosen number of cycles (up to 99), the applicator switches itself off and may, optionally, emit a buzzer signal.

This 'loop' version gives compact spots and excellent reproducibility (C.V. below 3%) with aqueous or organic-solvent solutions, provided that the initial careful alignment of the loops as set by jig is not disturbed [1]. In the newer version now described the need for alignment checks is obviated by the use of glass capillaries instead of loops (Fig. 2). There is the incidental benefit that the shape of the sample cups is less critical, although a conical bottom is still preferable if the sample volume is only ∼0.1 ml and a large proportion is to be transferred to the plate; a commercially available vessel as made for electron microscopy can be used if polypropylene can be tolerated. Success with capillaries, following early discouraging trials, has hinged on bevelling the ends of bought-in precision capillaries (Alpha Laboratories Ltd.) and rejecting any that performed poorly.

Fig. 2 — View of the delivery module with the capillaries mounted in the support bar (12-unit version).

Each socket in the detachable mounting bar is just wide enough to allow passage of the capillary with little play, the means of support being merely a silicone-rubber collar which is pre-fitted to the capillary and normally rests on the top of the bar. As in the original version, there is automatic cycling and, through a slipping clutch, provision for manual movement of the bar assembly (with power switched off) when, for example, the capillaries are to be checked for freedom of vertical movement with the bar *in situ*. If the foot of any capillary in the 12-unit (or, optionally, 8-unit) array is out-of-line, height adjustment is achieved merely by moving the collar. For a given TLC plate, the supporting platform is height-adjusted so that the aligned array rises by 0.5–1 mm when the capillary tips contact the plate. Drainage is usually quite quick and complete.

Cleaning of the capillaries when changing samples is achieved merely by replacing the sample-cup holder with a trough containing whichever solvent is used; the applicator is set to perform a few cycles, with a piece of thick filter paper on the platform. At the start of a day the capillaries should be cleaned with detergent solution, if not kept in detergent – this being a permissible overnight alternative to dry storage. Clean capillaries that have passed installation testing can usually be relied on to discharge efficiently at each descent onto the plate.

As for the original loops (still regarded as advantageous if the samples contain suspended solid that could block a capillary), test data for capillaries have been obtained with coloured solutes that were determined after 10 depositions onto a plate and elution for spectrophotometry [1]. For 5 sets of depositions with an aqueous solution in each of two tests (Table 1), the C.V. was usually below 2%,

Table 1   Mean delivery volume (μl) and (*italicized*) C. V. values for individual capillaries in an array, ascertained with an aqueous solution of tartrazine in glass sample cups. For supplementary details, *see text*. No correction was made for possible evaporation during any series of tests. The mean volume for the set, of interest only to a user who preferred to ignore capillary 'individuality' in respect of delivery volume, was 1.17 ± 0.025 (S.D.) μl.

Capillary no:	76	77	78	79	80	81	82	83	84	85	86	87
Position in the array:	A	B	C	D	E	F	G	H	I	J	K	L
Test 1:	1.15	1.19	1.16	1.16	1.14	1.17	1.19	1.17	1.16	1.17	1.12	1.21
	1.5%	*1.1%*	*0.7%*	*1.3%*	*0.5%*	*0.9%*	*0.4%*	*0.5%*	*0.5%*	*0.6%*	*0.5%*	*0.5%*
Test 2:	1.16	1.19	1.17	1.17	1.16	1.18	1.20	1.16	1.15	1.18	1.12	1.22
	2.2%	*1.6%*	*1.0%*	*2.2%*	*1.4%*	*1.2%*	*1.7%*	*2.1%*	*1.2%*	*1.3%*	*1.8%*	*1.9%*

each capillary having a characteristic mean delivery volume as already observed with loops [1]. Tests with different capillaries in a different applicator (Table 2) confirmed these findings with glass sample cups, even with dichloroethane instead of water, and gave equally good results with polypropylene cups. The applicator evidently meets the rigorous needs of quantitative TLC, even if the precaution of including an internal standard in each sample is not taken. A constraint not

Table 2 Tests as in Table 1 but with different capillaries and a different applicator, and with trial of plastic cups and of a non-aqueous medium. In the one instance where the C.V. was high (first entry for capillary no. 64), evaporation during the rather prolonged series of tests was quite noticeable, with a consequent progressive increase in the strength of the solution applied.

Capillary no:	64	67	72
Aqueous tartrazine	1.21	1.16	1.17
	2.8%	1.0%	1.1%
Aqueous tartrazine in polypropylene cups	1.19	1.16	1.17
	1.3%	1.2%	1.1%
Dimethyl yellow in dichloroethane	1.20	1.15	1.16
	1.0%	1.8%	1.7%

peculiar to this design of applicator is that with very volatile solvents such as acetone the dispensing element may become clogged with solute through drying out at the tip.

Acknowledgement is made to Mrs. L. L. Basarab for skilled and patient help to the Department of Health & Social Security for grant support towards developing the first version and to company analysts who carried out independent testing.

Reference
[1] Leppard, J. P., Harrison, A. D. R. & Nicholas, J. D. (1976), *J. Chromatog. Sci.*, **14**, 438–443.

Comments on #E-2, P. B. Stockwell — AUTOMATED SAMPLE
PRE-TREATMENT
Remark by Colin R. Jones, concerning automatic solvent extraction apparatus: the described design entails having each sample await its turn for extraction, whereas by manual methods batches of tubes can be shaken and centrifuged simultaneously. *Answer to* L. E. Martin. — Concerning the choice of centrifugal separation rather than phase separation by a mixing coil, suitable tubing for the mixing-coil technique was not available at the time the apparatus was developed.

Comment on #E-3, W. Pacha & H. Eckert —
AUTOMATED PHARMACOKINETIC ANALYSIS
Reply to J. Chamberlain. — The units were made for us by a company, and the total system is available for about 120,000 Sw. fr.

264   **Notes and Comments**   [NC(E)]

Comments on #E–4, M. M. Johnston *et al.* –
AUTOMATED CHLORTHALIDONE ASSAY
& #NC(E)-1, P. E. Grebow *et al.* – INDAPAMIDE ASSAY

Question by J. C. Swann. – In view of the non-competitive nature of the inhibition, has consideration been given to overcoming non-linearity at high chlorthalidone concentrations by calculating a parameter theoretically α inhibitor concn. whatever its level relative to the inhibition constant [the parameter being: % inhibition/(% inhibition + 100)] ? *Reply.* – Yes, we have indeed plotted logarithmic functions which increase the range; but the advantage does not outweigh the convenience of using the recorder reading directly as a measure of inhibitor concentration; it is easier and quicker to dilute the sample and re-assay it than to do the more complex data analysis. *Question on indapamide by* B. Scales. – In relation to assay specificity and to possible interference by urinary metabolites, are the structures known for the NaOH degradation product and for its reaction product with formaldehyde? *Reply.* – The structure of the fluorescent derivative is not known; it is not a simple hydrolysis product, but probably involves an oxidative mechanism. Metabolites do not interfere, being removed in the solvent extraction procedure.

Editor's citation: EXAMPLE OF USE OF AN AUTOMATIC EXTRACTOR

Phenobarbital, pyrimidone and phenytoin have been assayed by GC after extraction from serum by use of a microprocessor-controlled automatic extractor (Du Pont) [1]. Extraction entails passage through a cartridge containing a lipophilic resin which is evidently of XAD-2 type. An ingenious rotor allows 12 cartridges to be processed simultaneously, with *in situ* collection of the run-off and then of the desired eluate without manual intervention.

Editor's citation: AUTOMATION IN RESIDUE STUDIES

References are given in a review [2] to enzymic detection of cholinesterase inhibitors, e.g. in eluates from RP-HPLC columns, and to automation of chromatographic delipidization [cf. last entry in section #NC(C)].

[1] Onge, L. M. St., Dolar, E., Anglim, M. A. & Least, C. J. (1979) *Clin. Chem.,* **25**, 1373-1376.
[2] Schooley, D. A. & Quistad, G. B. (1979) *Progr. Drug Metabolism,* **3**, 1-113.

#F Ensuring Reliable Results, Especially for Drugs in Blood

#F-1 INFLUENCE OF BLOOD-SAMPLING PROCEDURES ON DRUG LEVELS IN PLASMA

OLOF BORGÅ, INGA PETTERS and RUNE DAHLQVIST,
Department of Clinical Pharmacology, Huddinge Hospital, Huddinge, Sweden

After withdrawal of blood samples, the plasma level is very sensitive to factors that influence drug distribution between red cells and plasma, especially factors such as inhibitors that affect binding. For hydrophilic drugs the plasma/red cell drug ratio in vivo *may change with storage time* in vitro. *Even slight haemolysis during sampling may raise the plasma level of a drug bound with high affinity to red cells. With the 'heparin-lock' technique, plasma propranol may fall, probably through a rise in FFA and thus altered drug binding to plasma α_1-acid glycoprotein; this indicates that heparinization for repeated venous blood sampling calls for wariness.*

There appear to be few systematic studies on the effect of blood-sampling procedures on observed drug levels in plasma. A summary is now attempted of some published results and of two studies performed by ourselves.

Drugs that are sufficiently lipophilic may penetrate the red blood cell (RBC) membrane. In general they will equilibrate rapidly between plasma and RBCs. The plasma/RBC ratio at equilibrium is determined by binding processes in plasma as well as in RBCs. Then unbound drug levels inside and outside the cell will be equal. Various factors may be critical for the binding processes, e.g. temperature, pH and binding inhibitors. The commonly used procedures for blood sampling and subsequent separation of plasma or serum from RBCs by centrifugation lead to a decrease in temperature, and hence to disturbance of *in vivo* binding equilibria between RBC and plasma, resulting in an altered plasma/RBC concentration ratio. In general, one would expect the most notable changes to occur with drugs that are extensively bound to plasma proteins and/or RBCs, whereas drugs which are not distributed into RBCs will not be affected, Table 1 gives examples of drugs known to bind to RBCs and, in the rare cases where it has been identified, the molecular species responsible for the binding.

Table 1 Examples of drugs bound to erythrocytes.

Drug	Intracellular binding component
acetazolamide	carbonic anhydrase
various primary sulphonamides	carbonic anhydrase
chlorthalidone	carbonic anhydrase
minocycline, tetracycline	carbonic anhydrase
penicillin G	haemoglobin
dicloxacillin	haemoglobin
pentazocine	unknown
tricyclic antidepressants (e.g. imipramine)	unknown
phenytoin	unknown

High-affinity binding of various drugs has, for example, been shown with the intracellular enzyme carbonic anhydrase. Examples are now given of various factors involved in the blood sampling and how they may affect the measured plasma level.

EFFECT OF TEMPERATURE

Plasma-protein binding of phenytoin increases with decreasing temperature, while RBC binding is little affected. It was thus predicted that plasma separation with refrigerated blood samples would lead to higher phenytoin levels than separation at room temperature [1]. In a small study performed in connection with a study on the kinetics of phenytoin in healthy subjects, we were able to confirm this hypothesis (Table 2). Two blood samples were drawn from four

Table 2 Phenytoin levels (μg/ml) in plasma samples isolated at two different temperatures.

Subject	a 24°	b 4°	b/a
1	3.07	3.50	1.14
2	2.92	3.05	1.04
3	2.97	3.43	1.15
4	5.30	5.65	1.07
		Mean	1.10

subjects 20 h after a 300 mg oral dose of phenytoin sodium salt (Dilantin). The samples were allowed to equilibrate for 1 h at either 24° or 4° and centrifuged at the respective temperatures. Plasma levels were determined in duplicate using a precise mass-fragmentographic procedure [2]. Samples isolated at the lower temperature gave approximately 10% higher levels. It should be noted that the true *in vivo* drug level was not determined at either temperature.

EFFECT OF DISPLACING AGENTS

Binding of a drug to a plasma protein may be inhibited by compounds bound to the same protein. With some evacuated tubes (Vacutainer, Becton & Dickinson), plasma levels of propanolol were lower than those obtained using all-glass equipment [3]. With Vacutainers, a higher RBC/plasma ratio was found, and it was concluded that some contamination in the Vacutainer butyl rubber stopper was causing a lower degree of plasma binding of propranolol and consequently a re-distribution to RBCs *in vitro*. Later the displacing agent was identified as tris(butoxyethyl) phosphate (TBEP) [4]. Displacement selectively involved the plasma protein α_1-acid glycoprotein [4]. Several lipophilic bases appear to be sensitive to this mechanism, while acidic drugs, which normally bind to albumin, are generally unaffected. A general formula for *in vitro* concentrations has been derived:

$$C_{\text{p}} = \frac{f_{\text{RBC}} \cdot C_{\text{B}}}{Hcr \cdot f_{\text{p}} + (1 - Hcr) \cdot f_{\text{RBC}}}$$

where C_{p}, C_{B} = drug concentrations in plasma and whole blood; Hcr = haematocrit; f_{p}, f_{RBC} = unbound fractions in plasma and RBC.

The formula may be used to calculate the alteration in $C_{\text{p}}/C_{\text{B}}$ ratio that will occur if, e.g., the unbound fraction of the drug in plasma is changed by virtue of a displacing agent (such with propranolol), altered temperature (phenytoin) or pH.

In this context it should be noted that certain cannulas for repeated blood sampling may cause a displacing effect similar to that noted above with Vacutainers [5].

TIME DEPENDENCE

In the previous discussion we have assumed distribution equilibrium between RBC and plasma. However, for some less lipophilic drugs the diffusion through the cell membrane occurs slowly. This fact may be of concern in studies of rapid distribution or elimination processes of a drug. Thus RBC uptake and release of penicillin G have approximate half-lives of 50–60 min, while that of elimination from the body is 30 min [6]. Therefore, when blood levels rise or fall, RBC concentrations will always be lagging behind the plasma concentrations, and equilibration will not occur *in vivo*. If, however, the blood samples are allowed to stay for varying lengths of time on the bench-top, as usually done to obtain serum samples, equilibration will be approached to a varying extent in the samples, producing very irregular pharmacokinetic results.

EFFECTS OF HAEMOLYSIS

Slight haemolysis is of little concern unless the RBC/plasma ratio is high, as in the case of chlorthalidone. At equilibrium (which is reached fairly slowly with this drug) the RBC level is approximately 40 times that of plasma [7, 8]. In experiments using ^{14}C-labelled drug we were able to demonstrate a predictable and quite considerable effect of haemolysis on the plasma chlorthalidone level. The haemolysis was brought about by adding increasing amounts of hypotonic sodium chloride solution, and determined by a spectrophotometric assay of haemoglobin, sensitive down to a haemolysis of 0.1% (P. Collste, unpublished work in author's laboratory). A 25% increase in the apparent plasma level was found at a haemolysis of 1% (Fig. 1), which, owing to the inherent yellow

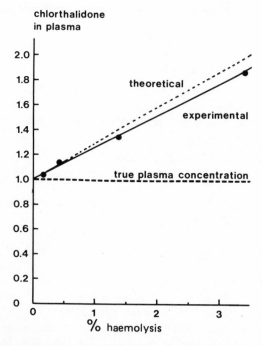

Fig. 1 – Chlorthalidone plasma concentration in relative units as a function of degree of haemolysis. The true plasma level is set to 1.0. Theoretical (*broken*) and experimental (*solid*) curves agree well.

colour of plasma, was impossible to detect by the naked eye. This value was close to that predicted (Fig. 1) by use of the following formula:

$$c_{hem} = \frac{(1 - Hcr) \cdot c + f \cdot Hcr \cdot 40c}{! - Hcr + f \cdot Hcr}$$

where f = fraction of RBCs destroyed; c = true plasma concentration in non-haemolysed sample; Hcr = haematocrit.

The factor 40 is the actual RBC/plasma ratio for chlorthalidone.

USE OF IN-DWELLING HEPARINIZED CANNULAS

This technique, the so-called heparin lock, is common practice when repeated blood samples are wanted during a relatively short time period. Usually the cannula is flushed with heparinized saline after each blood sample to prevent clot formation. Considerable amounts of heparin may thus be administered to the subject, causing free fatty acid (FFA) levels in plasma to rise. For propranol [5] and quinidine [9] the unbound fraction will rise owing to displacement of protein binding. Total plasma levels on the other hand will fall owing to re-distribution of displaced drug to tissues, and elimination processes may be profoundly affected.

Thus the last is an example of how the blood sampling procedure may not only introduce artifacts in the plasma level determination, but also interfere with the entire biological process being studied.

References

[1] Ehrnebo, M. & Odar-Cederlöf, I. (1977) *Eur. J. Clin. Pharmacol.*, **11**, 37–42.

[2] Hoppel, C., Garle, M. & Elander, M. (1976) *J. Chromatog.*, **116**, 53–61.

[3] Cotham, R. H. & Shand, D. (1975) *Clin. Pharmacol. Ther.*, **18**, 535–538.

[4] Borgå, O., Piaksky, K. M. & Nilsen, B. G. (1977) *Clin. Pharmacol. Ther.*, **22**, 539–544.

[5] Wood, M., Shand, D. G. & Wood, A. J. J. (1979) *Clin. Pharmacol. Ther.*, **25**, 103–107.

[6] Kornguth, M. L. & Kunin, C. M. (1976) *J. Infect. Dis.*, **133**, 175–184.

[7] Collste, P., Garle, M., Rawlins, M. D. & Sjöqvist, F. (1976) *Eur. J. Clin. Pharmacol.*, **9**, 319–325.

[8] Beerman, B., Hellström, K., Lindström, B. & Rosén, A. (1975) *Clin. Pharmacol. Ther.*, **17**, 424–432.

[9] Kessler, K. M., Leech, R. C. & Spann, J. F. (1979) *Clin. Pharmacol. Ther.*, **25**, 204–210.

#F-2   SOME MATRIX-EFFECTS ON THE EXTRACTION OF HYDROPHOBIC AMINES AND A QUATERNARY AMMONIUM COMPOUND FROM SERUM AND PLASMA

D. WESTERLUND and L. B. NILSSON, Astra Läkemedel AB, Research & Development Laboratories, Analytical Chemistry, S-151 85 Södertälje, Sweden

The extraction of protriptyline and zimelidine has been particularly studied in relation to the influence on solvent-extractability that may be exerted by certain serum and plasma components and by a plasticizer, MEHP, that may be present in blood products. Observed influences are attributable to ion-pairing and/or adduct formation.

The biological matrix may have the following features:
—The compounds to be determined often occur in very low concentrations together with many endogenous compounds in a much higher concentration range.
—Endogenous macromolecules often give complexes with the compounds of interest and/or the reagents.
—Ionized components may cause unintended ion-pair extractions into an organic phase or ion-pair formation in the aqueous phase.
—Hydrophobic endogenous compounds may enhance extraction by adduct-formation.

Protein binding of drugs is often strongly pH-dependent. The order of affinity to the most important macromolecule in this respect, albumin, is usually such that anions are more extensively bound than neutral and cationic forms [e.g. 1]. The conformation and charge of endogenous macromolecules alter with pH, and for albumin very large structural changes appear at pH values outside the range 4-10 [2]; the saturation of an aqueous phase with an organic solvent, as is the case in extractions, may also affect the conformation of the proteins [e.g. 3] and hence their binding properties.

Macromolecules often have many binding sites with similar or different affinity towards a compound; but in the analysis of drugs that often are present in very low concentrations a $1 + 1$ complex will dominate because of large excess of the protein. The influence of a protein P on extraction of an amine A at alkaline pH may then be described by the simple relationship:

$$D_A = k_{d(A)} \times (1 + K_{PA} \times [P])^{-1} \qquad (1)$$

where D_A and $k_{d(A)}$ are the distribution ratio and distribution constant for the amine, respectively, K_{PA} is equilibrium constant for a 1 + 1 complex between the amine and the protein, and $[P]$ is the protein molarity. The concentration range of albumin in blood is normally between 0.41 and 0.65 mM, and K_{PA} must consequently be >150 in order to reduce the amine-extraction by more than 10%.

Amongst the numerous anions in blood [e.g. 4], chloride and bicarbonate are present in concentrations such that they may give rise to ion-pair extraction of hydrophobic amines and quaternary ammonium compounds into suitable organic phases [5]. The fatty acids are notably hydrophobic, and give rather high ion-pair extraction constants with cations. However, the effectiveness of the ion-pair extraction is reduced by a competing distribution of the neutral form of the fatty acids themselves, and it can be shown [6] that the extraction ability of all hydrophobic ($>C_{12}$) fatty acids is the same provided that the pH is less than 8.4. Ion-pair extractions with fatty acids will furthermore be decreased by protein-binding [7] and also by adsorption of the surface-active fatty acid anions onto interfaces between the liquids [8].

The two main effects that have been discussed above, protein binding and ion-pair extraction with anions, will evidently have opposite influences on the extraction of amines from plasma or serum.

Some years ago we studied the influence of the addition of albumin and serum on the extraction of the antidepressant drug protriptyline (PT), a secondary amine, and of its tertiary (NPT) and quaternary (MPT) analogues [5, 6]. As those studies were designed to reveal the influence of the biological material, the conditions were chosen to give, theoretically, low distribution ratios, and no attempts were made to get quantitative extractions. It appeared that (1) on addition of albumin, the distribution ratios for low concentrations of both PT and MPT increased while the extraction of NPT was lowered; (2) the addition of serum gave an even higher extraction increase for MPT, dependent on both pH and the serum concentration. The respective conclusions were (1) NPT is more strongly bound to proteins than PT and MPT; (2) the extraction increase is due to the presence of anionic compounds that give extractable ion-pairs with the ammonium compounds and/or other compounds capable of giving extractable adducts. Since the extraction increases could not be satisfactorily explained by actual chloride and fatty acid concentration, other as yet unidentified components must be responsible for the effects.

We have recently started an investigation where we will systematically study the influence of selected endogenous compounds on the extraction of hydrophobic amines, the model compounds being a new antidepressant drug, zimelidine — a tertiary amine, and especially its monodesmethylated active metabolite, norzimelidine, a secondary amine.

Zimelidine

Endogenous compounds may, as mentioned above, act as counter-ions in ion-pair extractions and/or as adduct-formers. Palmitic acid as a representative fatty acid, cholesterol and bilirubin, all present in concentrations comparable to those in human blood, were found to have a negligible influence on the extraction of norzimelidine into dichloromethane. With lecithin [L-α-lecithin (dipalmitoyl)] in albumin solutions containing 0.1 M NaCl, however, the extraction of norzimelidine at pH 4.75 into methylene chloride increases linearly with lecithin concentration: at 2 mg/ml about 42% is extracted, compared with 10% in the absence of lecithin [9]. Since lecithin is a zwitterion, the extracted species is assumed to be a double ion-pair; one of the anionic components in extractions as mentioned above and from biological fluids is likely to be chloride. Since human blood contains also other phosphatides (e.g. cephalins, sphingomyelin and plasmalogens) for which the normal physiological levels may vary 5-fold or more, ion-pair extractions of drugs and endogenous molecules with such compounds may be of significance and require a careful planning of extraction procedures used in the assay from such a matrix in order to get accurate results.

Through the widespread use of plastic materials rich in additives, exogenous compounds may also participate in extraction equilibria in assay procedures for biological material. Di-(2-ethylhexyl)phthalate (DEHP) is frequently used as a plasticizer, and large amounts of this compound, e.g. 16–120 mg/l [10], and its degradation product mono-2-ethylhexylphthalate (MEHP), e.g. 4–56 mg/l [11], have been found in human plasma stored in PVC bags. MEHP is a rather hydrophobic acid that in a buffer-dichloromethane phase system mainly will be present in the organic phase over a wide pH-range. Its ability to influence the extraction of norzimelidine was tested [12]. MEHP was found to give adducts with norzimelidine, and from scrutiny of the data the best fit was obtained if it were assumed that 2 molecules of norzimelidine associate with 3 of MEHP.

In those extractions MEHP present at 1.4 mM increased the extraction of norzimelidine from 20% to 73% at pH 5.7, and the effect was still significant with one-tenth of this concentration. When plasma was present the effect was less: in two experiments the extraction increased from 20% to 36% and from 28% to 49% respectively with 1.5 mM MEHP. The smaller effect must then be due to reactions competing with the phthalate, possibly protein binding and/or extraction with endogenous compounds.

Similar findings on the extraction of zimelidine and norzimelidine, as perchlorate ion-pairs, were recently reported [13]: the addition of pooled plasma gave a 2- to 3-fold enhancement of extraction into dichloromethane. The opposite effect was, however, observed for corresponding 2-naphthalene-sulphonate ion pairs, when the extraction into the organic phase fell to about one-third on addition of plasma, probably because of an extensive protein binding of this counter-ion.

References

[1] Klotz, I. M., Gelewiz, E. W., & Urquhart, J. M. (1952) *J. Am. Chem. Soc.,* **74,** 209–211.

[2] Tanford, C., Swanson, S. A. & Shore, W. S. (1955) *J. Am. Chem. Soc.,* **77,** 6414–6421.

[3] Sun, S. F. (1969) *Biochim. Biophys. Acta,* **181,** 473–476.

[4] White, A., Handler, P. & Smith, E. L. (1964) *Principles of Biochemistry,* (3rd edn.), McGraw-Hill, New York, *vide* p. 628.

[5] Westerlund, D. & Theodorsen, A. (1975) *Acta Pharm. Suec.,* **12,** 127–148.

[6] Westerlund, D. (1975) Studies on the fluorimetric determination of sub-nanomole amounts of ammonium compounds by ion-pair extraction with application to the bioanalysis. Thesis, *Acta Universitatis Upsaliensis,* **4,** *vide* p. 18.

[7] Spector, A. A., Fletcher, J. E. & Ashbrook, J. D. (1971) *Biochemistry,* **10,** 3229–3232.

[8] Westerlund, D. & Söderqvist, A. (1975) *Acta Pharm. Suec.,* **12,** 277–284.

[9] Westerlund, D. & Nilsson, L. B. (1979) to be published.

[10] Vessman, J. & Rietz, G. (1974) *J. Chromatog.* **100,** 153–163.

[11] Vessman, J. & Rietz, G. (1978) *Vox Sang.,* **35,** 75–80.

[12] Nilsson, L. B., Vessman, J. & Westerlund, D. (1979) to be published.

[13] Westerlund, D., Nilsson, L. B. & Jaksch, Y. (1979) *J. Liq. Chromatog.,* **2,** 373–405.

#F-3 FATTY ACIDS AND PLASTICIZERS AS POTENTIAL INTERFERENCES IN THE BIOANALYSIS OF DRUGS

ROKUS A. de ZEEUW and JAN E. GREVING, Department of Toxicology, Laboratory for Pharmaceutical and Analytical Chemistry, State University, 9713 AW Groningen, The Netherlands

Interferences with plasma, serum and whole blood frequently arise from fatty acids, which are present at concentrations in the order of 10^{-3}-10^{-4} M and can easily be extracted with organic solvents, even from alkaline aqueous solutions, because of their strongly lipophilic character. Other sources are contact of the human skin with glassware and the occurrence of traces of fatty acids in organic solvents. Avoidance of fatty interference needs a thorough knowledge of properties of the drug as well of the interfering substance, e.g. extraction and chromatography behaviour, stability constants of complexes, and reactivity towards derivatizing or masking agents. In general, the interfering substance is present in much larger quantities than the drug (a factor 10^2 or 10^3 is not unusual) and the problems of obviating interference increase with decreasing drug concentration.

Notable interferences also arise from phthalate esters, which extraction solvents contain in only ppb or ppm amounts. The enrichment obtained by evaporating the extract and reconstituting in a small volume can lead to drug-like concentrations. Distillation of the solvent usually provides adequate purification. Other sources of contamination are contact of the biological fluid or an extract thereof with plastic syringes, plastic tubing, plastic containers and glass vials or containers with cap-inlays or septa consisting of some kind of polymeric material.

Although powerful separation techniques are at our disposal, (especially chromatographic) and detectors, the complex composition of the biomatrix as well as the analytical procedure itself still represent sources of interference. This becomes especially noticeable when trying to meet the ongoing demands to increase sensitivity to ng/ml or pg/ml level. The detection limit may then be set by the presence of interferences and not by the capabilities of the instrumentation [1-3].

FATTY ACIDS

Fatty acids, and their esters are encountered not only as endogenous compounds but also in the 'laboratory environment'. In the biomatrix they occur as free fatty acids and as esters in triglycerides, cholesterol, phospholipids, etc. Palmitic, oleic, stearic and linoleic acids make up about 80% of the free, long-chain fatty acids, and are thus most frequently encountered as interfering compounds. In serum and plasma these fatty acids may reach concentrations up to 350 μg/ml ($\sim 10^{-3}$M), mostly complexed to albumin and lipoproteins. In urine and saliva, their concentrations are lower, though urine may be quite rich in other acids such as hippuric acid.

Because of their acid character, fatty acids will interfere especially in the analysis of acidic and neutral drugs. Yet, because of their strongly lipophilic character, they can also be extracted from aqueous alkaline solutions with solvents such as chloroform or dichloromethane [4]. Thus (Fig. 1), in the methanol-dissolved residue from a chloroform extract of alkaline (pH 9.5) stomach contents as analysed by GC/MS in a case of trimipramine poisoning, the major peak is methyl stearate, followed by a combined peak of methyl

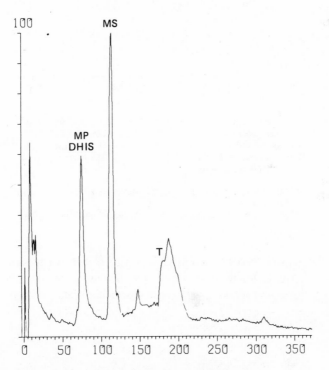

Fig. 1 – GC/MS analysis of a trimipramine intoxication specimen.
A chloroform extract of alkalinized stomach contents was evaporated to dryness and the residue re-constituted in a small amount of methanol for injection into a Finnigan GC/MS/COM instrument, model 3300, operated in the CI mode with methane as reactant gas. GC column: 3% OV-1 on Chromosorb W, at at 225°. MP = methyl palmitate, DHIS = dihydroiminostilbene, MS = methyl stearate [*not*, in this context, mass spectrometry], T = trimipramine.

palmitate and dihydroiminostilbene, a thermal degradation product of trimi-pramine. Peak T was due to trimipramine itself and was present as a shoulder on a third, unidentified, endogenous compound. The methyl fatty acid esters are the result of flash methylation in the injection port. Fatty acid interference is also shown in Fig. 2. Here, we were asked to analyze post-mortem blood for the presence of any drugs that may have been involved, without any hint about their identity, concentrations, etc. Obviously this requires analytical techniques known to give minimal recovery losses, and high sensitivity. The choice was GC/MS analysis, on a XAD-2 eluate of whole blood.

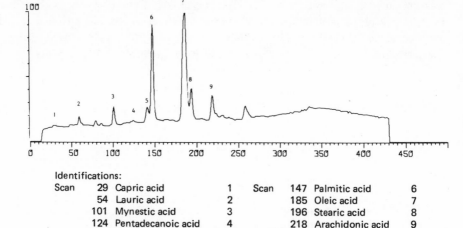

Identifications:

Scan				Scan			
	29	Capric acid	1		147	Palmitic acid	6
	54	Lauric acid	2		185	Oleic acid	7
	101	Mynestic acid	3		196	Stearic acid	8
	124	Pentadecanoic acid	4		218	Arachidonic acid	9
	141	Hexadecenoic acid	5				

Fig. 2 – GC/MS analysis in the screening for the presence of an unknown poison in post-mortem blood.
Methanol eluate of an XAD-2 extract of whole blood, analyzed as for Fig. 1 but at 200°.

Furthermore, fatty acid interference may arise if there is a derivatization step affecting all acidic functions in the sample, including the fatty acids, as in extractive alkylation procedures [1], or in crown ether-catalysed derivatizations [5].

OXYPHENONIUM
BROMIDE

BENACTYZINE
METHOIODIDE
(internal standard)

A powerful tool to avoid or to circumvent interference by fatty acids is to increase the selectivity of the extraction procedure. If we are dealing with ionizable, hydrophilic or moderately lipophilic drugs, back-extraction steps can be quite effective, provided we pay adequate attention to the fundamental aspects involved. Thus, in the analysis of plasma for the anticholinergic drug oxyphenonium bromide (OxBr), a quaternary ammonium compound with an ester function (*see panel*), whose plasma levels are in the lower ng/ml range, the potentially interfering material is present in a 100,000-fold excess. The analytical approach is summarized in Scheme 1 [3]. Ox^+ is first extracted into dichloro-ethane (DCE) by ion pairing with perchlorate as counter-ion. As fatty acids co-extracted in large amount, we decided to apply a back-extraction. We therefore applied a displacement procedure by adding tetraphenylammonium (TPeA) phosphate. In combination with ClO_4^-, $TPeA^+$ has an extraction constant E_{Qx} which is several times higher than that of Ox^+ with perchlorate, which results in an exchange of Ox^+ to the aqueous phase. Fatty acids are retained in the DCE phase. The next step is a clean-up to further remove lipophilic substances.

Thereafter (Scheme 1) Ox is hydrolysed to give an acid which is derivatized after extraction into dichloromethane as the $TPeA^+$ ion pair. The selectivity of the latter process is highly dependent on the nature and the concentration of quaternary counter-ion and the pH of the aqueous phase. Extractive alkylation will result in complete derivatization of *all* acidic compounds including fatty acids, as will be the case with catalytic procedures such as crown ether addition (Fig. 3).

Thus in order to design an optimized procedure for the determination of a drug in biological materials, the analyst must know, both for the drug and for potential interferences in the biomatrix, the key properties such as extraction behaviour, stability constants of complexes, reactivity toward derivatizing agents and chromatographic characteristics.

Besides the examples given in the oxyphenonium approach, there are, of course, other possibilities:

—column chromatographic clean-up procedures, e.g. in the analysis of chlorinated hydrocarbon pesticides or various catecholamines,

—increased chromatographic resolution as in capillary GC, though it remains to

be seen whether the capacity of those columns will be adequate, especially after prolonged routine exposure to endogenous material,
—selective detection, e.g. UV and fluorescence techniques in HPLC, AFID and ECD in GC, and MS.

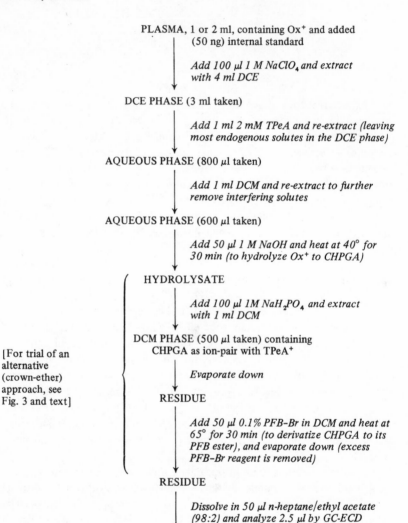

Scheme 1. Ion-pairing and other steps in the determination of oxyphenonium bromide (OxBr) in plasma, with benactyzine methiodide as internal standard.

Abbreviations: DCE, dichloroethane; DCM, dichloromethane; TPeA, tetraphenylammonium (phosphate); CHPGA, cyclohexylphenylglycollic acid; PFB, pentafluorobenzyl.

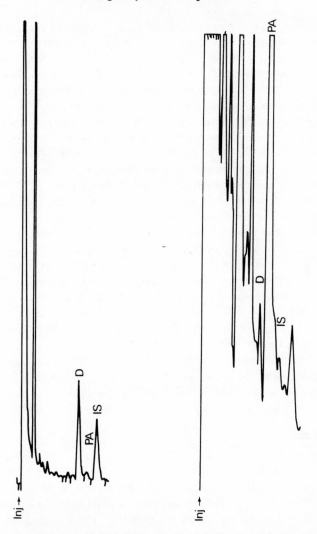

Fig. 3 – GC-ECD runs, showing extent of interference by fatty acids in the determination of oxyphenonium bromide as added to plasma along with the internal standard (each 10 ng/ml).
For abbreviations, see legend to Scheme 1.
Left: Extraction of CHPGA as ion pair with TPeA$^+$ followed by derivatization with PFB-Br as in Scheme 1.
Retention times: CHPGA-derivative (denoted 'D'), 4.34 min; internal standard derivative ('IS'), 5.62 min; palmitic acid derivative ('^{6}PA'; interference) 4.96 min.
Right: Variation of Scheme 1 (as indicated), with crown ether-catalyzed derivatization as described by Durst *et al.* [5].
Retention times *(abbreviations as above)*: D, 4.74 min; I.S., 5.83 min; PA, 5.05 min; other fatty acid derivatives (interferences), 2.19, 2.84, 3.49, 3.79 and 6.75 min.

Though each technique helps in particular ways to improve selectivity and achieve very high sensitivity, emphatically it still remains necessary to minimize the amounts of interfering substances. Although the interference may go undetected, its mere presence may easily pollute the analytical systems. Moreover, each additional step in the analytical procedure entails the risk of recovery losses that becomes more pronounced with decreasing drug concentrations.

Though fatty acid contamination may come largely from endogenous material, the possibilities for exogenous contaminations should not be underestimated. The presence of fatty acids on the human skin can easily result in contaminations. Contact of glassware with the skin will leave a fingerprint of fatty material which can creep to the inside fairly easily if the glassware is wetted, even with solvents like water [6]. Furthermore, trace amounts of fatty acids may be present in chemicals and solvents, especially those which, as for ethanol and ethyl acetate, have been produced by fermentation processes. Because of the concentration factors reached in the evaporation of organic extracts and re-dissolution in a very small amount of solvent, and because of build-up when glassware is being used repeatedly [6], these trace amounts can easily develop into noticeable interferences. Therefore, cleaning of glassware to avoid the build-up of these and other potential interferences (see below) is critical. The procedure we now use routinely is as follows. Leave to stand overnight in a freshly prepared mixture of 80 parts hydrogen peroxide 30%, 300 parts HCl 36% and 120 parts distilled water. The next morning the glassware is rinsed with distilled water and dried in an oven with a fan for 1 h at 150°. This cleaned glassware should be handled by a pair of tongs and must not be touched with plastic or rubber gloves or come into contact with skin.

PLASTICIZERS

The widespread use of plasticizers in the polymer industry and the fact that these compounds can easily migrate out of the plastic by extraction or diffusion has been a cause for concern with regard to environmental contamination [8]. The most commonly used plasticizers are phthalic acid esters, adipic acid esters, sebacic acid esters and organic phosphate esters. These compounds are easily soluble in the organic solvents used for the extraction of drugs from aqueous samples and they are frequently encountered as exogenous interferences in drug analyses [9, 10]. The major source for this contamination was found to be the cap liners of the bottles or containers in which the high-quality solvents for extraction were delivered [10]. Phthalate esters were most frequently found in these solvents, their concentrations ranging from 0.1 to 5 ppm. In general, the quality indication on the label of the solvents was no guarantee of the degree of contamination. In contrast, it was found that cheaper solvent qualities which had cork inlays wrapped in aluminium foil had not detectable contamination. Even a contamination of a few ppm may be serious in view of the above-mentioned concentration factors. Figure 4 exemplifies this for a high-quality dichloroethane

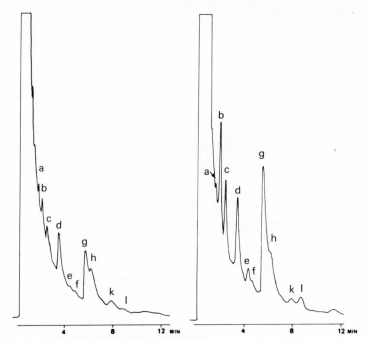

Fig. 4 – GC-FCD runs to reveal plasticizers.
Left: Residue from drying down 50 ml of high grade dichlorethane; taken up in
150 μl of chloroform, and 2 μl injected. Column: 3% OV-1 on Chromosorb G-HP,
220°.
Right: Extract of the cap inlay of high-grade dichloroethane. The cap inlay was
cut into small pieces and extracted with 50 ml of plasticizer-free dichloroethane;
further treatment as for the solvent study.

and a cap inlay of white polymer material. The major peaks are due to dipentylph-
thalate and benzylbutylphthalate. Phthalate esters respond strongly in GC-FID
and GC-ECD and, having absorption maxima around 230 and 275 nm, will also
interfere when using a UV-absorption detector. Fortunately, most plasticizers can
be removed by distillation of the extraction solvent prior to use.

The Biological fluids such as whole blood, plasma or serum can become contaminated with plasticizers when brought into contact with or stored in plastic
containers, bags or tubing [7, 8, 11]. Stoppers or septa used to close blood
collection tubes can also be a source of contamination, notably with tri-2-
butoxyethyl phosphate [9]. Thus contact with such materials must be avoided.

The extent to which plasticizers may interfere with drug bioanalysis is
demonstrated (Fig. 5) by GC/MS scans of a urine extract and a blood extract
from a subject who had ingested phencyclidine [9]. It is interesting to note that
in the blood sample fatty acid and cholesterol interference occur as well. MS is
evidently an almost indispensable tool in such systematic toxicological analyses.
Moreover, inclusion of fatty acids and plasticizers in data collections for the most
common analytical techniques, e.g. GC, HPLC, TLC and MS, is very important.

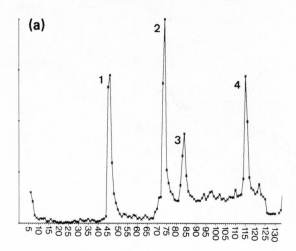

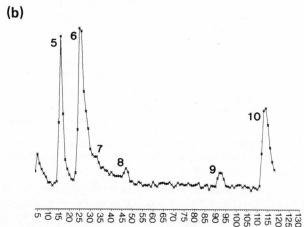

Fig. 5 – GC/MS scans of (a) urine and (b) blood extracts, in a systematic toxicological analysis in which phencyclidine was found [9]. Substance identification was done by computer-assisted comparison of the mass spectrum with a general library of about 800 spectra: 1 = phenyl cyclohexane, 2 = phenyclidine, 3 = dibutylphthalate, 4 = dioctyladipate, 5 = dibutylphthalate, 6 = palmitic acid, 7 = oleic acid, 8 = dioctyladipate, 9 = unidentified, 10 = cholesterol. *From ref. [9] with permission of the publisher.*

References

[1] Gyllenhaal, O., Brötell, H. & Hartvig, P. (1976), *J. Chromatog.*, **129**, 295–301.

[2] Greving, J. E., Jonkman, J. H. G. & De Zeeuw, R. A. (1978) *J. Chromatog.*, **148**, 389–395.

[3] Greving, J. E., Jonkman, J. H. G., Fiks, F., De Zeeuw, R. A., Van Bork, L. E. & Orie, N. G. M. (1977) *J. Chromatog.*, **142**, 611–619.

[4] Westerlund, D. & Söderqvist, A. (1975) *Acta Pharm. Suec.*, **12**, 277–284.
[5] Durst, H. D., Milano, M., Kikta, E. J., Conelly, S. A. & Grushka, E. (1975) *Anal. Chem.*, **47**, 1787–1801.
[6] Greving, J. E., unpublished observations.
[7] Vessman, J. & Rietz, G. (1974) *J. Chromatog.*, **100**, 153–163.
[8] Environmental Health Perspectives (1973) *Exptl. Issue No. 3*, U.S. Dept. of Health, Education and Welfare, Washington DC.
[9] Costello, C. E., Sakari, I. & Bieman, K. (1973) in *Techniques of Combined Gas Chromatography/Mass Spectrometry, Application to Organic Analysis*, (McFadden, W., ed.), Wiley, New York, pp. 377–384.
[10] De Zeeuw, R. A., Jonkman, J. H. G. & Van Mansvelt, F. J. W. (1975) *Anal. Biochem.*, **67**, 339–341.
[11] Jaeger, R. J. & Rubin, R. J. (1972) *New Engl. J. Med.*, **287**, 1114–1117.

#F–4 **MODIFICATIONS IN THE WORK-UP PROCEDURE FOR DRUGS DUE TO THE BIOLOGICAL MATRIX**

JÖRGEN VESSMAN, AB Hässle, S-431 20 Mölndal, Sweden

For drugs in a plasma matrix, extraction into an organic solvent may be facilitated by pre-dilution with water, yet may still be slower than if in pure aqueous solution. It may be advantageous to use a one-phase extraction system which is finally rendered two-phase by adding water. An explanation is suggested for the difficulty in extracting some lipophilic compounds.

The process of analysing drugs in biological samples comprises three major steps: (a) isolation, (b) separation, (c) quantitation. The latter two steps have been studied extensively for more than a decade, whereas the first step, isolation or the coarse separation from the biological matrix, has had less attention. Especially for blood plasma this first step has not infrequently been treated as if one was dealing with essentially an aqueous solution. In the following some examples will show that this might be an over-simplification.

Terodiline

The secondary amine, terodiline (*see panel*), was found to behave quite differently in the presence of plasma components compared to pure aqueous solutions [1]. Figure 1 illustrates results with samples spiked with ^{14}C-labelled material. Two points are of interest:
(1) There was a considerable delay in the achievement of quantitative extraction,
(2) The dilution of the plasma sample 5-fold with water led to more rapid extraction.

Although *dilution of plasma samples* often is stipulated in the literature,

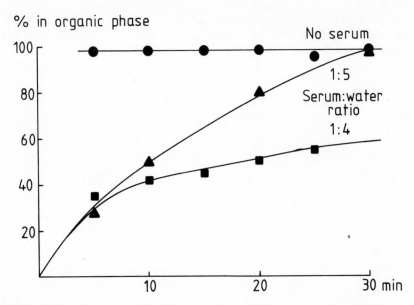

Fig. 1 – Extraction of ^{14}C-terodiline into heptane. *From ref. [1], with acknowledgement.*

there are only few examples describing the positive effect of the dilution. Two examples concern the quaternary ammonium compound emepronium [2] and the secondary amine, methindione [3].

The other effect, pronounced *time-dependent extraction,* has also been found in a few studies. For chloroimipramine (*see panel*), a rather lipophilic

tertiary amine, slow extraction into diethyl ether from spiked porcine plasma
was observed (Fig. 2). The secondary amine, a metabolite, was extracted with
less pronounced time-dependence [4].

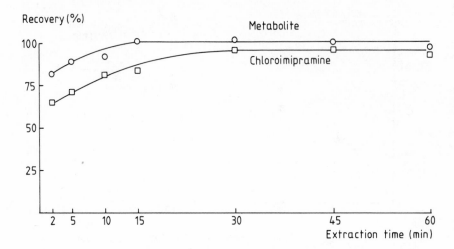

Fig. 2 – Extraction of chloroimipramine into diethyl ether.

Aprinidine

A more lipophilic compound is aprinidine. The partition equilibria for this
amine indicate that a high pH value is not necessary. Figure 3 shows that extraction into dichloromethane was especially slow when the sample was taken to pH
12.5 [5]. Possibly too high pH values can initiate some kind of denaturation,
which retards the extraction. Incipient precipitation of proteins was in this case
observed only at pH 12.5. In contrast to terodiline, dilution of the aqueous
phase gave *decreased* extraction yields.

These observations of time-dependent extractions might reflect the presence
of plasma components acting as a third phase in addition to plasma water and

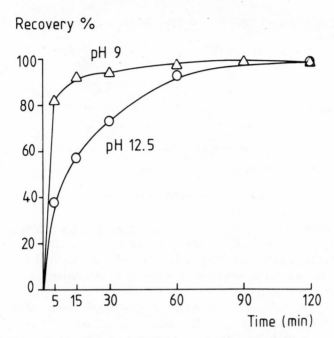

Fig. 3 – Extraction of aprinidine into dichloromethane.

the organic solvent. The afore-mentioned substances are all amines and can be present both in charged and uncharged form, which will facilitate diffusion in both polar and non-polar media. What will happen if a compound does not have protolytic properties?

Some experience was gained in the analysis of diethylhexyl phthalate (DEHP) in blood plasma [6]. Although the solubility in water is only around 1μg/ml, direct extraction from plasma into organic solvents failed to give reasonable recoveries. Instead a modification was investigated, which can best be described as a type of homogeneous extraction. The partially dried-down sample was soaked with methanol containing 10% of toluene. This single phase was after 15 min admixed with so much water as to result in a two-phase system. The DEHP was found in the small organic layer consisting of toluene.

When an experimental organosilicon compound (*overleaf*) with the same strong lipophilic character as DEHP came to be assayed, this procedure could be evaluated further [7, 8]. The organosilicon compound, tersely termed 2,6-*cis*, has a water solubility of around 3μg/ml (3 ppm). In partition studies it is entirely distributed into the organic phase. Yet it was soon realised that the extraction from plasma or serum was markedly time-dependent. Thus, with heptane it was shown that after 24 h 25% still remained in the biological matrix (a patient sample, not a spiked one). This observation indicates more strongly than data from any of the above-mentioned amines that some time-dependent process is of importance.

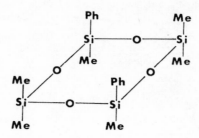

2,6-*cis*-diphenylhexamethlcyclotetranloxane (2,6-*cis*)

The problem was approached with a homogeneous extraction variant, in this case using sample/methanol/heptane in the proportions 1:10:3 to give a single homogeneous phase. After 15 min this was admixed with 10 vol. of water, whereby heptane was recovered as an upper phase that contained essentially all of the 2,6-*cis*.

The afore-mentioned two-phase extraction was made with heptane. Experiments were performed with some other solvents of varying polarity. The extraction yields were somewhat different from that with heptane (Table 1). Possibly the more a solvent takes up water and dissolves in water, the better the initial extraction.

Table 1 Extraction yield of '2,6-*cis*' from human serum with various solvents. (The value obtained by homogeneous extraction is taken as 100%.)

Solvent	% recovered after extraction for:	
	5 min	30 min
Diethyl ether	14.2	37.2
n-Heptane	2.8	14.5
Chloroform	2.1	12.7
Di*iso*butyl ketone	22.9	23.7
n-Pentanol	10.7	13.7

DEHP has been shown to be present mainly in lipoproteins. Presumably 2,6-*cis* likewise resides in lipophilic components in plasma, e.g. lipoproteins. If so, the compound is more or less shielded in a two-phase extraction from direct contact with the organic solvent. This relates to the structure of lipoproteins. Figure 4 shows schematically the two-phase nature of the lipid-containing particle, with the lipophilic core and the hydrophilic envelope or coating. Where a compound has extremely low water-solubility, as with 2,6-*cis*, it is most probable that in a two-phase extraction the rate of extraction will be controlled by diffusion, leading to a very slow process. On the other hand, in homogeneous extraction, the entire sample is brought into solution, including lipophilic and hydrophilic compartments. Upon formation of a new two-phase system the compound is distributed into the phase of similar nature.

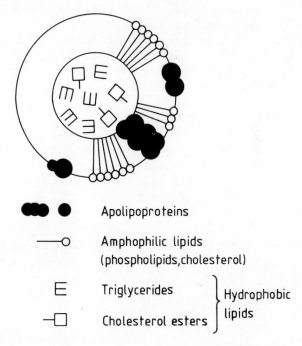

Fig. 4 – Schematic picture of a lipoprotein particle. *Reproduced by kind permission of Dr. Peter Nilsson-Ehle, University Hospital, Lund, Sweden.*

This approach is not in fact new. In lipid literature, there are procedures using the so-called Folch mixture, where methanol/chloroform (2:1) is used to create a homogeneous mixture with the sample. This is eventually divided into two phases, with the least polar lipids in the chloroform layer.

From the above-mentioned observations it is clear that especially for lipophilic compounds one has to pay much more attention to the composition of the biological sample in relation to the character of the substance of interest.

References

[1] Hartvig, P., Freij, G. & Vessman, J. (1974) *Acta Pharm. Suecica,* **11,** 97-108.

[2] Vessman, J., Strömberg, S. & Rietz, G. (1970) *Acta Pharm. Suecica,* 7, 363-372.

[3] Vessman, J., Strömberg, S. & Freij, G. (1974) *J. Chromatog.,* 94, 239-244.

[4] Lagerström, P.-O., Carlsson, I. & Persson, B.-A. (1976) *Acta Pharm. Suecica,* 13, 157-166.

[5] Lagerström, P.-O. & Persson, B.-A. (1978) *J. Chromatog.,* 149, 331-340.

[6] Vessman, J. & Rietz, G. (1974) *J. Chromatog.,* 100, 153-163.

[7] Hammar, C.-G., Freij, G., Strömberg, S. & Vessman, J. (1975) *Acta Pharmacol. Toxicol.,* 36, Suppl. 3, 33-39.

[8] Vessman, J., Hammar, C.-G., Lindeke, B., Strömberg, S., LeVier, R., Robinson, R., Spielvogel, D. & Hannemann, L. (1978) in *Biochemistry of Silicon and Related Problems* (Bendz, G. & Lindquist, I., eds,) Plenum Press, New York, pp. 535-558.

#F–5   STABILITY OF DRUGS IN STORED BIOLOGICAL SAMPLES

H. K. ADAM, Pharmacokinetics Section, Safety of Medicines Department, ICI Pharmaceuticals Division, Alderley Park, Macclesfield, SK10 4TG, U.K.

Despite its undesirability, storage of samples is frequently forced by circumstances upon analysts involved in drug measurement. A procedure is proposed which allows prospective examination of the stability of drugs in stored biological samples. It permits statistical evaluation and quantification of the limits of the statement 'no significant change'. Problems which may be encountered in stability studies and ways of getting round them are discussed.

From the point of view of the analyst involved in dealing with biological samples, Utopia would consist of a land where:

(a) analytical procedures performed perfectly at all times with sufficient sensitivity and accuracy;
(b) staff and equipment functioned faultlessly all the time;
(c) the throughput of samples was geared to match, or at least not exceed, the resources available at any given time;
(d) the laboratory in which the analyses were carried out was next-door to the source of the samples.

Unfortunately, we live in an imperfect world. One of the consequences of inhabiting the real world is that samples received into an analytical laboratory frequently have to be transported to the laboratory or, for various reasons, have to be stored before analyses can be commenced. Detailed reasons as surveyed by B. Scales in Vol. 7 (this series) do not concern us here, other than to point out that, on occasions, methodology may run into problems, equipment breaks down or staff have vacations, large-scale studies can overwhelm us with samples, or studies may be carried out in a centre some distance from the analytical laboratory.

Since it is virtually certain that one or several of these factors will occur at some stage in every analyst's career, the starting point for the discussion is the assumption that some samples will have to be stored before analyses can commence. This assumption automatically leads us to consider two factors. Firstly, under what conditions should samples be stored? Secondly, what effect will the

storage have on the analytical results generated from these samples? This latter
question must be answered before any results generated from stored samples can
be evaluated.

Not surprisingly, there have been several studies [e.g. 1–4] concerning the
effect of transport and/or storage of samples for clinico-chemical or pathological
evaluation. Indeed in standard clinical chemistry texts [e.g. 5] there are statements on the effect of storage on the reliability of the determinations. For the
assay of exogenous substances the literature is fragmentary and usually consists
of asides in papers on methodology. Indeed, amongst 22 assay descriptions for
exogenous substances in a recent volume [6], none made any explicit mention
of sample stability.

The problem indeed exists. The recovery of indomethacin from plasma
decreased with duration of deep-freeze storage [7], and the concentration of
clonazepam in serum, as determined by immunoassay, decreased by 63% when
samples were stored for 48 h at room temperature [8].

This presentation will consider:
(a) How should we try to ascertain whether storage affects the result?
(b) What problems can occur in evaluating the stability of samples?
(c) If there is sample instability, can we (i) prevent it or (ii) correct for it in our
    results?

IS THERE A PROBLEM?

It is possible to detect gross instability by careful planning and evaluation of
replicate analyses within a routine quality control programme. Thus, recently in
our own laboratory, with replicate analyses carried out two weeks after the
initial estimation, with storage at 4° between analyses, showed an average
decrease of 33 ± 6% (S.E. $n = 10$). This was tracked down to deacetylation
by serum esterases during storage, confirmed by a rise in the corresponding
deacetylated product.

An alternative approach is to prepare a calibration series in the expected
range, store it and analyze as unknowns along with the stored samples. This
option on occasions may be dictated by circumstances but it allows retrospective
evaluation only.

To clearly establish if a problem exists it is essential that a formal stability
programme be set up. Two approaches to this are possible. Firstly one can use
real samples whose volume is sufficiently large to allow several replicate analyses
over the desired period of time. This procedure has the advantage of examining
samples which contain any metabolic products and allows evaluation of any
problems, e.g. deconjugation, which this may cause. However, this method has
the major disadvantage that the accuracy of the 'true' value cannot be established
and measures of precision only can be obtained.

The best procedure is to carry out a *prospective* stability study to try to
anticipate any problems. This can be done by examining a series of known

standards, stored under exactly the same conditions as the real samples may have to be stored. With the proviso that problems from metabolites will not be seen, this is the recommended method because (i) the baseline or 'true' value is known and (ii) the range of concentrations to be examined can be set by the analyst (it should of course cover the anticipated range in real samples).

The method we have used is to add known amounts of the substance (with a check on the stock solution concentration, e.g. by UV) to a bulk sample of, say, plasma and then split into aliquots (e.g. 2 ml) and analyze after storage for suitable times. The storage is in the vessels in which real samples would be kept, e.g. plastic tubes for blood, plasma and serum and glass or plastic for urine. We choose two storage conditions, (1) room temperature for 2 weeks (in darkness) and (2) −20° for 1, 3 and 6 months, to provide an indication of any problems.

Our procedure is to store a calibration series (6 standards including zero) for the intervals described above and then compare the results obtained from the stored samples with those from a freshly prepared standard series. With modern equipment it is unusual not to obtain a linear calibration graph. Hence statistical evaluation can be applied to any differences in intercept or slope of the regression lines obtained. By this procedure not only can differences be detected but confidence intervals can also be set for any differences observed.

The procedure to evaluate such studies is illustrated in Fig. 1 and Tables 1 and 2. Figure 1 shows a set of data obtained for ICI 100,795 with the freshly

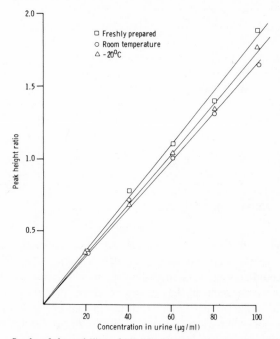

Fig. 1 − Study of the stability of ICI 100,795 in urine stored for 2 weeks.

prepared standards compared with those stored at room temperature for 2 weeks and −20° for 2 weeks. (The drug is a methyl ester.)

Table 1 shows the regression parameters from the data. Note that there is a suggestion of matrix instability (intercept a) for the room temperature samples but its magnitude is not significant. By t-test one can then calculate the significance of differences between slopes and the 95% confidence interval for these differences. These are shown in Table 2, where the differences are expressed as % of the slope of the freshly prepared calibration series. None of the differences was statistically significant. Also given in Table 2 are the 95% confidence limits of the differences. From these we can see that we would have detected decreases of 14%, 11%, 9% and 16% respectively in the samples stored at −20° for increasing times.

Table 1 Regression parameters from calibration series (2 week's storage).

	Freshly prepared	Room temperature	−20°
r^2	.996	.999	.998
a (intercept)	.0043	.0276	.0038
b (slope)	.0183	.0163	.0173
S.E.$_b$	.0006	.0003	.0004

Table 2 Differences in slopes as percentage of that from freshly prepared standards

	Room temperature	95% C.L.	−20°	95% C.L.
Week 2	−10.9*	±8.6	−5.5	± 8.8
Week 5	-	-	−3.3	± 7.6
Week 12	-	-	−3.1	± 5.9
Week 22	-	-	−5.9	±10.1

* Difference significant at $P = 0.05$

Given these latter statements we can safely conclude that ICI 100,795 is stable in urine if stored at −20° in glass bottles. Note that without the confidence intervals we could only have stated that no significant decrease occurred.

By a procedure such as that described above it is possible to anticipate storage and, assuming results similar to those in Table 2, to have confidence in the results on any samples which have to be stored for periods covered by the stability study.

PROBLEM AREAS

(1) The assay is sufficiently sensitive to determine small differences as significant, as exemplified in Table 2. – A decrease of 10.9% is significant. This suggests that care should be taken to ensure that samples are kept at room temperature for the minimum time before freezing. However, assuming first order decay and allowing the worst possible loss, i.e. – 19.5% at 2 weeks (outer limit of 95% confidence interval) storage for 3 days should cause < 5% decrease. Thus no correction could realistically be made.

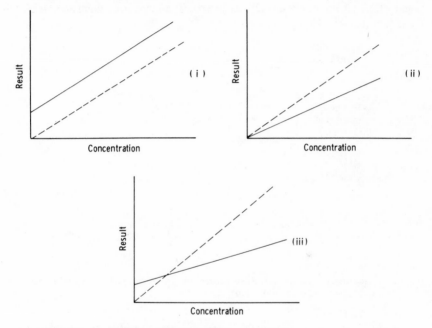

Fig. 2 – Possible results from prospective stability studies: *solid line*, stored sample; *dashed line*, freshly prepared sample.

(2) Three types of plot (Fig. 2) may be obtained by the recommended procedure:
 (i) Matrix instability but no compound instability – straightforward, allow background subtraction.
(ii) Compound unstable but no increase in background – again straightforward (t-test, etc., on slopes).
(iii) Matrix and compound instability – this area is problematical and expert statistical advice should be sought.

IF INSTABILITY DOES OCCUR

If a major problem is identified the optimal solution is to prevent it, or at least circumvent it, e.g. by increasing the polarity of the solvent [7], or by addition of antioxidants [8]. In our own laboratories we have found that serum esterases can be almost totally inhibited by the use of 5 mg/ml fluoride as anticoagulant instead of the commercially available ~1 mg/ml. On occasions the use of anti-bacterial agents (especially in animal urine) or freeze-drying [9] may solve the problem. If instability occurs its degree may vary with the storage medium. Atenolol is unstable in both blood and plasma at 4°, but Fig. 3 shows that the rate of decline in plasma is about three times that in blood (W. Bastain and J. McAinsh, personal communication). Thus in this case haemolyzed whole blood offers a distinct advantage over plasma if samples containing atenolol have to be stored at 4°.

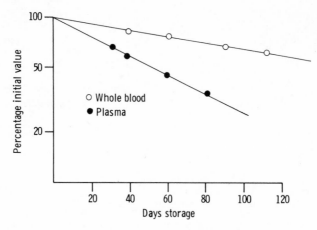

Fig. 3 – Stability of atenolol when stored at 4°. Results of W. Bastain and J. McAinsh (personal communication).

In the final analysis the advice is 'do not store samples'. If you must, then store at −20°. If storage conditions to prevent instability cannot be found, then correction factors based on a plot of percentage decrease with time (usually first order) represent the last available option.

Acknowledgements
The author thanks Mrs. H. Houghton for expert technical assistance and laboratory colleagues of Pharmacokinetics Section for helpful discussions.

References
[1] McClellan, E. K., Nakamura, R. M., Haas, W., Moyer, D. L. & Kunitake, G. M. (1964) *Am. J. Clin. Path.*, **42**, 152-155.
[2] Aimes, C. R. & Corpas, A. (1971) *J. Med. Microbiol.*, **4**, 362-365.

[3] Jefferson, H., Dalton, H. P., Escobar, M. R. & Allison, M. J. (1975) *Am. J. Clin. Path.,* **64**, 689-693.
[4] Greendyke, R. M., Banzhar, J. C., Pelysko, S. & Bauman, B. (1977) *Am. J. Clin. Path,,* **68**, 508-510.
[5] Katchmar, J. F. & Moss, D. W. in *Fundamentals of Clinical Chemistry* (Teitz, N., ed.), W. B. Saunders, Philadelphia, pp. 565-698.
[6] Various authors (1978) *J. Chromatog.,* (Biomed. Applications), **146**.
[7] Ferry, D. G., Ferry, D. M., Moller, P. W. & McQueen, E. G. (1974) *J. Chromatog.,* **89**, 110-112.
[8] Spiegel, H. E., Symington, J. & Savulich, A. (1978) *Res. Comm. in Chem. Pathol. & Pharmacol.,* **19**, 271-279.
[9] Proksch, G. J. & Bonderman, D. P. (1976) *Clin. Chem.,* **22**, 456-460.

#F-6 GLP IN BIOANALYTICAL METHOD VALIDATION

J. A. F. de SILVA, Department of Pharmacokinetics and Biopharmaceutics, Hoffmann-La Roche Inc., Nutley, N.J., 07110, U.S.A.

In the framework of 'Good Laboratory Practices' (GLP) and the particular content of analyzing drugs in biological fluids, recommendations are put forward for designing and validating assay procedures, especially chromatographic and radioimmunological (RIA). Aspects considered include the role of processed 'determinate' and 'internal' standards, as exemplified by an HPLC assay.

The mandatory Good Laboratory Practices (GLP) regulations of *Non-clinical Laboratory Studies* as initiated by the U.S. Food & Drugs Administration (FDA), were issued in the Federal Register [(43-FR-59986) Vol. 43, No. 247, Part II] on December 22, 1978, and became effective on June 20, 1979. Compliance with these regulations required the establishment of a set of Standard Operating Procedures (SOPs) for all analytical methods used in the conduct and evaluation of preclinical toxicity/safety evaluation studies covered by the regulations including those used in pertinent bioavailability/bioequivalence studies. The impact of these regulations on the analytical chemist and the practice of his speciality has been discussed in several publications of general interest [1–5].

Bioanalytical method development is emphasized since it is essential for the above studies. Statistical validation of an assay with respect to accuracy, precision and reproducibility (intra- and inter-assay) is required. The sensitivity limit of accurate quantitation and the linear dynamic range of the detection system must be defined. Specificity of the assay for the parent drug and/or its major metabolites should be demonstrated by selective extraction and/or a chromatographic separation technique coupled to a specific detection system where possible (e.g. electron capture, mass-spectrometric, electrochemical). Any compromise between sensitivity and specificity of an assay will have to be probed [6].

The overall recovery and reproducibility should be statistically evaluated by replicate analysis of spiked control samples covering the *in vivo* concentration range concerned, i.e. from the limit of sensitivity to the upper end of linearity of the detection system. The calibration curve should be defined by linear regression analysis. The accuracy of a chemical assay can be further verified using the radioisotopically labelled compound both *in vitro* and especially during *in vivo* drug-metabolism studies.

Chromatographic assays (GC, HPLC, GC–MS) should include a suitable internal standard, i.e. a chemical analogue of the compound to be quantitated, added to the original biological specimen and *processed through the method* to compensate for any procedural or adventitious variations incurred in sample handling.

The stability of the drug in the biological specimen under prolonged storage conditions, and under the conditions of analysis (e.g. enzymatic hydrolysis of esters, thermal rearrangement during GC analysis, protein precipitation by the addition of ethanolic standard solutions, and photochemical instability) must be determined, since any losses incurred can significantly affect the overall recovery, precision and reproducibility of an assay, and hence the validity of the data.

Most bioanalytical methods developed may be categorized into *I. Physicochemical assays using non-radioactive chemical compounds* (e.g. GC, HPLC, GC–MS, luminescence, electrochemical, spectrophotometric), and *II. Physicochemical assays using radiolabelled compounds* (e.g. labelled drug or derivative, and RIA). The following Standard Operating Procedures seem warranted for the validation of a bioanalytical method before it is deemed ready for use in preclinical pharmacokinetic or bioavailability studies [7] as complying with the GLP regulations for non-clinical laboratory studies.

SUGGESTED POLICIES FOR ASSAY VALIDATION IN RELATION TO GLP

[Editor's note: Since the text represents 'in-house' drafting, only minimal editing has now been done.]

The fundamental criteria required for the validation of any analytical method are (a) *accuracy* of the measurement, (b) *precision* of the measurement in replicate analyses on any given run (intra-assay) and in replicate analyses on consecutive runs on different occasions (inter-assay), and (c) *specificity* of the assay for the component being determined, especially in biological specimens taken from *in vivo* experiments. The statistical definitions and analytical interpretation of the above criteria have been adequately reviewed [8]. *Accuracy* is a function of the *absolute purity* of the analytical standard (authentic compound) used, the correctness of the gravimetric step in weighing the required amount of standard, quantitative transfer, solubility, and volumetric dilutions made to yield the working standard solutions. Systematic errors can jeopardize accuracy.

Precision is a function of the proper calibration, setting up and operation of the analytical instrument used in the final determinate step, i.e. sensitivity limit (signal-to-noise ratio) and linearity of the system, and the meticulous care used in every step (manual or automated) involved in the overall sample processing in the method (i.e. volumetric transfer of a representative aliquot of the total sample for analysis, pH adjustment, selective extraction, clean-up/derivatization, preconcentration, solubilization of residue, and analysis of an aliquot in the final determinate step). Precision can be adversely affected by indeterminate or random errors incurred in sample manipulation.

I. Physico-chemical assays using non-radioactive chemical compounds

(a) *Authentic standard compound.* – Although method development, *in vitro*, may be initiated with unreleased material, assay validation with *in vivo* samples requires released material, i.e., authentic chemical compound from a bulk lot certified [by Quality Control (Assurance)] as to percentage purity and chemical analysis.

Since the free base or acid of a compound (if ionizable) is generally used as the basis for quantitative analysis, appropriate corrections in the amount of the compound weighed, if a salt, should be made to yield unit weight of the free base or acid for quantitation of *in vivo* levels.

(b) *Choice of the biological specimen* to be selected for analysis (whole bood, plasma, serum or urine) is influenced by the pKa of the compound, its ionic behaviour, i.e. whether it is either acidic, neutral, basic or amphoteric in nature, and its protein-binding characteristics.

(c) *Quantitative extraction* of the parent drug and/or its major metabolites into a suitable solvent is determined by its pKa, pH-partition behaviour and solubility in the solvent. The absolute accuracy and ultimate sensitivity of the method is greatly influenced by the percentage recovery of the compounds to be determined.

(d) *Sample clean-up* (cf. this author's amplification, *this volume*) is governed by the tolerance and/or susceptibility of the detector system to contamination by endogenous and/or exogenous impurities present in the final residue presented for quantitative analysis. The precision and reproducibility of an analysis can be seriously impaired by idiosyncrasies of the specimen *per se,* by contaminants, and by physical losses incurred in sample work-up, especially when performed at or near the limit of detection.

(e) *Stability of the drug* (and its metabolites) during sample storage (temperature and duration) and during sample work-up must be determined. The long-term stability of a compound in a biological specimen under either freezer room ($-17°$ to $-20°$) or deep-freeze (-70 to $-90°$) conditions for at least 3–6 months should be determined (Table 1). Special precautions to be observed must be defined where needed to ensure stability during storage and/or sample processing (photochemical, chemical, thermal instability), especially if the compound has inherent stability problems. This factor is particularly important if the breakdown products formed *in vitro* also exhibits pharmacological activity and/or toxic side-effects, as opposed to similar effects seen from its biotransformation products *in vivo*.

(f) *Choice of the analytical method* to be used is governed by the physico-chemical properties of the drug compound and/or its analytically useful derivatives amenable to instrumental analysis (GC, HPLC, TLC, GC-MS, luminescence, electrochemical).

 i. All analytical instruments must be maintained in proper operating condition with periodic servicing, calibration, and with documentation thereof. The calibration and performance characteristics of a given instrument must be verified

Table 1 Stability studies.
A. HPLC assay of a compound **I**, and estimation of a presumed decomposition product **II**, in plasma stored at −17° (n.m. = non-measurable; n = no. of observations).

Day of study	Spiking into pooled plasma (μg/ml)	Amount of **I** found, ± S.D.	n	% of **I** lost	% of **II** estimated
0	5.00	4.90 ±0.19	5	n.m.	n.m.
1	5.00	5.00 ±0.12	4	n.m.	n.m.
7	5.00	5.10 ±0.14	5	n.m.	n.m.
24	5.00	4.78 ±0.09	5	4.4	−
49	5.00	4.07 ±0.13	5	18.6	18.4
70	5.00	4.04 ±0.06	5	19.2	16.5

B. HPLC assay of **I** and estimation of **II** in urine.

Storage temp.	Spiking into pooled urine (μg/0.1 m)	Day of study	Amount found, ± S.D.	n	% of **I** lost	% of **II** estimated
−17°	0.35	0	0.35 ±0.016	5	−	−
	0.35	1	0.33 ±0.007	4	(5.7)	n.m.
	0.35	7	0.25 ±0.012	5	28.6	n.m.
	0.35	14	0.21 ±0.030	5	40.0	10.9
	0.35	35	0.17 ±0.020	5	51.4	15.1
	0.35	43	0.17 ±0.017	5	52.0	16.9
−90°	0.30	0	0.30 ±0.021	5	−	−
	0.30	1	0.32 ±0.014	5	−	−
	0.30	13	0.30 ±0.013	5	−	−
	0.30	27	0.34 ±0.032	5	−	−
	0.30	48	0.32 ±0.03	4	−	−
	0.30	68	0.29 ±0.02	5	−	−

by each analyst with respect to *sensitivity* and *linearity* of the detection system for each day of operation.

ii. All solutions of reagents and analytical standards used in an assay must be properly labelled with chemical name, concentration, and an expiry date, after which they should be discarded.

iii. The analytical instrumental system should be calibrated using a standard curve of authentic standards over the concentration range used for quantitation, i.e. the sensitivity limit for accurate quantitation and the linear dynamic range of the detection system must be defined. There should be no data reported that are below the statistically validated sensitivity limit of the assay (i.e. generally not less than twice the blank value) for spectrophotometric, luminescence or electrochemical methods, or below a 5:1 signal-to-noise ratio or 1 cm peak height for chromatographic methods entailing measurement of gaussian-shaped peaks.

iv. Where possible, assays with chromatography as the final step should include an analogous chemical compound as *internal standard* which matches, in recovery and stability, although not in chromatographic position, the *determinate standard* (i.e. the compound to be quantitated), so as to monitor the intra- and inter-assay variability caused by procedural variations incurred in overall sample handling. Where analysis is by GC-MS, instrument calibration for sensitivity and linearity, and the determination of accuracy, precision and reproducibility, should be made with standard solutions containing graded amounts of pure authentic drug containing the internal standard [preferably the actual drug with a stable isotope label (^{2}H, ^{15}N, ^{13}C)], added in a fixed amount. The overall recovery from biological specimens should be determined at concentrations encompassing those expected in the unknown, e.g. 100 pg to 2000 ng/ml of plasma. (The statistical aspects of quantitative GC-MS analysis are presented in an excellent monograph by Millard [9]; see also J. de Ridder in Vol. 7 of this series.)

v. Spiking into the biological matrix for the Determinate Standard Curve should encompass the highest and lowest concentrations expected in the unknowns. During *method development* the % recovery and the S.D. of each concentration on the calibration curve must be determined by replicate analysis ($n \geq 3$) to define the statistical characteristics of the curve (slope, intercept and correlation coefficient) using linear regression analysis. The role of internal standards in this connection is discussed in detail by Curry & Whelpton [10].

vi. *In processing any batch of unknown specimens,* spiked samples containing graded amounts of authentic compound determinate standards must be processed in parallel, enabling the absolute recovery to be determined. With use of a processed *internal standard* in a chromatographic assay (see iv.) determination of the unknowns can be by a ratio method based on either peak height (for very sharp narrow peaks) or peak area (for broader peaks with a width at half-height > 4 mm). Such compensation for handling variations ensures greater confidence in singlicate determinations.

vii. *The linearity of the assay* must be demonstrated by the analysis of replicates of known concentrations of the authentic compound added to control specimen (blood, plasma, etc.) in the expected concentration range of the 'unknowns'. [*Editor's comment:* This is arguably the preferable approach (cf. v.) to establishing the calibration curve.]

viii. *The sensitivity limit of the assay* (e.g. '1.00 ± 0.05 ng/ml of blood,), below which no data should be reported, must be statistically defined and validated [cf. McAinsh *et al., later in this vol.*]. Values below the limit may be denoted either as non-measurable (n.m.) or, e.g., as less than 1 ng/ml. The volume of sample needed to achieve the stated sensitivity should also be given.

ix. *Overall recovery* of authentic standards of a given drug added to 'control' initially (see vi.) is determined by comparing the slope value (determined by linear regression analysis) of the Determinate (Processed) Standard Curve to that of the Instrument Calibration Curve generated for the analysis.

x. *Statistical evaluation* must furnish the accuracy, precision, reproducibility and % overall recovery, in replicate analyses of known concentrations added to control biological sample (intra- and inter-assay variability) [11–13]. The use of a radioisotopically labelled compound as an absolute standard for both *in vitro* and *in vivo* evaluation of the above statistical parameters ensures greater confidence with respect to the accuracy and specificity of the assay. Analysis of unknowns containing a radiolabelled compound is an excellent means of validating the accuracy and specificity of a newly developed chemical assay against an older assay by using the radiometric data as the absolute standard. Correlation of the two methods against each other by linear regression analysis ($r \geq 0.98$) is the most acceptable form of validation, as shown in Fig. 1 for clonazepam

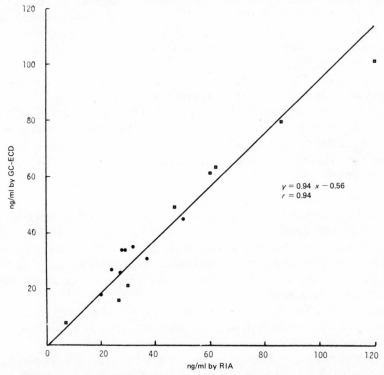

Fig. 1 – Correlation between plasma clonazepam concentrations determined by RIA and GC-ECD in a series of patients receiving clonazepam alone (●) and with concomitant anticonvulsants (■).

plasma concentrations determined by GC-ECD and RIA in patients undergoing anticonvulsant therapy with clonazepam alone or concomitant therapy with other anticonvulsant drugs (phenobarbital, primidone, diphenylhydantoin).

Assay modification. – Any modification of the original assay procedure (e.g. inclusion of newly elucidated metabolites, shortening the time of analysis, change of extraction solvents, automation, etc.) will require re-validation of the overall recovery and all the statistical parameters of the original assay.

Assay automation. – Assay methods developed initially as manual procedures may be automated by the use of suitable hardware and/or computerized data-handling systems. All the statistical parameters defined for the manual operation must be re-defined for the automated process. Standards should be introduced periodically during the automated run to check the precision and accuracy of the automated system, and re-calibration should be performed if necessary.

II. Physico-chemical assays using radioactive compounds

(A) *Radiometric analysis (concluding with liquid scintillation spectrometry)*

Accuracy checks on all spectrometers are performed monthly by assaying a series of quenched standards (^{3}H and/or ^{14}C) containing a known amount of radioactivity. The disintegrations per min (dis/min) determined experimentally are compared to the dis/min known to be present in each standard. In the event that the instrument calibration has drifted, a new calibration curve is determined for subsequent use. The print-out for each of these monthly calibrations should be displayed beside the instrument for ready reference.

The raw data is fed into the computer on punch tape using a data editor. The computer prints out raw data (sample number, total counts obtained, net counts per minute and the Automatic External Standardization (AES) or External Standard Pulse (ESP) as an instrument calibration check, and calculated results (efficiency and dis/min).

(B) *Radioimmunoassay (RIA) development*

1. *Specificity.* – Whenever possible, specificity will be validated by comparison with an established, specific physico-chemical procedure, e.g. GC-ECD, HPLC, or GC-MS. Should no alternative procedure exist, specificity will be evaluated on the basis of cross-reactivity studies with known metabolites and/or potential competitors. The specificity of each batch of antiserum used for a particular study must be determined.

2. *Radioligand.* – The specific acitivity and radiochemical purity of the radioligand must be reported for each assay, and the tracer re-purified if evidence of decomposition with time exists.

3. *Calibration curves.* – There is no best or standard procedure for plotting RIA dose-response data, but the logit-log plot of B/B_0 *versus* dose should be used where applicable. After manual evaluation of a logit-log plot, the data will be analyzed by means of a weighted linear regression analysis of the curve

essentially as described by Rodbard [14]. In those instances in which the data cannot be fitted by logit-log transformation, a manual best-fit by eye will be plotted on semi-log graph paper.

4.   *Limit of sensitivity.* – The sensitivity of a calibration curve will not exceed 90% of the mean counts bound at zero standard, i.e. at least a 10% inhibition of binding of the radioligand to the antibody following addition of unlabelled drug.

5.   *Precision.* – The response variable in logit-log calibration curves must be calculated for each run by reading the standards as unknowns against the computer-calculated regression line.

6.   *Duplicates* must be run for all samples, including standards and unknowns.

7.   During *method development* the intra- and inter-assay C.V.s must be determined by analysis of at least 6 replicates of samples which are representative of high, medium and low concentrations of the drug in the unknown samples.

8.   *Quality control* – the following RIA variables must be recorded for each run: (a) a total count of radioligand added to each tube; (b) non-specific binding as % of total count; (c) specific binding of radioligand to antibody as % of total count after correction for non-specific binding; (d) the slope and intercept of the regression line, when using the logit-log program; (e) the 10, 50 and 90% B/B_0 intercepts of the calibration curve; (f) the response variable for the standards; (g) the values for standards randomly placed among the unknowns, as calibration checks to ensure accuracy.

An excellent discussion on the development and validation of RIA procedures for anticonvulsant drugs is presented by Cook *et al.* [15].

EXAMPLE OF A CHROMATOGRAPHIC ASSAY

In an HPLC assay for an imidazo-1,4-benzodiazepine ('Z'; analogous to midazolam [16]) and internal standard, gaussian-shaped peaks (254 nm) that were completely resolved enabled accurate quantitation using either peak-height or peak-area ratio techniques. A Hewlett-Packard Laboratory Data System, Model 3352-B, was programmed to compute the essential statistical parameters.

Intra-assay variability was determined in quadruplicate at 100, 400, 800 and 1000 ng of Z added/ml of plasma, using the peak-area ratio technique (Table 2). The linear regression line for the equation $y = mX + b$ gave a correlation coefficient of 0.9994, validating the linearity of the calibration curve. The values for the recovered determinate standards, when calculated as unknowns against the computer-generated linear regression line, gave an average % deviation of 2.42, indicating the closeness of fit of the recovery data about the line.

Inter-assay variability for Z was determined in singlicate at the same 4 concentrations added/ml of plasma, in three replicate runs using the peak-height ratio technique (Table 3). Linear regression analysis gave a correlation coefficient ranging from 0.9992 to 0.9999, verifying the linearity of the calibration curve, with an average % deviation of 0.654 to 4.57, verifying the long-term reproducibility of the assay.

Table 2

Intra-assay variability for compound Z
at 4 concentrations (n=4)

```
BASIC
◆RUN
THE MAX. ENTRY IS 40 DATA POINTS. USE TSS-DATARED FOR MORE

ENTER THE NUMBER OF CALIBRATION POINTS
?◆16
ENTER DATA POINTS AS: CONC.,STD.-RESPONSE,REF.-RESPONSE
?◆100,20577,225978
?◆100,21581,226388
?◆100,20996,232505
?◆100,21810,231189
?◆400,72878,227297
?◆400,75193,230463
?◆400,74402,227404
?◆400,75064,228489
?◆800,146146,231566
?◆800,149497,232440
?◆800,146989,230044
?◆800,146376,232164
?◆1000,175059,230644
?◆1000,180931,233543
?◆1000,172240,222226
?◆1000,183223,241870
```

$$Y = 7.54529E\text{-}04 \quad \bullet \ X + \ 2.13756E\text{-}02$$

COEFFICIENT OF DETERMINATION (R^2): .998773
COEFFICIENT OF CORRELATION (R): .999386

GIVEN CONC.	CALC'D CONC.	RATIO	% DEVIATION
100	92.3515	9.10575E-02	-7.64845
100	98.0107	9.53275E-02	-1.98935
100	91.3521	9.03034E-02	-8.64789
100	96.6998	9.43384E-02	-3.30023
400	396.609	.320629	-.847626
400	404.085	.326269	1.02119
400	405.292	.32718	1.32291
400	407.072	.328523	1.76808
800	803.113	.63112	1.01414
800	824.075	.643164	3.00934
800	818.504	.63896	2.31297
800	807.272	.630485	.903951
1000	977.597	.759001	-2.24028
1000	998.433	.774722	-.156653
1000	998.89	.775067	-.111011
1000	975.643	.757527	-2.43566

THE AVERAGE % DEVIATION = 2.42092

Table 3

Inter-assay variability for compound **Z**
at 4 concentrations for 3 replicate runs

```
RUN 1  ENTER THE NUMBER OF CALIBRATION POINTS
       ?←4
       ENTER DATA POINTS AS: CONC.,STD.-RESPONSE,REF.-RESPONSE
       ?◆100,2.3,20.9
       ?◆400,9.1,20.1
       ?◆800,19.3,22.3
       ?◆1000,21.2,19.8

              Y= 1.04119E-03  ◆ X +  3.27749E-02
       COEFFICIENT OF DETERMINATION (R^2): .999926
       COEFFICIENT OF CORRELATION (R): .999963

       GIVEN CONC.    CALC'D CONC.    RATIO      % DEVIATION
         100            97.1931      .133971      -2.80688
         400           403.348       .452736       .836975
         800           802.588       .868421       .323517
        1000           996.872      1.07071        -.31283

                       THE AVERAGE % DEVIATION = 1.07005
```

```
RUN 2  ENTER THE NUMBER OF CALIBRATION POINTS
       ?◆4
       ENTER DATA POINTS AS: CONC.,STD.-RESPONSE,REF.-RESPONSE
       ?◆100,2.6,19.9
       ?◆400,7.9,18.9
       ?◆800,17.7,20.0
       ?◆1000,23.5,21.8

              Y= 1.07039E-03  ◆ X + 1.21441E-02
       COEFFICIENT OF DETERMINATION (R^2): .998351
       COEFFICIENT OF CORRELATION (R): .999175

       GIVEN CONC.    CALC'D CONC.    RATIO      % DEVIATION
         100           110.664       .130653      10.6642
         400           378.979       .417989      -5.25513
         800           815.075       .885          1.38441
        1000           995.282      1.07798        -.471777

                       THE AVERAGE % DEVIATION = 4.56898
```

```
RUN 3  ENTER THE NUMBER OF CALIBRATION POINTS
       ?◆4
       ENTER DATA POINTS AS: CONC.,STD.-RESPONSE,REF.-RESPONSE
       ?◆100,2.5,18.5
       ?◆400,9.1,19.1
       ?◆800,16.8,18.2
       ?◆1000,20.5,17.9

              Y= 1.12176E-03  ◆ X + 2.49641E-02
       COEFFICIENT OF DETERMINATION (R^2): .999976
       COEFFICIENT OF CORRELATION (R): .999988

       GIVEN CONC.    CALC'D CONC.    RATIO      % DEVIATION
         100           98.2127       .135135      -1.78728
         400          402.471        .47644        .617767
         800          800.629        .923077      7.85829E-02
        1000          998.688       1.14525        -.131239

                       THE AVERAGE % DEVIATION = .653716
```

Table 4 indicates a mean *intra-assay* coefficient of variation of 1.5%, with a range of 2.9% at 100 ng/ml, improving to 1.1% at 1000 ng/ml. The mean *inter-assay* coefficient of variation for 3 replicate runs is 2.5% with a range of 6.0% at 100 ng/ml, improving to 0.1% at 1000 ng/ml, validating the accuracy and precision of the assay.

Table 4 Statistical evaluation of the precision and reproducibility of an assay for compound **Z** in plasma (n = no. of replicates).

Amount added ng/ml	Amount found ng/ml	n	C.V., %
Intra-assay variability			
100	94.6 ± 2.8	4	2.9
400	403.3 ± 3.9	4	0.9
800	814.5 ± 7.1	4	0.9
1000	987.6 ± 11.1	4	1.1
			Mean = 1.5
Inter-assay variability			
100	102.0 ± 6.1	3	6.0
400	394.9 ± 11.3	3	2.9
800	806.1 ± 6.4	3	0.8
1000	996.9 ± 1.4	3	0.1
			Mean = 2.5

The concentration of **Z** in the unknowns is based on the linear regression line of the recovered determinate standards processed along with the unknowns (Table 5). The computer-generated equation of the line gave a correlation coefficient of 0.9994 and an average % deviation of 1.94% over the calibration range. This indicates excellent closeness of fit of the actual data about the linear regression line.

The concentration in the unknowns is determined by interpolation from the calibration curve of the recovered determinate standards, using the peak-height ratio technique, and printed in concentration (ng/ml) by sample number. Hard copy of computer-data print-outs is a very acceptable form of presentation of analytical data, since all the essential details are on each report. Data printed on thermal writers should be either Xerox-copied or photocopied for archival storage, since the original will fade with time. The regulatory agencies prefer this type of hard copy for data reports, and it may well be the required way of the future.

Table 5

Calculation of unknown concentrations of **Z**

based on least-squares regression line

```
ENTER DATA POINTS AS: CONC.,STD.-RESPONSE,REF.-RESPONSE
?♦100,2.3,20.9
?♦400,9.1,20.1
?♦800,19.8,22.1
?♦1000,21.2,19.8
          Y=  1.05388E-03   ♦ X +  3.23523E-02
COEFFICIENT OF DETERMINATION (R^2): .998754
COEFFICIENT OF CORRELATION (R): .999377
```

GIVEN CONC.	CALC'D CONC.	RATIO	% DEVIATION
100	96.4233	.133971	-3.57672
400	398.89	.452736	-.277496
800	819.421	.895928	2.42764
1000	985.264	1.07071	-1.47357

```
          THE AVERAGE % DEVIATION = 1.93886

ENTER THE NUMBER OF UNKNOWNS TO BE FOUND
?♦18
ENTER: SAMPLE #, UNK. RESPONSE, REF. RESPONSE
?♦5651,3.7,19.9
?♦5653,2.2,21.8
?♦5656,1.0,22.7
?♦5661,2.8,20.7
?♦5663,2.1,19.9
?♦5666,1.8,21.5
?♦5671,3.6,21.4
?♦5673,2.5,22.7
?♦5676,1.9,23.9
?♦5681,3.7,21.8
?♦5683,2.5,22.0
?♦5686,0.7,22.4
?♦5691,6.9,23.3
?♦5693,5.0,22.9
?♦5696,1.9,21.9
?♦5701,3.6,23.2
?♦5703,2.0,20.9
?♦5706,0.7,21.6
```

SAMPLE #	RATIO	CONCENTRATION
5651	.18593	145.725
5653	.100917	65.0594
5656	4.40529E-02	11.1023
5661	.135266	97.6515
5663	.105528	69.4339
5666	8.37209E-02	48.7422
5671	.168224	128.925
5673	.110132	73.803
5676	7.94979E-02	44.7351
5681	.169725	130.349
5683	.113636	77.1281
5686	3.12500E-02	-1.04595
5691	.296137	250.298
5693	.218341	176.479
5696	8.67580E-02	51.6239
5701	.155172	116.54
5703	9.56938E-02	60.1029
5706	3.24074E-02	5.22870E-02

Acknowledgements
The author is indebted to Dr. W. R. Dixon for Fig. 1, Dr. M. A. Brooks and Mrs. T. L. Lee for Table 1, Mr. C. V. Puglisi for Tables 2-5, Mr. M. R. Hockman for the computer programs, and especially to Ms. V. Waddell for her untiring assistance in the preparation of this manuscript.

References
 [1] Smith, R. V. (1976) *American Laboratory*, 8 (8), 47-53.
 [2] Spencer, E. Y. (1977) *Chem. Tech.*, 7, 576-579.
 [3] Horwitz, W. (1978) *Anal. Chem.*, 50, 521A-524A.
 [4] Hodges, R. M. (1978) *Anal. Chem.*, 50, 531A-540A.
 [5] Libby, R. A. (1978) *Anal. Chem.*, 50, 975A-976A.
 [6] de Silva, J. A. F. (1978) in *Blood Drugs and Other Analytical Challenges* [Methodological Surveys (A), Vol. 7] (Reid, E., ed.), Horwood, Chichester, pp. 7-28.
 [7] Schwartz, M. A. & de Silva, J. A. F. (1979) in *Principles and Perspectives in Drug Bioavailability* (Blanchard, J., Sawchuk, R. J. & Brodie, B. B., eds.), S. Karger, Basle, pp. 90-119.
 [8] Connors, K. A. (1975) *A Textbook of Pharmaceutical Analysis* (2nd edn.), Wiley, (1978), New York (see pp. 554-567).
 [9] Millard, B. J. (1978) *Quantitative Mass Spectrometry,* Heydon, London, pp. 39-90.
 [10] Curry, S. H. & Whelpton, R. (1978), as for [6], pp. 29-41.
 [11] Scales, B. (1978), as for [6], pp. 55-59.
 [12] Chamberlain, J. (1978), as for [6], pp. 55-59.
 [13] Fingl, E. (1978) in *Antiepileptic Drugs: Quantitative Analysis and Interpretation* (Pippenger, C. E., Penry, J. K. & Kutt, H., eds.), Raven Press, New York, pp. 209-218.
 [14] Rodbard, D. (1974) *Clin. Chem.,* 20, 1255-1270.
 [15] Cook, C. E., Christensen, H. D., Amerson, E. W., Kepler, J. A., Tallent, C. R. & Taylor, G. F. (1976) in *Quantitative Analytic Studies in Epilepsy* (Kellaway, P. & Peterson, I., eds.), Raven Press, New York, pp. 39-58.
 [16] Puglisi, C. V., Meyer, J. C., D'Arconte, L., Brooks, M. A. & de Silva, J. A. F. (1978), *J. Chromatog.,* 145, 81-96.

#F–7 LIMITS OF DETECTION

J. McAINSH, R. A. FERGUSON and B. F. HOLMES,
ICI Pharmaceutical Division, Safety of Medicines Department, Alderley Park, Macclesfield, SK10 4TG, U.K.

In determining the limit of detection of any analytical procedure, one problem which constantly arises is the separation of background noise peaks from those attributable to the drug under investigation. Many approaches have been used, e.g. half the bottom standard or twice background, but all are qualitative and do not reflect the inherent scatter in any calibration line. A confidence-limit approach based on a statistical analysis of a calibration plot is proposed which allows an estimation of the experimental error in any one observation. Where the error is such that the confidence interval for the estimated concentration includes zero, the measured concentration is considered non-detectable. Yet this approach is not without its difficulties, and it is felt there is still some way to go before the limit of detection problem is finally resolved.

The determination of the limit of detection of any analytical procedure is of importance to anyone carrying out pharmacokinetic studies in animals or man. In particular, for any drug there will be a time point after dosing at which the measured concentration approaches the instrumental background noise, and the question arises as to how these data should be handled. Moreover, prior to the first dose there may appear to be a positive concentration, which could vitiate a study if thought to be due to the presence of drug in the systemic blood. For these and many other situations, it is important to quantify the concentration which cannot be handled with any degree of accuracy i.e. to determine the limit of detection of the analytical procedure. It is now sought to identify the problem further and outline some of the more qualitative approaches to its solution, and then to propose a more quantitative statistical approach and deal with some of the conclusions to which it leads.

EXAMPLE BASED ON A GC-ECD ASSAY

The GC-ECD traces in Fig. 1, for a drug and internal standard spiked into a biological fluid at different concentrations, show peaks at about 15 and 5 min

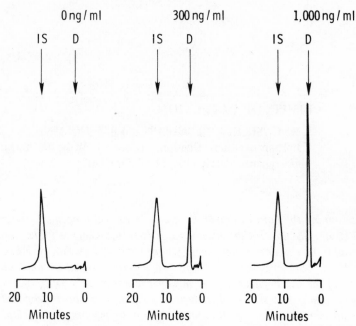

Fig. 1 – Example of GC (electron capture) traces for a drug in a biological fluid. The drug (D) was spiked into the fluid to give the concentration indicated. An internal standard (IS) was spiked in at a constant level.

respectively. For the two levels of drug spike, 300 and 1000 ng/ml, the peak is clearly identifiable, but for the blank only the internal standard is clearly identifiable, although at the retention time of the drug there is evidence of a peak. Since no drug was spiked into this sample, this peak must be due to background noise arising from many possible sources such as impurities in the carrier gas, the state of the column, the state of the detector, impurities in the organic solvents used in the chemical extraction procedure, or the biological fluid itself.

The calibration line resulting from these and other data is plotted in Fig. 2 as the peak height ratio of drug to internal standard *vs.* the known spiked concentration. The first question which arises is whether to treat the data by a statistical regression analysis or by a best-eye-fit. If the former is preferred then one has to decide between a linear regression analysis or a quadratic analysis to account for any non-linearity. In the present instance a linear regression analysis was carried out which produced the line shown in Fig. 2; the observation that the intercept from the linear regression analysis is not the same as the measured blank value will be considered later. My personal view is that a regression analysis should be undertaken wherever possible because a best-eye-fit is in itself qualitative and that is what I am attempting to replace by a quantitative estimate of the limit of detection.

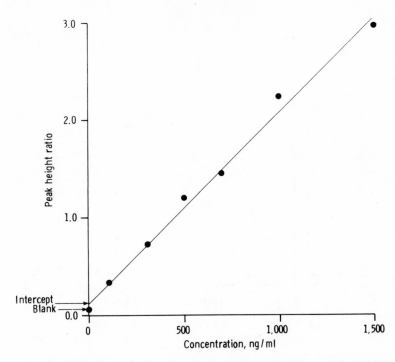

Fig. 2 – Example of a calibration curve for a drug spiked into a biological fluid, together with an internal standard so as to furnish a 'peak-height ratio'. The blank and two drug points correspond to the traces shown in Fig. 1.

A well-known additional complication which arises in this type of analytical work is the variability of the blank peak-height ratio as illustrated for some of our chlorhexidine GC data. The peak-height ratio on 28 March 1979 was just a little over 0.10 for the blank. By 4 April it had dropped to under 0.08 and by 10 April to about 0.01. However, on 18 April it was nearly 0.06 and on 11 May it was about 0.035. This variation makes it imperative that a standard curve be run with every set of unknowns on each and every day that analysis is carried out. It also implies that the limit of detection will vary on a day-to-day basis and must be assessed on each occasion. If the blank peak-height ratio did not vary and if the regression intercept was always the same, then perhaps a fixed limit of detection could be contemplated.

Now if a regression analysis has been performed on a curve for standards, the interpretation of the peak-height ratio for a biological sample of unknown concentration is straightforward (Fig. 3). The unknown peak-height ratio R1 will produce a concentration C1. Likewise R2 will produce C2. However, R3 will produce a concentration of zero and R4 would, in theory anyway, produce a negative concentration, C4. This method of computation called Direct Numerical Calculation therefore is associated with a major problem, and to get

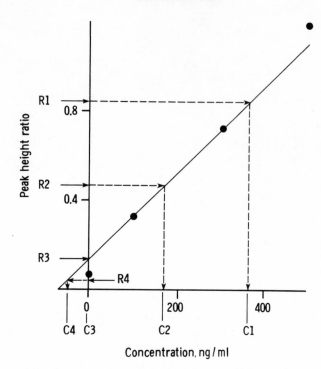

Fig. 3 – Application of the calibration curve shown in Fig. 2 to assess the drug concentrations in test samples, by 'direct numerical calculation'. For the observed peak-height ratios R1 and R2, the perpendiculars dropped from the calibration curve (only the lower part of which is shown) give the concentrations as C1 and C2. Note, however, that R3 gives a zero value and R4 a negative value, such that the term 'non-detectable' has to be used (*see text*).

round it any concentration which is negative or zero should be quoted as non-detectable. However, this method does not account for the variability in the blanks and it can produce small positive concentrations in which one has no faith. How to distinguish real concentrations from non-detectability is the key to the problem.

In principle it is a matter of establishing a 'non-detectable box'; any small positive concentrations, arising from peak-height ratios larger than the blank, that fall within the box would be quoted as 'non-detectable'. The problem now becomes how to define the boundary conditions for the box, since once this is achieved the separation of small positive concentrations and detectability is straightforward. Amongst the many qualitative methods available for achieving this end, one is shown in Fig. 4. Only the bottom two calibration points of the previous calibration line (Fig. 2) are represented. The bottom positive standard was 100 ng/ml. This Half-Bottom Standard approach can be tackled in two ways. One entails forming the box (Fig. 4) by halving the concentration to 50 ng/ml, and then projecting upwards to the line and then horizontal to the *y*-axis. Then

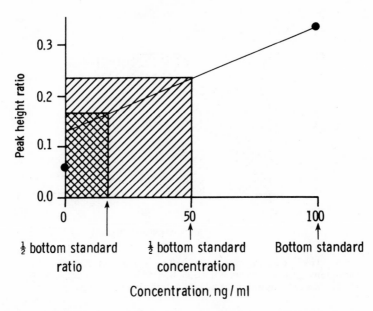

Fig. 4 – The 'half-bottom standard' approach to assessing detectability. Only the two lowest points in Fig. 2 have been shown. For explanation of hatching and for interpretation, *see text*.

any measured peak height ratio between zero and 0.024 would produce a non-detectable concentration. The alternative approach is to start with the peak-height ratio of the bottom standard, this being halved to produce the double-hatched box shown in Fig. 4. This approach would result in a much lower limit of detection. However, either way suffers from the arbitrary choice of the bottom standard and of the proportion thereof.

Another qualitative approach to the problem is known as the twice-background method. Fig. 5 illustrates the same two calibration points and line as in Fig. 4. In this method the first decision is whether to use the regression intercept or the measured blank as the representation of the background. The advantage of the former is that it takes into account all points in the calibration line whereas the advantage of the latter is that it is a direct measure. Not withstanding the problem that this decision poses, I have depicted the box formed using twice the intercept.

In the present example the twice blank would in fact produce a negative concentration for the limit of detection. Fig. 5 indicates other drawbacks to this method. If the blank is in fact used, the limit of detection depends on a single measured quantity, but if anything is wrong with that unique measurement then the error will be transmitted throughout the analytical determinations. Another type of drawback arises if the regression intercept is zero or negative, both likely possibilities dependent on the scatter of the complete calibration run. Finally

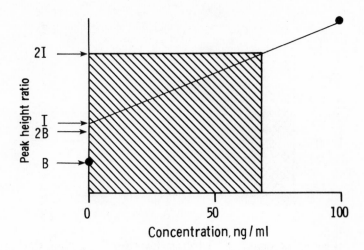

Fig. 5 – The 'twice-background' approach to assessing detectability. The observed blank intercept is denoted **B**, and the regression intercept **I**. The box has been arbitrarily based on twice the latter.

this method suffers the same drawback as the other, that is the arbitrary choice of twice. One might well choose three times or any other multipliers.

Statistical approach

At this point statistics come to the rescue, through furnishing an estimate of the error and confidence which can be placed on the analytical value for an unknown concentration. The interrupted sloping lines shown in Fig. 6 depict the 90% confidence interval for a predicted concentration. Thus, when a peak-height ratio is observed to have a value of 2.0, the horizontal line drawn from the y-axis cuts the calibration line at the estimated concentration and the interrupted lines at the two confidence limits. If this procedure were to be repeated many times, on 90% of the occasions the interval would enclose the true concentration that is in the sample.

The limit of detection is a one-sided limit below which we are 95% sure a new blank observation would fall. Because this is a one-sided statement, i.e. we are giving a limit *below* which the values will almost certainly fall, it is comparable with a 90% two-sided limit for the estimated concentration. In practice this means that the boundary conditions for the formation of the non-detectable box occurs when the upper interrupted line cuts the y-axis, the line then being taken horizontally to the calibration line and a perpendicular dropped. Thus any measured concentration with a lower confidence limit equal to or less than zero will be quoted as non-detectable.

The 95% level for the limit of detection is admittedly to some extent arbitrary. However, the value has a well-defined meaning and obvious

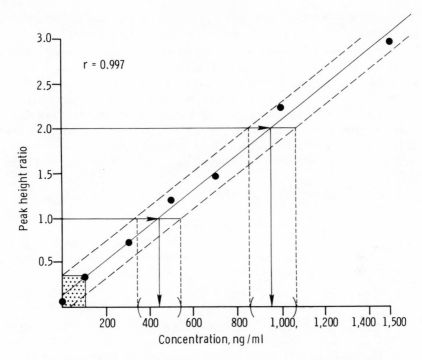

Fig. 6 – The 'confidence-limit' approach to assessing detectability, for the curve shown in Fig. 2 (regression coefficient 0.997). For construction of the 'non-detectable box' (speckled rectangle; note difference in scale from Figs. 3–5), *see text*. The vertical interrupted lines denote the 90% confidence intervals for two notional test results.

properties, viz. we are 95% certain that a new observation will fall below the limit of detection if it is a blank. So although it is in a sense arbitrary its properties are apparent whereas those of, say, twice background are not.

It is relevant to this particular approach that we carry out most of our regression analysis for calibrations with the Hewlett Packard HP90 system, and the requisite programmes can be put onto the standard magnetic card input, thus allowing a print-out of the limit of detection and unknown concentrations in a conventional manner with the minimum of intervention. This would seem as an obligatory feature of any method to be used to assess the limit of detection.

CONCLUSIONS AND IMPLICATIONS

The various techniques show considerable variation for the illustrative data.
The *half-bottom standard* approach gives 50 or 15 ng/ml.
The *twice-background approach* gives 70 or −9 ng/ml.

The *confidence-limit approach* gives 110 ng/ml; the fact that it gives a higher limit than the other two has been observed a number of times in our work.

One question which the results in the example pose, and to which there is no ready answer, is how do you estimate the mean when you have a non-detectable value (ND) in a column of otherwise positive results? Does one use zero or one of the above-zero results or the values computed by the regression analysis?

Concerning the consequences of replicate analysis of the uncertainty associated with a determination, for a peak-height ratio of 1.0 the estimated concentration on the basis of the above illustrative calibration is 443 ng/ml. For a single analysis only the limits seem rather wide (Table 1), reflecting the inherent scatter in any

Table 1 Replicate analysis and uncertainty. The values relate to the foregoing example of an assay.

Peak-height ratio	Estimated concentration, ng/ml	90% confidence interval		
		Single	Duplicate	Triplicate
1.0	443	337 – 548	363 – 522	374 – 511
2.0	956	849 – 1066	874 – 1041	884 – 1031

one determination. Duplicate analysis would tighten these somewhat, but triplicate analysis has little further effect (Table 1). At a peak-height ratio of 2.0 the estimated concentration is 956 ng/ml with the same absolute range for the confidence limits in the case of a single determination. Once again duplicate analysis would be of some help, but triplicate analysis shows little advantage over duplicate as was likewise concluded by Curry [1] earlier in this series.

In the illustrative assay the line has a regression coefficient of 0.997. Table 2

Table 2 Illustration of effect of increased variability on confidence interval (observed peak-height ratio = 1, concentration 443 ng/ml)

Correlation coefficient (r)	Standard deviation	90% confidence interval
0.997	0.095	337 – 548
0.989	0.170	250 – 631
0.980	0.232	176 – 700
0.970	0.283	113 – 758

illustrates what happens to the confidence limits as the regression coefficient begins to fall. Clearly the standard deviation will rise and the confidence r limits on the estimated concentration become very large, notably so for an r value as high as 0.97 although still reasonable.

Thus the solving of an actual problem in trying to determine the limits of detection, through the quantitative confidence limit approach, has disclosed the higher than anticipated error associated with an individual determination unless the regression coefficients are very, very good. This, I conclude, merely emphasizes the very high degree of precision requisite for producing meaningful data for subsequent pharmacokinetic analysis.

Reference

1.   Curry, S. H. and Whelpton, R. (1978) In *Blood Drugs and Other Analytical Challenges* [Vol. 7 of Methodological Surveys (A)] (Reid, E., ed.), Horwood, Chichester, pp. 29–41. [*See also* #NC(F)-3, *this vol. –Ed.*].

#F–8   ASSESSMENT OF ASSAY PRECISION FROM LINEAR
       CALIBRATION GRAPHS

J. P. LEPPARD, Wolfson Bioanalytical Unit, Institute of Industrial and Environmental Health and Safety, University of Surrey, Guildford GU2 5XH, U.K.

Lack of published guidance, even in pertinent contributions to Vol. 7 of this series [1], has prompted the following consideration of precision assessment in the context of assaying biological samples for drugs on the basis of the data for 'processed' (spiked-in, 'determinate') standards. Misgivings are expressed about the risk of unjustified obsessiveness. An example is given of application of the recommended approach to the calculation of confidence limits.

Many analyses, particularly chromatographic analyses for drugs in blood, are performed batchwise. It is customary [J. A. F. de Silva, article #F-6, *this vol.*] to include within each batch of samples a set of standards of different known concentrations for the purpose of producing a (hopefully linear) calibration graph to aid in the determination of the unknown samples. It is also desirable to analyze at least one sample in replicate to determine the precision of the assay. However, for a given batch there may have to be a strict size limit, so that it becomes imperative to make the most economical use of the known samples (here termed *assessment samples*) in order to be able to assay the maximum number of unknown samples in the batch. Statistical theory suggests that linear regression analysis may be helpful in achieving this end, and the following discussion attempts to summarize the relevant theory. More detail is given in standard statistical texts [e.g. 2–4]. The symbols used largely follow those in Snedecor & Cochran [4], and are summarized later in this article.

THE REGRESSION LINE

If X denotes the concentration of the compound being analysed, and Y denotes the measured value (e.g. chromatographic peak height or area) resulting from the analysis of a sample containing that concentration of compound, the set of (X, Y) points obtained from the analysis of the calibration standards is assumed to be a randomly drawn sample from a hypothetical, infinitely large 'population' of points making up the 'true' calibration line.

This true line is assumed to have the form $Y = \alpha + \beta X$. The values of α and β cannot be determined in practice, but regression analysis enables any given set of experimental calibration points to provide *estimates* of them (a and b respectively). The equation of the experimental line, $Y = a + bX$, is then an estimate of the equation for the population line (Fig. 1).

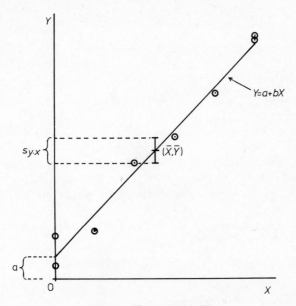

Fig. 1 – The regression line.

Thus, if n is the number of points on the graph, the slope of the line, b, is given by

$$b = \frac{\Sigma(XY) - \dfrac{\Sigma X \Sigma Y}{n}}{\Sigma X^2 - \dfrac{(\Sigma X)^2}{n}} = \frac{\Sigma xy}{\Sigma x^2} \qquad (1)$$

(N.B. Lower-case x and y refer to deviations from mean values and are used in the following formulae since they simplify the mathematical form. The actual mathematical relationships are given in the symbol summary later).

The intercept, a, is given by

$$a = \bar{Y} - b\bar{X} \qquad (2)$$

where $\bar{X}$, $\bar{Y}$ are the means of all the X and Y values respectively. Note that, in general, the point $(\bar{X}, \bar{Y})$ will lie on the regression line.

ASSAY PRECISION

The regression line is the 'best' one through the calibration points in the sense that it minimizes the sum of the squares of the deviations of the points vertically about the line. This vertical scatter gives an indication of the assay precision, and can be quantified. Thus

$$s_{y.x} = \sqrt{\frac{\Sigma d_{y.x}^2}{n-2}} \tag{3}$$

[where
$$\Sigma d_{y.x}^2 = \Sigma y^2 - \frac{(\Sigma xy)^2}{\Sigma x^2} \tag{4}]$$

is the standard deviation of the experimental points vertically about the regression line, and is analogous to the convential standard deviation for a set of replicate measurements about their mean value. (The term $s_{y.x}$ is an experimentally determined *estimate* of the scatter of the points making up the population line, in the same way that a and b are estimates of the population line parameters α and β).

It is also possible to calculate the 'standard error'* of the slope

$$s_b = \frac{s_{y.x}}{\sqrt{\Sigma x^2}} \tag{5}$$

and the standard error of the intercept

$$s_a = s_{y.x} \sqrt{\frac{\Sigma X^2}{n \Sigma x^2}} \tag{6}$$

These can be used in conjunction with a t-test to determine confidence limits for b and a, or whether the slope or intercept respectively differs from some chosen value, such as the slope of an earlier calibration graph, or, especially, zero in the case of an intercept (*cf*. J. McAinsh *et al*., #F-7, *this vol.*).

---

* The terms 'standard deviation' and 'standard error' are synonymous (p. 50 in [4]). Standard deviation is usually reserved to express the distribution or 'spread' of a series of individual measurements, while standard error is used to express the distribution of parameters derived from them. Thus the well-known standard error of the mean (SEM) is the standard deviation that would be obtained for a set of mean values derived by taking repeated multiple samples from a population and calculating the mean for each sample: the SEM expresses the distribution of mean values that would be obtained (and which would in fact be a 'tighter' distribution than the individual measurements possessed). Likewise, the standard error of the slope expresses the distribution of values for the slope that would be obtained if the experimental process for plotting the calibration graph were repeated a number of times: effectively, it measures how precisely the value for the slope can be relied upon (which is why it can be used—like other standard errors—to determine confidence limits).

COMPLICATIONS

The derivation of the above equations assumes that the precision of the assay, i.e. the standard deviation (s_y) of replicated Y values for a given value of X, is the same throughout the range of X values. If this is so, then $s_{y.x}$ is a direct indication of this standard deviation (Fig. 2a). However, as S. H. Curry & R. Whelpton indicate [1], in drug assays the standard deviation of the Y values is not independent of X, but proportional (or at least near-proportional) to X, in which case some of the formulae quoted above are inappropriate. (Other types of precision behaviour can occur, but are fortunately less common in analytical chemistry.) It is vital to note that the regression analysis does not of itself give any indication of the behaviour of the precision with change of concentration, and in fact, if the wrong behaviour is assumed, totally misleading results may be obtained.

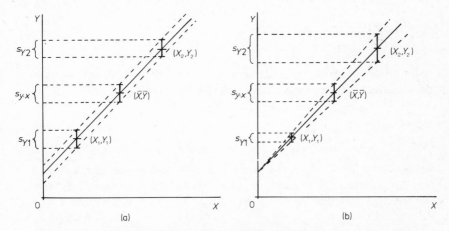

Fig. 2 – (a) Assay precision constant throughout concentration range. The precision at Y_1 and Y_2 (s_{Y_1}, s_{Y_2}) is estimated in each case by $s_{y.x}$.
(b) Assay precision proportional to concentration. The absolute precision is indicated by $S_{y.x}$ approximately only in the region of the point ($\bar{X}$, $\bar{Y}$).

In the situation where the precision is proportional to X (Fig. 2b) equations (1) and (2) for b and a give inaccurate estimates of β and α. It is better to use weighted regression analysis, in which allowance is made for the increased variability at large values of X. This leads to equations (7) and (8) for calculation of b and a [cf. 5].

$$b = \frac{\left(\sum \frac{1}{X^2}\right)\left(\sum \frac{Y}{X}\right) - \left(\sum \frac{1}{X}\right)\left(\sum \frac{Y}{X^2}\right)}{n\left(\sum \frac{1}{X^2}\right) - \left(\sum \frac{1}{X}\right)^2} \qquad (7)$$

$$a = \frac{\sum \frac{Y}{X^2} - b \sum \frac{1}{X}}{\sum \frac{1}{X^2}} \tag{8}$$

Equation (3) similarly gives an inappropriate estimate for the analytical precision, and a weighted regression treatment should again be used. Unfortunately, no simple description of such a treatment seems to be readily available. If the intercept, a, is sufficiently small that the line can be assumed to pass through the origin, then the best estimate of the slope is given by the relatively simple expression in equation 9:

$$b = \frac{\sum \frac{Y}{X}}{n} \tag{9}$$

Under these circumstances it has been found empirically (author's unpublished experiments) that the *relative* standard deviation about the regression,

$$s_{y.x, \text{rel}} = \frac{s_{y.x}}{\bar{Y}} \tag{10}$$

analogous to the C.V. can be taken as approximately constant, and gives at least an indication of the assay precision throughout the range of X. From it, the precision in absolute terms can be calculated for any chosen value, Y' of Y as

$$s_{Y'} = s_{y.x, \text{rel}} \cdot Y' \tag{11}$$

It must be stressed that this is purely a rule-of-thumb approach which has been found to give reasonable estimates of assay precision in practice, particularly when the precision is good. Its theoretical validity is doubtful, however, and it is likely to meet with disapproval from a statistician.

Of course, the precision of the Y values, i.e. the measured values, is only of secondary interest to the analyst. Far more relevant is the precision of the concentration (i.e. the X value) determined from the calibration graph for an unknown sample. Here again, regression analysis can be of assistance. In the situation where the precision of the Y values ($s_{Y'}$) is independent of X, the standard deviation of the calculated X value, X' is given by

$$s_{X'} = \frac{s_{y.x}}{b} \sqrt{\frac{1}{m} + \frac{1}{n} + \frac{(Y' - \bar{Y})^2}{b^2 \Sigma x^2}} \tag{12}$$

where Y' is the particular value of Y under consideration, and m is the number of independent replicate determinations used in obtaining it (m may of course be 1).

Where $s_{y.x}$ is proportional to X, $s_{y.x,\text{rel}}$ calculated as in equation (10) can be taken, with even more emphasis on the provisos mentioned above, as approximately equal to $\dfrac{s_{x'}}{X'}$, so that $s_{x'}$ can be calculated to give at least a ball-park figure for the precision of the X' value under consideration.

CONCLUSIONS

Application of the foregoing equations would imply that, if the behaviour of the precision with concentration were known, a calibration graph need only consist of three points—two to define the line, and one other to indicate the precision. This would be admirable for the original purpose of minimizing the number of assessment standards in an assay batch; but unfortunately the parameters determined from only three points would be very unreliable, as reflected in wide confidence limits. [The confidence limits determined from regression analysis involve Student's t, with $n - 2$ degrees of freedom (d.f.)—i.e. in this case, one d.f. only. An examination of the t-table will show that for significance levels less than 20% — probability values greater than 80% — t is very large.]

In any case, the behaviour of the assay precision with concentration is very often *not* known, at least in any specific instance, and it becomes advisable to devote some assessment standards to determining the precision behaviour. Werner *et al.* [6] describe a method for assessing the precision behaviour where it is made up of constant and concentration-dependent components, but it evidently requires very large numbers of replicate analyses (of the order of 100) in order to have any confidence in the results. Even if this seems to be a case of 'over-kill' it would appear to be necessary to perform replicate analyses at no less than two extreme concentration levels to determine whether the precision is concentration-dependent or not.

The worked example below shows the general advisability of analyzing unknowns in duplicate, as S. H. Curry & R. Whelpton [in 1] advocate, and it may be possible to use these results in the elucidation of the precision behaviour. Whatever approach is adopted, however, it seems inescapable that relatively large numbers of assessment samples (perhaps, even, not less than 10) will be needed in each analysis batch, if any reliance is to be placed on the results. Where the maximum batch size is, as often happens, of the order of 30, and considering the advisability of analyzing unknowns in duplicate, we are left with a very poor return for our effort—30 time-consuming assays for perhaps 10 results!

There appears, in fact, to be an incompatibility between the statistician's requirements for large numbers, and the analyst's desire for results. The incompatibility hinges upon the degree of confidence which can be placed in the results obtained. Elsewhere in this volume, J. McAinsh gives an example of 90% confidence limits of 337 to 548 ng/ml for an assay result of 443 ng/ml, estimated from a 7-point calibration graph which appears to an analyst to be a creditable straight line: the confidence limits seem intuitively to be too wide. The example below leads to similar hesitancy. If 90% confidence limits are sought for an assay result, they can quite easily cover half the concentration range investigated. Thus, whereas an analyst might have felt he could quote his results as being accurate within a few percentage points, the statistician tells him he can only quote within an order of magnitude!

What is to be made of this situation? A number of points are worth making. Firstly, however poor our analytical methods might appear in the light of statistical analysis, there is no doubt that they are in general more specific and more precise now than in the past, so that there is no immediate cause for paranoia in the analytical fraternity.

Secondly, there does seem to be a greater awareness among analysts of some of the statistical techniques available to them, probably because the ready availability of computers and advanced calculators with built-in statistical programs has made the techniques more accessible and less tedious. However, a whole battery of techniques familiar to statisticians could be used with advantage in analytical chemistry to make assay results more meaningful quantitatively; yet the existence of such techniques is largely unknown to the chemists. Greater dialogue between the two fraternities seems an essential requirement for rectifying the situation, and since statisticians are generally very willing to help when called upon, the onus would appear to be on the chemists to make the first approaches.

Thirdly, the situation should be rectified soon. With current developments in safety legislation it is vital that, having performed an analysis, the analyst should be able to state quantitatively how reliable his results are, and whether they meet any relevant legal requirements. At least one court case recently hinged upon the fact that the maximum permitted body content of a radio-isotope gave a count rate in the specified test which could not be distinguished from the background rate. If the legal requirements themselves are unrealistic in such a respect, it is to the analyst's advantage to know this before he presents his results.

There is thus a need for a realistic assessment of the meaningfulness of current analytical methods. No doubt on most occasions they will be found satisfactory, but as demands are made for the measurement of ever lower concentrations there is a danger of reaching a point where a given analytical method generates merely a series of meaningless numbers. We should be in a position to recognize this point so that we can take appropriate action — either by improving the analytical methodology or, in the last resort, by abandoning what will have become a futile exercise in favour of a more profitable occupation.

FURTHER REFLECTIONS

Most chromatographic assays for drugs in biological fluids involve several extraction, aliquotting, and pipetting steps, and theoretical considerations suggest that each of these steps is a potential source of concentration-proportional variability. Often they will be the main source of analytical error, and the overall precision for an assay at any specific concentration will often be proportional to concentration. It is re-emphasized that under these circumstances, which statisticians and chemists seldom consider, standard regression analysis leads to an unreliable calibration line, and wrong confidence limits for a concentration predicted from it corresponding to a subsequent Y-value. Amongst many textbooks now consulted, a few give the weighted expressions for a and b (equations 7, 8 and 9), but *not one* specifically gives confidence limits for calculated X (or concentration) values.

For what it is worth, I *believe* the weighted standard deviation about regression for the general line, $Y = a + bX$ is given by

$$s_w = \left(\frac{\sum \frac{Y^2}{X^2} + a^2 \sum \frac{1}{X^2} + nb^2 + 2ab \sum \frac{1}{X} - 2b \sum \frac{Y}{X} - 2a \sum \frac{Y}{X^2}}{n-2} \right)^{\frac{1}{2}} \tag{13}$$

When the line is forced through the origin ($a = 0$, $b = \dfrac{\sum \frac{Y}{X}}{n}$),

$$s_w = \left(\frac{\sum \frac{Y^2}{X^2} + nb^2 - 2b \sum \frac{Y}{X}}{n-1} \right)^{\frac{1}{2}} \tag{14}$$

$$= \left(\frac{\sum \frac{Y^2}{X^2} - \frac{\left(\sum \frac{Y}{X}\right)^2}{n}}{n-1} \right)^{\frac{1}{2}} \tag{14a}$$

The precision of the Y-values at any given X is then found from $X.s_w$.

Confidence limits for a calculated X' in the general case have so far defeated me. For the case of a line forced through the origin, I believe the limits are given by

$$\mathrm{CL}(X') = \frac{Y'}{b \pm t.s_w \sqrt{\frac{1}{m} + \frac{1}{n}}} \tag{15}$$

where t is Student's t for $(n-1)$ degrees of freedom at the chosen probability level. (N.B. These limits are not symmetrical about the most likely value of X', which must be calculated separately from $X' = \dfrac{Y'}{b}$).

I stress that equations (13) – (15) have been derived by myself unaided; if incorrect, they may provoke a statistician to respond with the correct versions. The situation where weighted regression should be used is both important enough and common enough to warrant authoritative guidance.

SYMBOLS AND FORMULAE

X Independent variable in calibration line.

Y Dependent variable in calibration line.

n Number of points in calibration line.

Σ Denotes 'sum of all terms of the form immediately following'.
Thus, ΣX is the sum of all the X values in the calibration line. ΣX^2 is the sum of all values of X^2. ΣXY is the sum of all the products of X and its associated Y.

$\bar{X} = \dfrac{\Sigma X}{n}$ Mean of the X values.

$\bar{Y} = \dfrac{\Sigma Y}{n}$ Mean of the Y values.

Y' A specific mean value of Y measured for an unknown sample.

X' The value of X corresponding to Y' as determined from the regression line.

m Number of independent replicate samples used for the determination of Y'.

$$x = X - \bar{X} = X - \frac{\Sigma X}{n}$$

$$y = Y - \bar{Y} = Y - \frac{\Sigma Y}{n}$$

$$\Sigma x = \Sigma X - n\bar{X}$$

$$\Sigma y = \Sigma Y - n\bar{Y}$$

$$\Sigma x^2 = \Sigma (X - \bar{X})^2 = \Sigma X^2 - \frac{(\Sigma X)^2}{n}$$

$$\Sigma y^2 = \Sigma(Y - \bar{Y})^2 = \Sigma Y^2 - \frac{(\Sigma Y)^2}{n}$$

$$\Sigma xy = \Sigma(X - \bar{X})(Y - \bar{Y}) = \Sigma XY - \frac{\Sigma X \Sigma Y}{n}$$

$$\Sigma d_{y.x}^2 = \Sigma y^2 - \frac{(\Sigma xy)^2}{\Sigma x^2}$$

$s_Y = \sqrt{\dfrac{\Sigma y^2}{n-1}}$ Standard deviation of independent replicate determinations of Y for a given analytical sample.

Equations relating to the regression line

(a) Precision constant throughout the range of X

$Y = a + bX$ Equation of regression line.

$b = \dfrac{\Sigma xy}{\Sigma x^2}$ Regression coefficient (= slope of regression line).

$a = \bar{Y} - b\bar{X}$ Intercept on Y-axis.

$r = \dfrac{\Sigma xy}{\sqrt{(\Sigma x^2)(\Sigma y^2)}}$ Correlation coefficient.

$s_{y.x} = \sqrt{\dfrac{\Sigma d_{y.x}^2}{n-2}}$ Standard deviation (or error) about the regression line.

$s_b = \dfrac{s_{y.x}}{\sqrt{\Sigma x^2}}$ Standard error of slope.

$s_a = s_{y.x}\sqrt{\dfrac{\Sigma X^2}{n\Sigma x^2}}$ Standard error of intercept.

$s_{X'} = \dfrac{s_{y.x}}{b}\sqrt{\dfrac{1}{m} + \dfrac{1}{n} + \dfrac{(Y' - \bar{Y})^2}{b^2\Sigma x^2}}$ Standard deviation of X', found from Y'.

(b) Standard deviation proportional to X

$$s_{y.x,\mathrm{rel}} = \frac{s_{y.x}}{\bar{Y}}$$

Relative standard deviation about regression line.
(If multiplied by 100 it may be quoted as a percentage, analogous to a coefficient of variation.)

$$b = \frac{\left(\sum \frac{1}{X^2}\right)\left(\sum \frac{Y}{X}\right) - \left(\sum \frac{1}{X}\right)\left(\sum \frac{Y}{X^2}\right)}{n \sum \frac{1}{X^2} - \left(\sum \frac{1}{X}\right)^2}$$

$$a = \frac{\sum \frac{Y}{X^2} - b \sum \frac{1}{X}}{\sum \frac{1}{X^2}}$$

If the line can be assumed to pass through the origin ($a = 0$),

$$b = \frac{\sum \frac{Y}{X}}{n}$$

ADDENDUM: AN EXAMPLE

The data in Table 1 summarize the calculations for a hypothetical GC analysis in which X represents the concentration of a drug being analyzed, and Y represents the peak height ratio for the drug compared to an internal standard. Fig. 3 shows a plot of the original data points, with the regression line drawn in.

Confidence limits for b and a

In general, confidence limits for a given value are determined from the product $t \times$ ('standard error' of the value) where t is Student's t (two-tailed) for the chosen level of probability and the appropriate number of degrees of freedom (d.f.). In the case of regression lines, d.f. $= n - 2$. Thus, in this example, for 90% confidence limits for b, d.f. $= 6$, and t (found from tables) $= 1.943$.

$$b = 0.206 \pm (1.943 \times 0.0116) = 0.206 \pm 0.0226$$

Similarly, $a = 0.205 \pm (1.943 \times 0.0735) = 0.205 \pm 0.143$.

In a similar manner, the slope, b, can be compared with the slope determined in a previous experiment to ascertain whether the method sensitivity has changed significantly. More often of interest is a test to ascertain whether the intercept is significantly different for zero (or some other chosen value). Thus

$$t_{exp} = \frac{a - 0}{s_a} = \frac{0.205}{0.07345} = 2.791.$$

This is greater than t quoted in the tables (1.943), and so the intercept can be deduced to be different from zero at a probability level of 90% (or at a significance level of 0.10).

Table 1 Data from a hypothetical GC assay (*see text*).

X (ng/ml)	Y	X^2	Y^2	XY
0	0.41	0	0.1681	0.00
0	0.12	0	0.0144	0.00
2	0.46	4	0.2116	0.92
4	1.11	16	1.2321	4.44
6	1.36	36	1.8496	8.16
8	1.78	64	3.1684	14.24
10	2.29	100	5.2441	22.90
10	2.33	100	5.4289	23.30
Σ 40	9.86	320	17.3172	73.96

$n = 8$; $\bar{X} = 5$; $\bar{Y} = 1.2325$

$\Sigma x^2 = 120$; $\Sigma y^2 = 5.16475$; $\Sigma xy = 24.66$

$\Sigma d_{y.x}^2 = 0.09712$

$b = 0.2055$	$s_{y.x} = 0.1272$
$a = 0.2050$	$s_b = 0.01161$
$r = 0.9906$	$s_a = 0.07345$

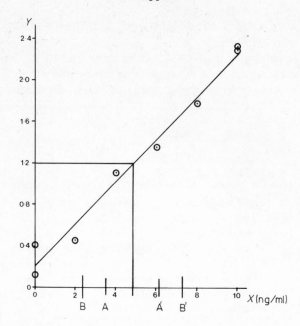

Fig. 3 – Estimation of confidence limits for a derived concentration X' of 4.84 ng/ml, corresponding to a measured peak height ratio of 1.2. The 90% confidence limits are indicated by the points A, A'., and the 99% limits by the points B, B'.

Assay precision

(a) *Precision independent of sample concentration*

The data presented do not give much indication of the influence of concentration on precision, although the standard deviations for the duplicate blanks and 10.0 ng/ml samples (0.205 and 0.03 respectively) are not inconsistent with the precision being constant throughout the concentration range. If this is so, then $s_{y.x} = 0.127$ is an indication of the assay precision. Equation (12) can then be used to determine the precision for the concentration found for a measured chromatographic peak height ratio in the case of an unknown sample. Table 2 gives, for three hypothetical peak height ratios, Y', the corresponding concentration (X'), found from the regression equation, and the precision $(s_{X'})$ depending on whether the peak height ratio was a single value, or the mean of two or three replicates $(m = 1,2,3)$. Note that in this context, 'replicate' means replicate samples worked up through the whole assay procedure, not simply replicate GC injections.

It is apparent from these values that the precision is better in the middle of the curve than at the extremities; moreover, as stated by Curry & Whelpton [in 1] for test samples, duplicates give a worthwhile improvement in precision over single measurements, and triplicates do not give a worthwhile further increase. These observations are generally true.

Table 2 Precision of calculated test values: illustrative results.

GC peak height ratio Y'	Concentration X'	$s_{X'}$		
		$m=1$	$m=2$	$m=3$
0.4	0.95	0.70	0.54	0.48
1.2	4.84	0.66	0.49	0.42
2.2	9.71	0.71	0.56	0.50

(b) *Precision proportional to sample concentration*
If there were grounds for believing that the standard deviation (in the Y direction) for replicates is proportional to the concentration, then, subject to the strictures mentioned earlier, $\dfrac{s_{y.x}}{\bar{Y}}$ can be taken as approximately constant, and approximately equal to $\dfrac{s_{X'}}{X'}$ for any value of X' in the range examined. Thus, in this example.

$$\frac{s_{y.x}}{\bar{Y}} = \frac{0.127}{1.2325} = 0.103 \text{ or } 10.3\%.$$

Therefore   for $Y' = 0.4$, $X' = 0.95$, $s_{x'} = 0.103 \times 0.95 = 0.098$ ng/ml;
for $Y' = 1.2$, $X' = 4.84$, $s_{x'} = 0.103 \times 4.84 = 0.499$ ng/ml;
for $Y' = 2.2$, $X' = 9.71$, $s_{x'} = 0.103 \times 9.71 = 1.000$ ng/ml.

Confidence limits for X'
Having calculated $s_{x'}$ it is usually sufficient to calculate the confidence limits for X' as

$$X' \pm t\, s_{X'}$$

where t is Student's t for the chosen level of probability with $n - 2$ d.f.

Thus, using the values in Table 2, 90% confidence limits are as follows:

for $X' = 0.95$, $m = 1$:— $0.95 \pm (1.943 \times 0.70) = 0.95 \pm 1.36$ ng/ml;

for $m = 2$, $X' = 0.95 \pm (1.943 \times 0.54) = 0.95 \pm 1.05$ ng/ml;

for $X' = 4.84$, $m = 1$:— $4.84 \pm (1.943 \times 0.66) = 4.84 \pm 1.28$ ng/ml.

If 99% confidence limits are required, $t = 3.707$, and, for example for $X' = 4.84$, $m = 1$, the limits obtained are $4.84 \pm (3.707 \times 0.66) = 4.84 \pm 2.45$ ng/ml. The 90% and 99% confidence limits for $X' = 4.84$ ng/ml are shown in Fig. 3.

General comments

Although the line shown in Fig. 3 is not a particularly good one, it is not unusually bad for samples of concentration less than 10 ng/ml, representing as it does a C. V. of about 10% in the middle of the range. It therefore comes as something of a shock to see how wide are the consequent confidence limits. The implications of this have already been discussed.

A second observation concerns the blank point at $Y = 0.41$. Faced with the eight calibration points in this example, an analyst might be tempted to regard this apparently high blank as a rogue, and omit it from his calculations in deriving the line. Were he to do so, it would undoubtedly improve the linearity of his line (e.g. $s_{y.x}$ would fall from 0.13 to 0.08). However, unless he had stronger reasons than convenience for omitting this value, he would be in grave danger of distorting the results. It so happens that the data above were produced synthetically, by taking a line with the equation $Y = 0.20 + 0.20X$ and superimposing a 'constant' random error with vertical standard deviation of 0.10. Table 3 compares some of the principal features of the regression lines calculated with and without the inclusion of the doubtful point.

Table 3 Regression line characteristics.

	All eight points	Omitting $Y = 0.41$	'True' line
n	8	7	—
b	0.2055	0.2183	0.2000
a	0.2050	0.1025	0.2000
r	0.9906	0.9961	—
$s_{y.x}$	0.1272	0.0826	0.1000
s_b	0.01161	0.00863	—
s_a	0.07345	0.05838	—

In addition to improving the linearity and precision estimate as represented by r and $s_{y.x}$, s_b and s_a, omission of the offending point increases the slope of the line b, and reduces the intercept a, pushing the calculated line away from the 'true' line for which it is supposed to be an estimate. Although the issue is more complicated than this, there is no doubt that to omit the blank value in these circumstances would be an unjustifiable act, giving a spuriously optimistic estimate of the assay precision, a false estimate of the equation for the calibration line, and, in consequence, wrong analytical results for unknown samples.

Acknowledgements

I am grateful to Dr. B. Scales and Dr. R. A. Ferguson (ICI Pharmaceuticals Ltd.) and to Professor S. H. Curry (now at the University of Florida) for commenting on this paper. In particular, I am indebted to Dr. Ferguson for the time and effort he devoted to identifying some fundamental errors in my original draft, particularly concerning the situation where precision is proportional to concentration. I have endeavoured to rectify these mistakes, but the responsibility for any 'residual error' (to use a statistical term) must rest on my shoulders.

Note Added in Proof

Weighted linear regression is dealt with in detail (including calculation of confidence limits for predicted X-values) in the *first* edition of a book published 20 years ago [7]. This material was unfortunately removed from the second edition of the book.

References

1. Reid, E. (ed.) (1978) *Blood Drugs and Other Analytical Challenges* (Vol. 7, this series), Horwood, Chichester. One pertinent article is that by S. H. Curry & R. Whelpton (see also citations in #0 by E. Reid, *this vol.*).
2. Acton, F. S. (1966) *Analysis of Straight Line Data*, Dover, New York.
3. Saunders, L. & Fleming, R. (1971) *Mathematics and Statistics for Use in the Biological and Pharmaceutical Sciences*, 2nd edn., Pharmaceutical Press, London.
4. Snedecor, G. W. & Cochran, W. G. (1967) *Statistical methods*, 6th edn., Iowa State University Press, Ames, Iowa.
5. Davies, O. L. & Goldsmith, P. L. (1972) *Statistical Methods in Research and Production* Oliver & Boyd, Edinburgh (*see* Section 7.5).
6. Werner, M., Brooks, S. H. & Knott, L. B. (1978) *Clin. Chem.* **24**, 1895-1898.
7. Brownlee, K. A. (1960), *Statistical Theory and Methodology in Science and Engineering*, 1st edn., Wiley, New York (*see* pp. 306-317).

#NC(F) Notes and Comments related to Ensuring Reliable Results, Especially for Drugs in Blood

#NC(F)-1 *A Note on*
POLICIES FOR INTERNAL STANDARDS

KENNETH H. DUDLEY, Department of Pharmacology, University of North Carolina School of Medicine, Chapel Hill, North Carolina 27514, U.S.A.

The selection process for an internal standard* should include consideration of all aspects of the chromatographic assay. An internal standard should mimic closely the properties of the solute itself. One which has essentially the same chemical structure as the solute, differing only by a slight modification of a chemically insignificant substituent, will normally provide the desirable chromatographic properties and close resemblance in extraction and derivatization properties. Many of the common error sources and pitfalls in chromatographic assays can be eliminated by the use of appropriate internal standards [1].

In clinical laboratories, appropriate synthetic compounds are often unavailable for use as internal standards. It has then been a common practice to use drugs from within the same structurally related drug class and to manipulate the drug solutions to suit the needs of this assay. Examples include the use of chlorpromazine for the quantitation of thioridazine [2], and of amitriptyline for the quantitation of dothiepin and its metabolites [3]. In patient monitoring, however, such a practice is valid only so long as the analyst has access to a patient's complete history of drug therapy and an awareness of the pharmacokinetic principles of drug elimination. It is an obvious error, with potential risk to a patient's welfare, to use a drug as an internal standard if the patient had received that drug and if sufficient time had not been given for complete disappearance of the drug from the body.

In the selection of an internal standard from among drug derivatives, an awareness of general biotransformation reactions can be helpful in avoiding

---

* The term here connotes 'processed internal standard' as defined on p. 63 in Vol. 7 (this series), which includes an article where S. H. Curry (with R. Whelpton, pp. 29–41; *see also this vol.*) argues for caution in the use of internal standards. –*Editor.*

analytical pitfalls. Care should be taken not to select a drug derivative as an internal standard if there is evidence or suspicion that the derivative is an *in vivo* metabolite of the drug to be quantitated. *N*-Dealkylated or phenolic derivatives of drugs are examples of common drug derivatives that should be rejected as candidate internal standards. As a rule, such derivatives are normally encountered as the principal *in vivo* metabolites of drugs with *N*-alkyl and/or aromatic ring substituents. A unique example surfaced during studies of the excretion of 5-(*p*-hydroxyphenyl)-5-phenylhydantoin (*p*-HPPH, *see panel*), the principal metabolite of phenytoin (5,5-diphenylhydantoin) in man and most animals. The synthetic *meta*-hydroxy analogue, *m*-HPPH, was used as the internal standard in the early GC determinations of *p*-HPPH, but was later rejected when *m*-HPPH was discovered to be a normal urinary product of phenytoin metabolism in the dog [4]. Although *m*-HPPH is not normally found as a true metabolite in man and other species, the compound is indeed an artifact produced by acid-catalyzed decomposition of the phenytoin-dihydrodiol metabolite [4,5]. A urine specimen is normally incubated with either acid or enzyme (β-glucuronidase), the former being the more convenient, for fission of the *p*-HPPH-glucuronide and extraction of *p*-HPPH. 5-Phenyl-5-(*p*-tolyl)-hydantoin (MPPH), which is commercially available, has been recommended as an internal standard for *p*-HPPH determinations [6], and, more recently, 5-(*p*-hydroxyphenyl)-5-(*p*-tolyl)hydantoin (HMPPH) has been evaluated against MPPH and likewise recommended [7] (*see panel*).

p-HPPH   m-HPPH   MPPH   MHPPH

In a GC assay where the derivatization method is rigorous, if not brutal, it is essential that the structure of the internal standards mimic as closely as possible that of the drug. The on-column methylation technique (OCMT), as applied in simultaneously determining phenytoin (PHT) and phenobarbital (PB), illustrates the need to make proper selections of internal standards.

PHT and PB are cyclic imides with pK'_a for weak acids, but differ considerably in chemical behaviour, particularly in alkaline solutions. The 5,5-disubstituted hydantoin system is exceptionally stable in hot alkaline solution, whereas the barbiturate system is degraded to ring-opened products. This same chemical behaviour is frequently observed in determinations of PHT and PB by on-column methylation. Laboratories have reported the presence of multiple peaks appearing in the chromatogram after injection of methanolic tetramethylammonium hydroxide (TMAH) or trimethylphenylammonium hydroxide (TMPAH) solutions of PB [8-10]. As shown by GC-MS [10], the on-column methylation reaction with PB and TMPAH gives rise to a 'normal' derivative, N_1, N_3-dimethylphenobarbital, and two degradation products, N-methyl-α-phenylbutyramide and N,N-dimethyl-α-phenylbutyramide. However, the problem of PB degradation can be eliminated by the use of an appropriate internal standard for PB, one that shows essentially the same methylation behaviour. 5-Ethyl-5-(p-tolyl)barbituric acid ('p-methylphenobarbital', MPB) is an appropriate internal standard, and the data shown in Figs 1 & 2 illustrate its usefulness.

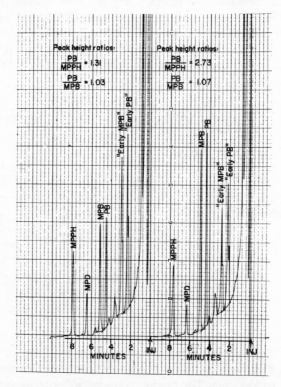

Fig. 1 – Gas chromatograms obtained by two consecutive injections of the final TMPAH extract (OCMT) from a patient plasma sample containing phenobarbital (cf. text for explanation of method and abbreviations).

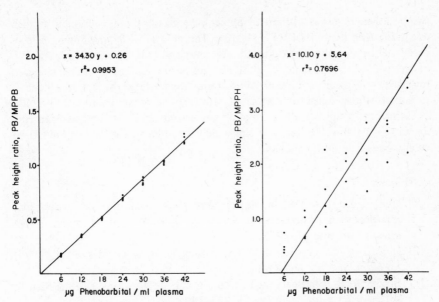

Fig. 2 – Comparison of calibration data for the determination of phenobarbital, as obtained by the use of MPPH and MPB as internal standards (cf. text for explanation of method and abbreviations).

Fig. 1 is a summary of chromatograms obtained from a patient plasma specimen which contained only PB. The chromatograms were obtained when two consecutive injections were made of the final TMPAH extract under the conditions of the OCMT used in our laboratory [11]. The peaks labelled as 'Early PB' and 'Early MPB' are due to degradation products of PB and MPB, respectively. MPD is the internal standard employed for primidone determinations; MPPH, the internal standard for PHT determinations [11]. By comparison of the peak height ratios of the PB/Early PB in the two chromatograms (Fig. 1), it can be seen how non-reproducible the degradation of PB can actually become from injection to injection. In one case, the PB/Early PB ratio is 0.63; in the other, 1.58. In the hypothetical situation that MPPH had been the internal standard for quantitating PB, the PB/MPPH peak heights ratio would have varied from 1.31 to 2.73. With MPB as the internal standard, the PB/MPB peak height ratio remains essentially constant, values of 1.03 and 1.07 being observed. The use of MPB as an internal standard in the OCMT for PB ensures the accuracy of the determination. At any given concentration of PB and MPB in the sample, the absolute value of the proportion of the degradation reaction may vary from injection to injection, but the proportions (or fractional amounts) of PB and MPB degraded are the same and, therefore, the ratio of their concentrations, as observed in terms of a peak height ratio, is a constant.

Another benefit from making the proper selection of an internal standard is exemplified by the data summarized in Fig. 2. These data were obtained when two dual column gas chromatographic instruments were calibrated for PB determinations by our OCMT [11, 12]. The data (Fig. 2) are a composite of peak height ratios obtained across all four columns in the calibration experiment. In the hypothetical situation that MPPH had been used as the internal standard for PB, it can be seen that the PB/MPPH ratios are scattered. In fact, it is difficult to discern which points are associated with any one of the columns. With MPB as the internal standard, the calibration curve from each column was essentially the same, enabling the use of a single calibration curve for PB determinations on any one of the four columns.

References

[1] Dudley, K. H. (1978) in, *Antiepileptic Drugs: Quantitative Analysis and Interpretation* (Pippenger, C. E., Penry, J. K. & Kutt, H., eds.), Raven Press, New York, pp. 19-34.

[2] Dinovo, E. C., Gottschalk, L. A., Nandi, B. R. & Geddes, P. G. (1976) *J. Pharm. Sci.*, **65**, 667-669.

[3] Brodie, R. R., Chasseaud, L. F., Crampton, E. L., Hawkins, D. R. & Risdall, P. C. (1977) *J. Int. Med. Res.*, **5**, 387-390.

[4] Glazko, A. J. (1972) in *Antiepileptic Drugs* (Woodbury, D. M., Penry, J. K. & Schmidt, R. P., eds.), Raven Press, New York, pp. 103-112.

[5] Maguire, J. H., Kraus, B. L., Butler, T. C. & Dudley, K. H. (1979) *Therap. Drug Monitor.*, **1**, 359-370.

[6] Kutt, H. (1978) in *Antiepileptic Drugs: Quantitative Analysis and Interpretation* (Pippenger, C. E., Penry, J. K. & Kutt, H., eds.), Raven Press, New York, pp. 307-314.

[7] Witkin, K. M., Bius, D. L., Teague, B. L., Wiese, L. S., Boyles, L. W. & Dudley, K. H. (1979) *Therap. Drug Monitor.*, **1**, 11-34.

[8] Perchalski, R. J., Scott, K. N., Wilder, B. J. & Hammer, R. H. (1973) *J. Pharm. Sci.*, **62**, 1735-1736.

[9] Van Meter, J. C. & Gillen, H. W. (1973) *Clin. Chem.*, **19**, 359-360.

[10] Osiewicz, R., Aggarwal, V., Young, R. M. & Sunshine, I. (1974) *J. Chromatog.*, **88**, 157-164.

[11] Dudley, K. H., Bius, D. L., Kraus, B. L. & Boyles, L. W. (1977) *Epilepsia*, **18**, 259-276.

[12] Dudley, K. H., Bius, D. L., Kraus, B. L. & Boyles, L. W. (1978) as for 6., pp. 35-41.

#NC(F)-2 *A note on*
THE USEFULNESS OF INTERNAL STANDARDS IN QUANTITATIVE BIOANALYTICAL METHODS

JÖRGEN VESSMAN, AB Hässle, S–431 20 Mölndal, Sweden

It is becoming more and more evident that the quantitative analysis of components in biological samples, especially if the series is large, is facilitated if a correctly chosen internal standard is added to the sample and then follows the compound to be analyzed right through to chromatographic quantitation. Numerous good examples are found in the analysis of drugs in pharmacokinetic studies, in therapy control, etc., although it does not follow that one can pay any less attention to any step in the analytical procedure. The physico-chemical properties of the internal standard should be close to, and as well-known as, those of the analyte in order to follow it in all steps in the analytical method, e.g. extraction, purification, derivatization, separation by chromatographic methods, and quantitation. The best choices are therefore found among homologues, analogues or isomers.

The internal standard is an important part of the *integrated* analytical procedure. It compensates for minor fluctuations in the conditions of some steps in the procedure, e.g.

(1) transfer of one phase in an extraction;

(2) non-stoichiometric reactions;

(3) quantitative derivatization reactions sensitive to certain reaction conditions;

(4) small but not fully controllable decomposition;

(5) variation in adsorptivity of surfaces.

Some experiences gained over fully 10 years are now discussed.

(1) Phase transfer is mainly of practical importance where rather than take an aliquot, one recovers as much as feasible of, e.g., an organic phase, as when the absolute amount of substance is limited; this applies especially when minute volumes are to be handled.

(2) An example is shown of a non-stoichiometric reaction. The quaternary ammonium compound emepronium is oxidized to benzophenone, which then is quantitated by GC-ECD [1]. The yield is around 65%, but only about 50% for the internal standard chosen, the *p*-chloro analogue. Yet this method has been used with a high degree of precision in the analysis of thousands of biological samples [2]. Some alternative internal standards are shown, of which the bromo, iodo and nitro compounds gave less volatile benzophenones and the fluoro and trifluoromethyl were too volatile [3]. The *o*-chloro analogue was less suitable as the reaction rate was too low. Other substituents could not be chosen as they would not withstand the oxidation conditions [4]. Another example of a

	R
Emepronium	H
Internal Standard	Cl
Alternative "	Br, I, NO_2 (F, CF_3)
Less suitable	o–Chloro

non-stoichiometric reaction where the *p*-chloro analogue is helpful concerns the demethylation of some tertiary amines [5]; thus Recipavrin is reacted with a chloroformate to form a carbamate in a yield of about 75%.

(3) An example of a quantitative derivatization sensitive to certain reaction conditions is the reaction of heptafluorobutyric anhydride, as used for secondary

Terodiline

amines [6], with terodiline [7], which being very strongly sterically hindered required a high concentration of the catalyst trimethylamine (TMA). Various internal standards were used, such as the n-pentyl homologue A; but it was not until compound B, with an exactly identical set-up around the secondary amino group, was used that reliable quantitative data could be obtained. Owing to the steric hindrance, terodiline and compound B were very sensitive to minor changes in reaction conditions, e.g. TMA concentration and the presence of water. This was not the case with A, which always reacted rapidly and quantitatively.

(4) These may be small but not fully controllable decomposition, e.g. on-column thermal degradation, as with meprobamate, which has been quite difficult to analyze by GC. The use of the illustrated internal standard with the same structural features did, however, improve the situation considerably [8].

$$O = C\,NH_2$$
$$|$$
$$O$$
$$|$$
$$CH_2$$
$$|$$
$$R-C-CH_2CH_2CH_3 \qquad\qquad R = CH_3 \qquad\qquad \text{Meprobamate}$$
$$|$$
$$CH_2$$
$$|$$
$$O$$
$$|$$
$$O = C\,NH_2 \qquad\qquad R = CH_2CH_2CH_3 \quad \text{Internal standard}$$

With the n-propyl homologue biological samples were analyzed with a relative S. D. of 1.6%, whereas with the often-proposed internal standard dibutylphthalate the value was 7%. This clearly shows that minor fluctuations in the performance of the instrument and especially the injector could be compensated for, even if it was not manifest that the compound was partly degraded upon injection into the column.

(5) With nanogram amounts, losses due to surface adsorption often occur. Emepronium, (as above) had been analyzed for some time with no indication of losses, by a procedure involving filtration of the quaternary ammonium/perchlorate

Table 1 Loss of ^{14}C-emepronium on silanized glass wool. The values are the % of recovered in three instances.

No filtration	Filtration through silanized glass wool	Filtration through non-silanized glass wool
108	79	105
109	43	101
108	101	104

ion pair through silanized glass wool [1]. Yet the latter, when ^{14}C labelled material became available, proved to cause varying and sometimes large losses (Table 1). The internal standard had sufficiently compensated for this loss. Use of non-silanized glass wool thereafter gave some improvement in the precision at the lowest levels that could be analyzed.

Another example concerns GC of metoprolol as the *bis*-trifluoroacetyl derivative [9]. Of the two internal standards shown, the homologue H93/47

Metoprolol

H 93 / 47

2 – methyl propranolol

Table 2 Analysis of metoprolol by GC-ECD as *bis*-TFA derivative.

Sample	Injection no.	Peak heights, arbitrary units			Peak-height ratio	
		2-Me propr.	H93/47	Metopr.	Metopr. / 2-Me propr.	Metopr. / H93/47
1	1	925	1030	1045	1.13	1.02
2	2	1164	1518	1525	1.31	1.00
3	3	1051	1439	1461	1.39	1.02
4	4	1146	1635	1639	1.43	1.00
5	5	1141	1633	1654	1.45	1.01
6	6	1205	1658	1615	1.37	0.99
7	30	1482	2304	2327	1.57	1.01

was superior to 2-methylpropranolol. The phenomenon was especially marked when an analytical series had been interrupted for 1-2 days and the column and the glass accessories had to be used again. Table 2 shows data for a series of runs in one working day, where the first samples exhibit considerably lower peak heights than the following ones. It is also clear that after 4 or 5 injections the increase in peak height is much smaller but that the 30th injection has quite another value than the 6th. In this situation, the peak height ratio is constant only with a homologous internal standard. The use of internal standard in such situations needs wariness; the analyte amounts should not differ too much from that of the internal standard.

The importance of structural resemblance of compound and internal standard has also been nicely shown in the analysis for one metabolite of metoprolol, B (*see panel*) [10]. Whereas metoprolol and metabolite A (H105/22) could be analyzed with mass fragmentography using tetradeuterated metoprolol as internal standard, metabolite B (H1 19/66) gave quite a large relative S.D., 10%, at a level of about 50 ng/ml. However, the use of pentadeuterated metabolite B (*see panel*) as internal standard improved the situation, the relative S.D.

Metabolite A of metoprolol

Metoprolol, deuterated
(*bis*-Tfaderivative)

Metabolite B

Metabolite B – d_5

decreasing from about 9% to 3.3% [10]. This may well be because the hydroxyl group in the benzylic position gives metabolite B properties other than metoprolol itself.

The previous examples have all been from the GC field. However, internal standards are now beginning to be used also in HPLC, e.g. in the analysis of catecholamines with electrochemical detection [11]. Recently it has been shown by Westerlund *et al.* [12] that naproxen and its dealkylated metabolite (*see panel*) can be favourably analyzed with the use of an internal standard. Table 3 shows that a considerable decrease in relative S.D. was obtained for naproxen when the same sample was evaluated with an internal standard. The dealkylated metabolite differs in being a phenol. This might explain why in this case the internal standard approach was not better.

Naproxen   Dealkylated metabolite

Internal standard

Table 3 Analysis of naproxen (A) and its dealkylated metabolite (B) by HPLC. The values represent relative standard deviation (peak-height measurements).

	Detector	Concn., μg/ml	No int. standard	With int. standard
A	Fluor.	5	8.81	1.44
		10	3.41	2.46
		20	2.64	0.81
	UV	25	2.13	1.46
B	Fluor.	3	1.56	1.33
		5	2.86	2.64

In conclusion, evidently the effort entailed in synthesizing a homologue or an analogue can often be justified by the considerable increase in precision that can be gained in bioanalytical methods.

Acknowledgements

Stimulating discussions with Magnar Ervik, Lars Johansson and Douglas Westerlund are gratefully acknowledged.

References

[1] Vessman, J., Stromberg, S. & Rietz, G. (1970) *Acta Pharm. Suecica,* 7, 363-372.
[2] Hallen, B., Sundwall, A., Elwin, C. E. & Vessman, J. (1979) *Acta Pharmacol. Toxicol.,* 44, 43-59.
[3] Vessman, J. & Hartvig, P. (1971) *Acta Pharm. Suecica,* 8, 235-250.
[4] Vessman, J. (1971) *Acta Pharm. Suecica,* 8, 251-260.
[5] Vessman, J., Hartvig, P. & Molander, M. (1973) *Anal. Lett.,* 6, 699-707.
[6] Walle, T. & Ehrsson, H. (1971) *Acta Pharm. Suecica,* 8, 27-38.
[7] Hartvig, P., Freij, G. & Vessman, J. (1974) *Acta Pharm. Seucica,* 11, 97-108.
[8] Arbin, A. & Ejderfjall, M-L., (1974) *Acta Pharm. Suecica,* 11, 439-446.
[9] Johansson, L, unpublished results at AB Hässle.
[10] Ervik, M., to be published.
[11] Refshauge, C., Kissinger, P. T., Dreiling, R., Blank, L., Freeman, R. & Adams, R. N. (1974) *Life Sciences,* 14, 311-322.
[12] Westerlund, D., Theodorsen, A. & Jaksch, Y. (1979) *J. Liq. Chromatog.* 3, 996-1001.

#NC(F)-3 *Notes on*
CRITICAL FACTORS IN DRUG ANALYSIS

STEPHEN H. CURRY* and **ROBIN WHELPTON**, Department of Pharmacology & Therapeutics, The London Hospital Medical College, London E1 2AD, U.K.

The objective behind any assay procedure for a trace organic compound in biological material is production of a defendable figure for the estimated concentration of the compound, in the biological system examined, at the time of sampling. The assay result must therefore be considered in the light of all stages in the assessment process—sample collection and storage, sample handling in the laboratory, analytical work, and data interpretation. Errors can arise at any of

* Present address: J. Hillis Miller Health Center, University of Florida, Gainesville, Florida 32610, U.S.A.

these stages. This paper summarizes some known or possible sources of error without attempting to provide advice on solving all possible problems. It was prepared after the conclusion of the Bioanalytical Forum which led to this book, and *account has been taken of some of the discussion remarks made at the Forum.*

SAMPLE COLLECTION AND STORAGE

Samples consist of blood or one of its fractions, urine, saliva, cerebrospinal fluid, or, occasionally, some other body material. The necessity for collection of an uncontaminated sample is obvious. Dilution during collection is permissible if its extent is known and so long as it does not affect any other feature of the procedure, such as extraction recovery. A similar guideline applies to additives. In this regard, it must be noted that while preservatives and antioxidants may prevent decomposition of the drug and/or prevent putrefaction of the sample, they can themselves cause chemical change. Particular problems can arise with antioxidants. These can cause reversion of drug metabolites to their parent compounds, if the metabolites are oxidation products. This has been shown to be a particular problem with the phenothiazine drugs, which can form both sulphoxides and N-oxides. These two types of oxidation product readily revert to sulphides and amines respectively [1, 2].

Blood sampling can lead to haemolysis. This will markedly affect assays of plasma concentrations for drugs localized in red cells [cf. Borgå *et al.*, #F-1]. Haemolysis can be caused by use of needles of excessively narrow diameter, over-rapid filling and emptying of syringes, and over-vigorous shaking of blood collecting tubes. Its likelihood probably varies between individuals, and is unquestionably strong with certain animal samples.

Samples should be obtained and retained sterile if possible, as microorganisms introduced during sampling could metabolize drugs. When drugs are administered intravenously, collection of blood samples through the injection needle or catheter must be strenuously resisted. Heroic control experiments are needed to validate such procedures. This problem was probed earlier in this series [3].

Blood samples clot to varying degrees, and this can affect drug concentrations in what is believed to be plasma but is partly serum, if the drug of interest is adsorbed onto the clot. It is wise to study either serum or plasma rather than a vague mixture, and so to use no anticoagulant, or enough anticoagulant, as appropriate. Anticoagulants themselves can interfere with assays, and care must be taken that the same type of collection tube is used in all samplings in any project. There are obvious problems in forensic science when a clean, identifiable sample is often unavailable.

Often there is the difficulty [3] that samples are transported from place to place in pharmocokinetic investigations, as the analytical laboratory is rarely adjacent to the human investigation centre. This sometimes involves posting of samples. Obviously, it is essential to know that the drug is chemically stable in all possible conditions. This is difficult to establish, and it is not always realized that a test of the system with a spiked sample may not be appropriate. Realistic test postings of standards are difficult to achieve. Furthermore, samples from the subjects will contain different materials, especially metabolites of the drug. This is important regardless of whether the metabolites are active or not, detectable or not, or of analytical interest in their own rights. Only a thorough and intelligent test of the system will suffice.

Samples are often centrifuged, and this provides an opportunity for decomposition, chemical change in plasma proteins, loss of volatile compounds, and further haemolysis, with the possibility that standards and unknowns may be influenced differently. Standards, controls and unknowns must all have similar treatment. J. Chamberlain has pointed out that in many double-blind studies control patients must be sampled in the same way as treated subjects, so that a complete series of blanks is made available.

For storage [*see also* #F-5, *this volume*] it is often assumed that the lowest available temperature is appropriate. However, examination of samples stored frozen, and comparison with aliquots of the same samples stored at 4°. can reveal protein changes in the frozen samples. If extraction recovery is affected by protein properties, then freezing may adversely affect the assay of a compound which is perfectly stable at 4°. Furthermore, much decomposition has more to do with microorganisms and light radiation than with storage temperature. Length of storage may be important, as a process of decomposition, or of reversion of metabolites to unchanged drug, occurring only slightly in one week can be a major problem if 3 months' storage is needed. This is especially important if samples arrive intermittently, in a project where assays are to be done as a batch. Batch handling is analytically desirable, but it entails between-sample differences in storage times.

Contamination from collection and storage tubes can be a problem. Protein binding of some drugs can be affected by plasticizers [#F-1], so if care is not taken that all storage tubes are similar, assay variations can result. This will be a problem only if the extraction is affected by protein. Chemical interference from contents of tubes must also be considered.

The question of whether one particular biological fluid should be assessed when there are alternative approaches available should be decided first and foremost on biological grounds. This particularly applies when considering plasma, serum and whole blood. Of course, there are times when only one of these media can be separately assessed; but where a choice is available, the needs of the experiment should dominate.

SAMPLE HANDLING IN THE LABORATORY

Much of the foregoing applies in the analytical laboratory as well as at the sampling site. Problems particular to the analytical worker start when he removes the sample from storage. Much can happen, in regard to decomposition, loss by volatilization, protein denaturation, and reversion of metabolites to parent drugs when a frozen sample is thawed, since during thawing the concentration of materials in the solutions around the ice-plugs are very different from those in the final thawed sample. Apart from this, it is incumbent on the analyst to ensure that his solvents and reagents are of constant and reproducible quality. Laboratory cleaning processes, especially where they involve detergents, can cause problems if the procedures are erratic.

Small samples, e.g. for micro assays or from paediatric patients, entail accentuation of problems concerning surface area, contamination, adsorption and evaporation.

ANALYTICAL WORK

The technique is here assumed to be adequate, in that it is specific and sufficiently sensitive. It is also assumed that the method is intelligently devised and used, without mysterious procedures such as indiscriminate use of isoamyl alcohol or buffer solutions of particular composition and pH, and that the endless problem of differentiation of drug and background material is under control. Variations between samples can, however, still arise as the result of non-specific factors.

A major problem is that of standards. As already implied, ideally standard samples are the same, and are handled in the same way, as the experimental samples, except in that the drug concentration is known, Unfortunately, this ideal is unattainable. Spiking is a particular problem. It cannot be assumed without evidence that the drug added to blank biological material will mix in the same way as that administered and mixed with a fraction of the same biological material *in vivo*. This is especially so if the drug sample is dissolved in an organic solvent before addition to the biological material, when there is a need either to dry down and re-dissolve, or to mix the solvent solution and the aqueous sample. This is not a severe problem with plasma, as it can be assumed that a solution is homogeneous. However, spiking of whole blood, and of tissues as unhomogenized material, is very difficult. Homogenates of tissues pose special problems. We can never be sure that drug material 'spiked' into a tissue homogenate will be bound within the homogenate in the same way as the drug material in a homogenate prepared from tissue collected from a drug-treated animal. However, it seems likely that when the tissue is reduced to a 'solution' by work-up procedures (especially, as M. D. Osselton remarked, if proteases are used—*cf.* #C-1), then drug dissolved will be equally extractable regardless of how it entered the tissue. Finally, in relation to spiking, spiking of serum is suspect, as no allowance is then made for changes during sample collection. Spiked samples can never contain a meaningful mix of metabolites. Storage of samples and standards together is a desirable habit.

Interfering materials cause difficulty. If fatty acids affect the accuracy of an assay [*cf.* #F-3], then variations between individuals in respect of free fatty acids in plasma will cause analytical variation. There has even been a suspicion at times that drug distribution within biological samples could be inhomogeneous.

Seemingly innocuous variations between laboratories must be noted with care. Different batches of samples, and different laboratory procedures, such as variations in tube shaking, vigour of mixing, tissue homogenization or sonification, dilution of samples before assay etc. can cause between-laboratory differences in assay results. Problems of this kind are dealt with by the intelligent use in each laboratory of full sets of control samples.

Of major importance is duplication. The general problem of duplication was dealt with by us in an earlier volume [4]. The question in relation to analytical handling concerns where duplication should start. It can be introduced at sample collection (two samples through the syringe, or one divided into two) or at any stage, perhaps even merely duplicate instrument readings on one extract. There seems to be a strong case for duplication at source of sample. This leads however to an important biological question. Is it better, in time-based experiments, to take two samples at each time point, or double the number of time points and take single samples at this double number?

A related question is the bracketing of samples with standards in chromatographic processes in which one sample or standard is investigated at a time. This leads to the question of how many standards are needed for a given number of samples. It seems unlikely that rigid rules could be devised to satisfy all comments on these matters. These matters have been considered by J. P. Leppard [#F-8] and J. A. F. de Silva [#F-6] in this volume (*see also* p. 370).

DATA INTERPRETATION

This was considered extensively in our previous publication [4]. Care must be taken to work within reasonable limits of claimed precision and accuracy. The limit of detection will be a function of variation in all of the influences now mentioned; but, above all, it must be known. Three matters, however, deserve further comment in the light of discussions at the Forum.

Firstly, internal standards have received much attention. One problem arises from the fact that the term is used differently by various authors. Sometimes a so-called 'internal standard' is really a solution of the actual compound of interest carried through the entire method*. An 'external standard' is then a pure solution. The procedure which is controversial is that of assessment of a known quantity of an analogue (or homologue) of the compound of interest, as an indicator of the performance of the assay for the test compound [K. H. Dudley, #NC(F)-1; J. Vessman, #NC(F)-2]. It is said that this is useful for testing extraction, purification, derivatization, chromatographic

---

*'determinate standard' (Editor's preferred term).

separation, and quantitation procedures. In our view, most of the arguments used in favour of the analogue internal standard idea can be used against it. It is fallacious to think that two analogues with similar extractability and derivatization or decomposition properties will be equally affected by a disturbance in the system. One or both will be affected, and there will be a 50% chance that it is the internal standard that is adversely affected. If this occurs, a correct result will be judged incorrect. This criticism should not be confused with our belief in the merits of an analogue used for checking a final chromatographic step, e.g. retention times, and the use of radioactive internal standards [4].

Secondly, extrapolation of confidence limits has been discussed by J. McAinsh [#F-7]. This may be fallacious. If 90% confidence limits are calculated from samples with high concentrations and then the same confidence limits are used for other samples to predict a limit of detection, no allowance is made for the near-certain greater variance at the lower levels [as considered by Leppard, #F-8].

Thirdly, the topic 'rogue' results ('outliers') was discussed at the Forum. There is a considerable literature on this [e.g. 5, 6]. A sensible approach would seem to be to search for the reason for an outlier occurring. This should not be frivolous or devious, such as blaming the outlier on distracting influences; but where extraction residues remain, they can be checked for pH variations, failure to add reagents, aliquot errors, etc. It seems that failing the discovery of such an error, and assuming that re-investigation of the biological sample is impossible, the data should be taken at face value. As such it should be recorded, although this does not imply that it must be interpreted.

CONCLUSION

The factors now surveyed are not intended to convey a pessimistic attitude to drug analysis. They do, however, exphasize the professional nature of participation therein. Rigid policies are needed in dealing with many problems; but professional supervision of laboratory work, and interpretation of data, will always be needed if the inevitable frequent deviations from the expected are to be properly handled.

References

[1] Cowan, D. A. (1976) in *Assay of Drugs and other Trace Compounds in Biological Fluids* (Reid, E. C. ed.), North Holland, Amsterdam, pp. 193-201.

[2] Curry, S. H. & Evans, S. (1976) *J. Pharm. Pharmacol.*, **28**, 467-468.

[3] Scales, B. (1978) in *Blood Drugs and Other Analytical Challenges,* (Vol. 7 this series) (Reid, E., ed.), Horwood, Chichester, pp. 43-54.

[4] Curry, S. H. & Whelpton, R. (1978) as for 3., pp. 29-41.

[5] Werner, M., Brooks, S. H. & Knott, L. B. (1978) *Clin. Chem.*, **24**, 1895.

[6] Grubbs, F. E. (1969) *Technometrics*, **11**, 1-21.

#NC(F)–4 PROBLEMS IN GETTING VALID RESULTS: AN OVERVIEW*

B. SCALES, Safety of Medicines Department, ICI Pharmaceuticals Division, Alderley Park, Macclesfield, SK 10 4TG, U.K.

There are two distinct phases in quantitative analysis: firstly, analytical development and method validation, and secondly, the routine use of the method and the interpretation of the results. No matter how these phases are managed, analytical problems will continually arise, which need solving in a most efficient manner.

Throughout the symposium, frequent reference was made to sample-handling problems which had been the cause of major difficulties in analytical development, or in performing routine methods reliably. In this article, an attempt is made to bring together some of these problems. It is convenient to collate the problems under the following headings, the order being the approximate chronological sequence of events occurring in the performance of an analytical technique.

1. THE BIOLOGICAL SAMPLE.

1.1 The choice of biological sample.
1.2 Artifacts associated with the presence of erythrocytes.
1.3 Chemical and biological instability.
1.4 Endogenous interferences.

2. SAMPLE WORK-UP.

2.1 Contamination from external sources.
2.2 Interference by components of organic polymers.
2.3 Loss of compound.
2.4 Solvent purity.
2.5 Extraction techniques.
2.6 Evaporation techniques.

3. QUANTITATION.

3.1 Standardization procedures.
3.2 Limits of detection.

4. CONCLUSIONS.

* S. H. Curry [#NC(F)–3] and, in turn, B. Scales have tackled a difficult assignment from the Editor—to knit together and reinforce Forum material including some remarks by discussants. The following participants (in order of text mention) are cited: J. A. F. de Silva, S. H. Curry, J. Vessman, P. F. G. Boon, K. H. Dudley, H. Borgå, H. K. Adam, A. P. Woodbridge, R. de Zeeuw, W. Dünges and L. E. Martin.
For other relevant points, the opening article [#0] and the index should be consulted.—*Ed.*

1. THE BIOLOGICAL SAMPLE

1.1 The choice of biological sample

It was suggested by de Silva (#D-4) that, although biological specimens for analysis are typically of certain types, the choice is dictated by both the chemical nature of the compound and pharmacokinetic principles. However, the analyst's preference for certain biological samples is frequently overridden by the type and purpose of the study and the species being investigated. Thus he may have a method of analysis which is perfectly satisfactory for a drug in plasma, but because, for example, the long-term study which he is monitoring is being performed in very small animals, it is inadvisable to take repeated blood samples. He may therefore have to make do with urine samples, which for analytical needs are much less satisfactory; thus the concentration of drug could be much lower and the interferences from contaminants much higher than in plasma.

The analyst may therefore have little choice in the type of biological sample that he has to analyze, and widely different samples such as blood, plasma, serum, urine, cerebrospinal fluid, bile, liver, muscle, kidney and fat feature prominently on his list. It is not uncommon that in order to optimize the analytical process a subtly different method has to be developed for each type of sample.

Much discussion centered on the actual reasons for the choice of whole blood, serum or plasma on which to do the analyses. Provided that the distribution of compound was not markedly in favour of the cellular components (in which case whole blood was preferred) there was a marked preference for using plasma rather than serum (Curry). However, J. Vessman noted that frozen and thawed plasma samples do throw a precipitate which is not seen with serum. This problem may be routinely overcome by ultrasonicating samples after thawing, to ensure homogeneity (Boon). It was suggested on theoretical grounds (Curry) that one would expect slightly higher concentrations of a compound in plasma than in serum. In practice, although this had been found for clonazepam (de Silva), the opposite had been found for phenytoin and phenobarbital (Dudley); for three compounds examined in this way by Vessman, no significant differences were seen.

1.2 Artifacts associated with the presence of erythrocytes

A difficulty in ensuring a representative sample is discussed by Borgå *et al*. [#F-1] in the context of ascertaining the actual *in vivo* plasma concentrations of drugs. Thus with the 'heparin-lock' technique there can be marked changes in apparent protein binding, and hence in distribution and elimination processes. With compounds which are highly bound in the erythrocyte, even slight haemolysis can seriously affect the measured plasma level. The authors cite examples where storage significantly altered the steady states, resulting in erroneous plasma values.

The common supposition that pharmacokinetic analysis can be performed using results for either whole blood or plasma is true in many cases, and whole blood has the advantage that sample deterioration on storage will have no effect on the actual total concentration of drug measured. However, Borgå *et al.* pointed out that in certain circumstances the steady-state equilibrium between erythrocytes and plasma may be virtually unattainable *in vivo*, and that the drug concentration/time profile for erythrocytes may lag behind the plasma profile. This illustrates the importance, in pharmacokinetic studies, of comparing like with like.

1.3 Chemical and biological instability

For many reasons, organic compounds decompose when in solution, and in aqueous solution the decomposition is often aggravated by the presence of a biological matrix. Although not desirable, biological samples do have to be stored, albeit briefly, for reasons such as transportation, analysis and re-analysis. It is therefore important to assess any instability and to be able to correct for it if necessary. Adam [#F-5] and de Silva [#F-6] indicate ways recommended, in their own laboratories, to check for instability and to minimize it. Storage at reduced temperatures (down to −70°) is generally successful. Problems highlighted in Forum discussions included protein denaturation, fractional crystallization and the resulting concentration of drug and metabolites near the sample surface, increased decomposition due to oxidation at the surface and ensuing problems when thawing to prepare a homogeneous solution. There was general agreement that the analyst must be alert to such problems. The addition of stabilizing agents, such as antioxidants and enzyme inhibitors, to the biological sample was advocated when long storage times are inescapable. Investigations covering many months are then necessary to prove the effectiveness of the measures taken. Adam [#F-5] considers how meaningful stability information can best be obtained.

It is whilst performing such checks for 'biological' instability of the drug, that matrix changes may become manifest. An unstable biological matrix can interfere with the analytical method, in a manner which is not obvious from studies with fresh biological samples.

1.4 Endogenous interferences

Attempts to quantify the compound of interest at low levels are invariably handicapped by the presence of interfering substances. The biological matrix is the most obvious source of these, and endogenous components have often been implicated as contributing towards irreproducible extraction efficiencies, high and unstable background noise and hence unacceptable limits of detection.

In practice, about 0.20 mg of heptane-extractable material is obtained from 1.0 ml of whole blood at any pH, and even larger amounts are extractable with more polar solvents,. This structurally undefined mixture may be present, therefore, at from 10^{-1} to 10^8 times the amount of the various compounds to be measured. As the search for compounds with increased biological activity continues, and the concentration for eliciting significant toxicity or therapeutic effect becomes less, it is not surprising that the direct applications of UV, fluorescence and electrochemical methods to extracts of biological samples is falling into disuse, and that most analytical methods now entail isolative and/or chromatographic steps (TLC, GC, HPTLC or HPLC) before the quantitative measurement.

An understanding of the endogenous components in the extract can help to ease many of the extraction and separation problems in analysis. Fatty acids such as palmitic, oleic, stearic and linoleic, and their derivatives, have been listed by de Zeeuw & Greving [#F-3] and Westerlund & Nilsson [#F-2] as being one major cause of interferences from serum and plasma. Amongst their effects are alterations to extraction efficiency by ion-pairing, significant reduction of reagent concentration thereby lowering the yield of the required derivative, and the production of interfering chromatographic peaks. These authors emphasize that many possibilities exist for eliminating the effects of interferences; thus one may attempt to separate and lose the interfering substances, but alternatively one may use selective derivatization and detection techniques as a cosmetic approach by which one could hope to ignore the interference. It was noted that the more clean-up steps the greater the risk also of exchanging endogenous for exogenous interferences. Some of the problems were overcome by the use of back-extraction procedures or by careful selection of solvents and pH. One example of a satisfactory clean-up procedure was given for oxyphenonium bromide in human plasma. Ion-pair extraction was followed by displacement of the base by competition with tetraphenylammonium ion; by this approach, followed by a further ion-pair extraction, fatty acid interference was almost eliminated. Final quantitation was by GC-ECD on the pentafluorobenzyl ester of a hydrolysis product; in all a rather lengthy procedure, but one which gave entirely satisfactory quantitation.

2.   SAMPLE WORK-UP

2.1 Contamination from external sources
Potential contamination of laboratory apparatus and chemicals was stressed by Woodbridge [#NC(C)-3]. He listed many materials conventionally found in the analytical laboratory, including bench polish, barrier creams, rubber bungs and plastic tubing, which could all lead to invalid results in pesticide residue analysis.

The presence of fatty acids inside glassware, through external contact with human skin, and from the traces present in some common solvents may be sufficient to cause interference. A routine cleaning and handling procedure to avoid accumulation of various interfering materials was described by de Zeeuw & Greving [#F-3]. When using micro scale work-up procedures [Dünges,

#NC(F)-6] most contamination of the micro glassware was conveniently removed by flaming in a bunsen flame or heating in a muffle furnace.

2.2 Interference by components of organic polymers
Modern technology has provided us with invaluable materials such as plastics which are a major cause of many analytical problems. Materials present in plastic containers can affect both the integrity of the sample and the subsequent analytical procedure. The most troublesome plasticizers include di-(2-ethylhexyl) phthalate [#F-2], tris(butoxyethyl) phosphate [#F-1] and additionally the esters of adipic and sebacic acid [#F-3].

The plasticizers are sufficiently non-volatile to concentrate during evaporation of solvent, yet sufficiently volatile to cause interfering peaks during chromatography. The trouble in the case of solvents was best remedied by distillation and further prevented by storage in all-glass vessels. Contamination of the biological samples with plasticizers is more difficult to overcome since plastic disposables for sampling and storage are now a seemingly necessary part of our equipment. The effect for the analyst can be disastrous; in the first place the plasticizers are frequently extractable and add to those already present in the organic solvent; secondly some plasticizers can cause a shift in protein binding [#F-1]; thirdly, and perhaps most importantly, some acidic metabolites, e.g. mono-(2-ethylhexyl) phthalate, are formed by the esterases in the blood and, because they are present at high concentrations, can markedly affect the extraction efficiency of basic drugs through the formation of ion-pairs [#F-2].

2.3 Loss of compound
Not only can sampling devices and containers introduce interfering materials into an analysis, they can also be responsible for the loss of material to be analyzed. de Silva [#D-4] has listed various physical and chemical factors which could contribute to the loss of compound from the sample. Most surfaces are capable of adsorbing appropriately structural organic molecules, and the severity of the loss from the biological matrix or the solvent extracts should normally become manifest during assessment of the overall recovery of the compound. Adsorption problems can be cured in some instances by silanization, although not invariably (Vessman).

Loss by diffusion though the plastic sampling devices and through the containers has been a known problem in the present author's laboratories for many years. Thus the analysis of 'Fluothane'* in biological samples by GC-ECD techniques can be done very successfully if care is taken to exclude the use of organic polymers at all stages. On the other hand, with slight changes in chemical structure of volatile lipophilic molecules, diffusion through some of these polymers may become insignificant. There is therefore no safe rule, and each compound must be examined on its own merits.

* 'Fluothane' (halothane) is a trade mark, the property of Imperial Chemical Industries Limited.

2.4 Solvent purity

There are few analytical techniques that do not require the use of organic solvents at some stage. It is seldom important that the solvent be pure, but it should be from impurities that can significantly interfere with the analysis. The volume of solvents used in analysis is large compared with that of the other chemicals, and although impurities may be present in low concentrations, the act of extracting these solvents, or evaporating to low volume, passing them through chromatographic columns or adding reagents, can result in a highly significant concentrating of these specific impurities which can then interfere with the analysis.

During analytical development work an attempt will have been made to select appropriate solvents or devise mixtures, with a view to optimizing their use for extraction of the test substance and the interfering biological components. At the same time, the solvents must be compatible with the other stages in the procedure.

Impurities in solvents may originate inadvertently, but inescapably, during the manufacturing process; or may be added intentionally during production, for example as stabilizers. It is well to be aware of the type of impurities which may be present, and to test whether variations in them can affect the analysis. This is readily done by incorporating the impurities, or concentrates of the solvents in question, into the appropriate stage of the analytical procedure. An acceptable analytical system would be hoped not to be influenced by small changes in the impurity level. However, even the most robust of methods can fail owing to changes in solvent impurities. It is often the unspecified impurities which by their very nature can cause the greatest problems. Thus the sulphur-containing impurities originating from aromatic solvents can be taken through extracting processes and seriously affect the performance of polarographic detectors, or give rise to major interfering GC peaks. Also, the amounts of unsaturated impurities present in aliphatic solvents may be a major factor in setting the limits of detection in a UV or a fluorescence assay. Trace impurities such as amines, aldehydes and alcohols can also react with the test compound, or compete with it for reaction with specific reagents, thereby producing unreliable results.

The situation is aggravated when the levels of these interfering impurities vary from batch to batch of solvent or reagent. This variability is a common source of the problems encountered when analytical methods are handed over from one person or laboratory to another. In the new situation, the reproducibility of the method may no longer be acceptable, and the linearity of the calibration plot, the limits of detection, and the selectivity may have changed. At the worst the interference may be such as to negate even the most valiant attempts to quantify the analyte level.

Should the level of impurity or batch-to-batch variability be a problem, in either analytical development or routine pharmacokinetic work, then batches of solvent sufficient for the total study or for significant segments of it ought

NC(F)-4]   **Problems in getting valid results: an overview**   359

to be either selected or prepared from purchased material, and earmarked for that analysis. The selection is often best done by trial-and-error, and suppliers of bulk solvents may let the customer select bulk batches prior to purchase. The preparation of large volumes of solvents, extensively purified according to standard methods in textbooks of practical organic chemistry, can be a lengthy procedure, especially if chemical modification, extraction, drying and distillation procedures are needed. Such purification is used only as a last resort; limited and selective purification is very often all that is needed, e.g. washing the solvents with aqueous solutions of the same pH as those used in the extraction procedure, or distillation to remove non-volatile residues. Acetone is a favoured solvent for a number of reactions, but trace impurities can cause problems, as in microscale work by W. Dünges (#D-9 in vol. 7, this series); condensation products and moisture have to be removed, and he warns against the injudicious use of molecular sieves.

Apart from preserving the stability of the solvent or a reagent solution, there may be little need to dry solvents used for extraction since they are invariably wetted during the extraction process. Indeed, many analysts recommend the use of water- or buffer-saturated solvents in the extraction process.

Purification of solvents may be of little value unless care is taken in their subsequent storage. Screw-capped glass bottles containing aluminium foil/cork inlays are favoured by de Zeeuw as the best way of preventing contamination by plasticizers. Oxidation can also be minimized by keeping the bottles at reduced temperature, in the dark, and as full as safety considerations permit.

2.5 Extraction techniques

In the execution of some analytical techniques it is desirable to transfer as much test compound as possible from the test sample to the final quantitative detection system. This helps ensure high reproducibility of the assay and achievement of the best limits of detection. However, in many contexts it is impossible, and unnecessary, to transfer the maximum amount of test compound, and in a complex multi-extraction and derivatization procedure, only 5–10% of the compound may be transferred. In fact, in some GC-ECD techniques only 0.001% of the original test compound may be transferred to the column, These transfers can be done quite reproducibly as proved by the good correlation coefficients for calibration plots which competent analysts usually obtain. The major problem in achieving this reproducibility lies not with the dispensing or injecting of aliquots but more usually with the ability to maximize other stages such as extraction, evaporation, dissolution and derivatization. Irreproducibility can occur in these processes for many reasons, but may not be observed, or even cured, until the analytical processes is handed over to someone with different experience.

Thus ethyl acetate would not be the preferred solvent if aqueous solutions had to be extracted at pH greater than 9.0. With care, when shaking is confined to short times at reduced temperatures ($2°$–$10°$), perfectly satisfactory results can be obtained; but at higher temperatures rapid hydrolysis can take place with a resultant decrease in pH of the alkaline solution, and a decrease in volume of the ethyl acetate due to both hydrolysis and to the solubilizing effect of the alcohol produced. Many solvents and especially their decomposition products can be quite reactive towards some chemical entities. Some test compounds can react quite unexpectedly during the extraction and drying-down process with ethers and halogenated aliphatic solvents.

The time-dependency of the extraction process has been discussed by Vessman (cf. #F-4). He showed that in some instances, doubling the extraction time of 15 min could improve the proportion extracted from serum into the organic phase by almost 50%. In one extreme case of an organosilicon compound, it was shown that in contrast to the rapid equilibrium reached between water and organic solvents, when plasma was used only 75% of the final equilibrium value was reached even after 24 h. Homogeneous extraction may offer a remedy.

The actual physical technique of optimally mixing immiscible liquids was mentioned in several papers and discussion sessions. No universally acceptable system was claimed, but gentle tumble-mixing (as in the Craig counter-current distribution process) or rolling would appear to be a very acceptable alternative to the reciprocating mixers in common use; the liquid surfaces are rapidly renewed, and there is the minimum production of troublesome emulsions which are so often produced by the more vigorous agitation techniques. On a smaller scale, extraction by swirling has some advantages, but it is of value only when equilibrium is rapidly achieved.

The extraction process as applied to biological samples is very complicated and poorly understood. The standardization of conditions such as time, temperature, batches of solvent and impurities, relative volumes of phases, and dilution of the biological sample is very important in order to achieve reproducible results.

2.6 Evaporation techniques

The analysis of drugs commonly entails the evaporation to dryness of organic extracts or solutions of reactants, and redissolving the residue in measured small volumes of solvent. Opportunities for losses are abundant; thus the compound may be lost by creeping over the walls of the vessels, by 'bumping' or by aerosol formation, or simply because it is volatile under the conditions used. If evaporation is too vigorous and prolonged, the compound may become firmly absorbed onto the glass surfaces or the traces of particulate material in the residues from the biological extract, and then be hard to redissolve.

Various suggestions about how to minimize these problems, and account for them, have been put forward. The swirling or centrifugal evaporation technique [Martin, #NC(F)-5] would appear to offer certain advantages in that bumping and aerosol formation are minimized. The careful use of well chosen internal standards [#NC(F)-1, #NC(F)-2], or the addition of tracer amounts of radio- or stable isotope-labelled compound, is a widely recommended means of checking and accounting for these losses, and also those which could occur on attempting to redissolve the residue.

Analysts have an innate conservatism when it comes to drastically changing their sample manipulation techniques. However, microscale pre-chromatographic procedures as advocated by Dünges [cf. #NC(F)-6] could help eliminate some of the above problems associated with clean-up on the conventional scale.

3.   QUANTITATION

3.1 Standardization procedures

Although not essential, it is common practice to produce calibration plots from which the concentration of compound in the unknowns can be derived. Ideally, these plots are produced from the analysis of standard samples, i.e. blank control samples to which known amounts of test compound have been added. Additionally, a further standard (an 'internal' or 'normalization' standard) is now frequently added in a fixed amount to all samples to be analyzed. To obtain valid results, the test and standard samples must be processed at the same time and in an identical manner, so that there is no distinction between the way that they are handled.

Recommendations for the use of internal standards have been discussed by de Silva [#D-4]. Vessman [#NC(F)-2] has illustrated their value in compensating for fluctuations such as phase transfers, non-stoichiometry of reactions, on-column stability and loss by adsorption onto glassware. The internal standard must be very carefully chosen, with physico-chemical properties as close as possible to those of the test compound, yet still allowing adequate separation from it by the final chromatographic and/or detection system. The use of the test compound labelled with a stable isotope as the internal standard has much to commend it, provided that a GC-MS facility for separating the isomers is available for routine use [cf. Martin & Tanner, #NC(D)-1]. Alternatively, when radio-labelled drug of high specific activity is available this can be used in trace amount.

The use of internal standards in no way obviates the need for careful analytical work. It has been argued (Curry & Whelpton in Vol. 7, this series) that the use of an internal standard can cause as many assay errors as it can correct. One fundamental assumption is that the losses of test compound and internal standards at the first and the subsequent stages of the analysis are independent of the mode of their addition to the biological sample. This has been questioned, especially when non-homogeneous samples have been spiked, but it may be true once the first step of extracting the compound from the bulk of the biological sample has been performed [Curry & Whelpton, #NC(F)-3].

3.2 Limits of detection

It is acknowledged that day-to-day changes in the recovery of compound and in instrument response do occur, and for this reason calibration plots are invariably prepared with each day's batch of unknown test samples. Should the sensitivity of the method change, for whatever reason, the availability of the calibration plot enables that change to be noted, yet still enables meaningful results to be obtained.

Problems exist when working close to the limits of detection because, other things being equal, the value of this limit is very dependent on the recovery of the compound and the sensitivity of the analytical instrument at the time the batch of samples was being analyzed. The arbitrary manner in which limits of detection are assessed has attracted some attention, but with little success in achieving a realistic practical approach which would consider this day-to-day variation in the analysis.

J. A. F. de Silva [#F-6] proposed various criteria which should be examined during the validation of an analytical method, prior to its routine use. One of these was that "the sensitivity limit of the assay must be statistically defined and validated". A major problem arises in that, although the method may be quite satisfactory in the hands of different analysts, when using it for concentrations well above the limits of detection, that very limit may be considered as a variable since it will depend on the daily care and attention to detail that the analyst takes at each analytical step and, for example, on the sensitivity of the detection system.

Although it may be valuable to set statistical values for the various criteria during assay validation, hopefully these would not be used to discount analytical results from a batch of samples which did not quite satisfy the set criteria; the analyst must be allowed some discretion.

The statistical treatment of data, and its use in assessing assay precision at various concentrations has been discussed by Leppard [#F-8]. Using similar statistical procedures McAinsh *et al.* [#F-7] suggested a confidence-limit approach to assessing the limits of detection, based on the statistical analysis of the calibration plot. These authors reject the conventional arbitrary quantitative assessment, and argue that the limit of detection is a variable which should be calculable for each batch of samples from the appropriate calibration plot, and not rigidly fixed at the time the method is validated. They perform a regression analysis on a desk-top computer (e.g. Hewlett Packard 97), with an additional programme to calculate and print out the limits of detection for each day's batch of samples. This could be a satisfactory approach, yet needs careful thought, especially in the selection of test samples and of the concentrations used for the standard calibration plot; obviously it is necessary to include a number of standards close to the anticipated limits of detection.

4. CONCLUSIONS

Staff doing analytical development often use quite different training and experience to those doing routine analysis, as is frequently reflected in the different ways that they will tackle the same problem. Without hindsight, one cannot be certain of the most rapid approach to problem-solving; but obviously a consideration of all the facts is essential. To this end, knowledge, experience and careful observation will enable the facts to be marshalled. Although a requirement for chemical training predominates in the initial analytical development work, the processes of trouble-shooting and the production of valid results frequently require a sound appreciation of relevant biochemical processes and pharmacokinetic principles. This, tempered by the application of a reasonable amount of intuition, is probably the most efficient and satisfying way of producing meaningful results.

#NC(F)-5 *A Note on*
PROBLEMS IN CONCENTRATION OF SOLVENT EXTRACTS FOR DRUG ANALYSIS

L. E. MARTIN, Glaxo Group Research Ltd., Ware, Herts, SG12 ODJ, U.K.

Most methods for determining drugs in biological fluids at the sub-μg concentration involve prior extraction of the sample with an organic solvent, e.g. 1–2 ml plasma is extracted with 0.5–5 ml of solvent. The solvent used for extraction should be chosen with care since, as B. Scales emphasizes [#NC(F)-4], impurities can be concentrated by the evaporation process; thus ephedrine reacts with aldehydes which may be present as impurities in diethylether (Beckett, Jones & Hollingsbee [1]. To overcome this type of problem, it is essential to use high-grade analytical solvents and often to distil them before use. The chemical nature of the solvents is also important; thus chlorinated hydrocarbons, e.g. dichlormethane, can alkylate basic compounds. For quantitative analysis by HPLC, GC or GC-MS, the volume of the solvent extract is usually concentrated by vacuum evaporation or in a nitrogen stream to 50–100 μl. During vacuum evaporation losses of drug may occur owing to violent boiling (bumping). One way to overcome this is to use anti-bumping granules. However, losses of drug can occur due to adsorption of material on the granules. In the author's laboratory, two designs of vacuum evaporators which prevent violent boiling are in use. These are Buchler Rotary Evapo-Mix (USA) and the Centrifugal Laboratory Dryer, described by Pepper [2]. In the former apparatus the tubes containing the solvent are placed in a carrier in a heated water bath and each tube is connected via a 'reflux head' to a vacuum manifold. The carrier containing the tubes is rocked back and forth by an oscillating motion which imparts a deep

Fig. 1 – Centrifugal evaporator.

swirling motion to the solvent in the test tubes. The centrifugal force of the swirl prevents 'bumping' during heating under vacuum.

In the second type of evaporator the tubes are placed in a centrifuge basket over which is placed a bell jar (Fig. 1). The centrifuge is started and the jar evacuated down to a pressure of 200 Pa. The evaporated solvent is trapped in Dewar flasks cooled in cardice. For solvents of higher b.p., e.g. ethylbutyl ketone, evaporation rate may be increased by placing a 275-watt internally silvered infrared lamp 3–12 inches from the bell jar.

The bottom of the tubes in which evaporation takes place should be drawn out into a conical tip. This allows the condensed solvent to reflux down the walls of the vessel and the last few μl of solvent are then collected in the tip of the tube. A problem with the evaporation of solvent extracts for biological material is that mg amounts of endogenous material may be precipitated during evaporation and the drug may be adsorbed on this precipitate. Ideally, evaporation should be stopped before precipitation occurs. This is often not possible when determining drugs at very low concentrations, and in this case the solution has to be taken to dryness. Complete re-dissolving of the precipitate in very small volumes of solvent or derivatizing agents can be a problem, and the drug can be trapped in undissolved material.

When the drug is available in a radiolabelled form a tracer amount should be added to the sample in order to assess losses due to evaporation, incomplete dissolution or surface adsorption of the drug during analysis. In some but not all cases, adsorption on the glass surface of the vessel can be prevented by silanization of glassware.

When carrying out evaporation in a flowing stream of nitrogen [#NC(F)-7, *this vol.*] losses of drugs as aerosol can occur, as should be checked with an internal standard.

References

[1] Beckett, A. H., Jones, G. R. & Hollingsbee, D. A. (1978) *J. Pharm. Pharmac.*, **30**, 15–19.
[2] Pepper, D. A. (1973) *Lab. Practice*, **22**, 538–539.

#NC(F)–6 *A Note on*
VOLUME REDUCTION UNDER PARTIAL REFLUX OF TRACE-ORGANIC EXTRACTS

W. DÜNGES*, K. KIESEL and K.-E. MÜLLER, Institut fur Pharmakologie der J. Gutenberg-Universitat, Mainz, W. Germany

With conventional methods for pre-GC concentration of dilute organic solutions severe losses of solutes occur in the sub-ml and nanomol range. Thus it has been claimed that compounds with a vapour pressure of more than 1 torr at 100° are lost with the widely used technique of concentrating with a nitrogen stream. Recently the technique of concentrating under partial reflux has been established as a routine laboratory method [1]. More than 60 compounds of different chemical classes have been investigated. All substances with b.p. 150° or higher in acetone or with b.p. of >170° in ethyl acetate can be concentrated with minimal losses. This was found with glass-capillary GC when a volume of 300 μl of such solutions containing 4–10 nanomol each was reduced to 7 μl.

The range of the method has been explored with a test mixture of phenols and anisoles. Ethyl acetate solutions of 4 nanomol in 3 ml or of 64 picomol in 250 μl could be concentrated to 7 μl with a recovery of 80–100%, showing only minimal deviations from GC calibration values. A cleaning procedure for micro glassware [2] and a new method for obtaining ultra-pure solvents in 500-ml batches are pre-requisites for this technique [1].

References
[1] Dünges, W. (1981) *Pre-chromatographic μl Techniques,* Hüthig-Verlag, Heidelberg.
[2] Dünges, W., Kiesel, K. & Müller, K.-E. (1978) in *Blood Drugs and Other Analytical Challenges* [*this series,* Vol. 7] (E. Reid, ed.), Horwood, Chichester, pp. 293–311.

* Contact address: Rilkeallee 14, D-65 Mainz 31, W. Germany.

#NC(F)–7 *A Note on*
A SIMPLE MULTI-TUBE SOLVENT EVAPORATOR

E. REID*, A. D. R. HARRISON† & J. P. LEPPARD, Wolfson Bioanalytical Unit, Institute of Industrial & Environmental Health & Safety, University of Surrey, Guildford GU2 5XH, U.K.

In multi-sample assays as performed, for example, on drugs or pesticides isolated from biological materials, there is often a need to reduce solvent extracts or GC derivatization media to a small volume or to dryness. A favoured method, especially

* To whom enquiries about the availability of the evaporator should be addressed.
† Present address: Albury Laboratories, Albury, Surrey.

in laboratories concerned with drug assay, is to expose the liquid surface in each of a set of tubes or vials in a heating bath to a stream of oxygen-free nitrogen, preferably adjustable to ensure escape of vapour but not of droplets that could entrain the drug. The past lack of commercial apparatus for multi-sample solvent removal by a simple means such as this has led many analysts, especially in the drug field, to improvise their own assembly based on a made-up gas distribution manifold [e.g. 1,2].

The design now described enables such a manifold to be moved easily between a raised idling position (Fig. 1) and a lowered working position (Fig. 2), with provision for exact height adjustment of the array as a whole or of individual gas inlets. For the rack which holds the vessels, different multi-hole plates are available to accommodate tubes of different diameters (suitably with conical bottoms) or micro-vials. To minimize local over-heating, the vessels rest on insulating pads fixed to the aluminium bottom plate of an air-bath ($\sim$150°max), underneath which are heating elements linked to a variable controller. One of

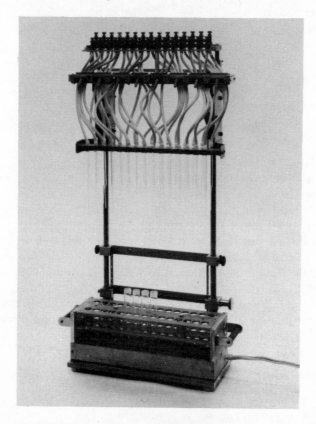

Fig. 1 – View of the evaporator in the raised (idling) position, showing the gas delivery system in its original form, now modified (*see text*).

Fig. 2 – View of the evaporator in the lowered (working) position, with merely some illustrative tubes in place rather than a full rack.

several advantages over an assembly that has recently come on the market (Techne Inc.) is the use of heat-resisting glass at the front and back of the rack, enabling individual vessels to be observed (with the aid, if desired, of a rear light) and removed as soon as dry; the back row is off-set to facilitate observation.

Disposable Pasteur pipettes serve as inlets. The manifold as illustrated, made up of plastic valve units as sold for aquaria, tends to distribute the gas unevenly. A conventional manifold is now favoured, with a valve in each side-arm that serves for shutting off the gas to a particular position or for supplementary regulation; the illustrated device for shutting off particular groups of tubes is thus superseded.

References

[1] Yeh, S. Y., Flanary, H. & Sloan, J. (1973) *Clin. Chem.*, **19**, 687–688.
[2] Maycock, K. (1977) *Lab. Practice*, **26**, 783.

Editor's explanation
Most of the points that came up in the Forum debates have been incorporated, with mention of the discussants concerned, in over-view articles [#NC(F)-3 and —4] written in retrospect. Some additional points are set down below, with category headings: THE SAMPLE; QUANTITATION; MATERIALS AND MANIPULATIONS.

THE SAMPLE with especial reference to #F-1 to #F-5
See the first part of #NC(F)-3, and part 1 of #NC(F)-4.
Additional points
It is because of better phase separation and higher recovery that whole blood is preferable to plasma or serum when dealing with basic drugs (J. A. F. de Silva, *answering* R. A. de Zeeuw; cf. #D-4). S. H. Curry, *replying to* R. A. de Zeeuw: information seems to be lacking on the possibility of uneven drug distribution or of evaporative losses when the sample is freeze-dried prior to shipping. [Concerning storage in the undried state, see (vi) in #0 as well as #NC(F)-3.— *Editor.*] *Remark by* J. Vessman relating to R. A. de Zeeuw's disinclination to pre-extract fatty acids lest the drug be pre-extracted too (ion-pairing may occur even with Cl^-): a solvent such as hexane, not conducive to ion-pair extraction, might be considered. Digitonin might offer a means of removing cholesterol (*remark by* J. Ramsey).

The distribution *in vitro* between RBC and plasma (cf. #F-1) has been shown to vary amongst a series of psychotropic drugs, and to be somewhat different for heparin compared with EDTA as anti-coagulant [1].

[1] Maguire, K. P., Burrows, G. D., Norman, T. R. & Scoggins, B. A. (1980) *Clin. Chem.*, **26**, 1624–1625.

QUANTITATION
with especial reference to #F-6 to -8
See the latter part of #NC(F)-3, and part 3 of #NC(F)-4; also (i) in #0
Additional points
Remarks by J. A. F. de Silva on 'rogue' results (outliers) [cf. part (i) in #0, and end of #NC(F)-3].—Our practice being to assay in singlicate, where there is an apparent rogue that cannot be attributed to (e.g.) interferences then we repeat the assay, possibly by another method; if confirmed, we have to accept it as the 'true' value even if it does not fall smoothly on the fall-off curve concerned. *Citation of* [1], by Editor.—Where duplicates are performed (in the RIA context) "an essentially arbitrary choice must be made of a limit for the ratio of observed to 'expected' differences between duplicates; a sample with a ratio above this limit may properly be rejected"; merely on statistical grounds, $\sim$11% of all pairs will differ by $\sim$3 times the S. D., a rough measure of which is $\sqrt{2} \times$ difference between duplicates in the assay. "Different people perform differently each

day, and it is bad practice to use the same criteria for every assay. On some assays it may be necessary to reject 1% of the distribution as outliers, but on others 50% may be more appropriate." The chosen limits may vary with the analyte concentration [1]. *Remark by* J. Chamberlain concerning the limit of detection: this, as a working policy, is to be the level at which the C. V. becomes greater than 30%.

[1] Ekins, R. P. & Malan, P. G. (1978) *Ann. Clin. Biochem.,* **15**, 125–126.

Comments by S. H. Curry on J. A. F. de Silva's 'GLP' (#F-6).—Notwithstanding some reservations, the suggested policies are valuable, and should not only govern our record-keeping but also be taught to students. *Comments by* the Editor are given in parts (i) & (viii) of #0. There should not be over-emphasis on daily checking of instrument linearity, and even for the 'determinate' standards spiked into the biological matrix, some day-to-day variation in slope may be tolerable. Nor should there be an obligation to adopt a 'processed' internal standard (a drug-like compound), in view of pitfalls that this may entail [S. H. Curry, #NC(F)-3; E. Reid, in Vol. 7, this series] ; duplication of all test samples may be preferable.

Editor's remarks on methods newly set up. — After statistical reassurance has been obtained, especially for linearity, from (say) three successful runs based on a full range of standards, it seems sufficient for routine assay to have merely 6–8 spiked tubes, say at only 2 levels differing at least 8-fold. This serves to establish within-run slope (assuming linearity) and check precision behaviour and sensitivity. Confirmation of precision is obtainable if some test samples are duplicated. However, most test samples may have to be singlicates for the sake of throughput, the risk of a lurking 'rogue' is reduced if there is a valid internal standard or if, in a bioavailability study, the blood curve is inspected for kinks that call for repeat assay. Confidence in singlicates is, however, weakened if it is decided to reject any calibration point as an inexplicable outlier. This unoriginal sketch is a 'holding operation' that may stimulate the publication [cf. #F-8] of deeper guidance by experts.

MATERIALS AND MANIPULATIONS
with especial reference to #F-3 and various other articles including #D-4
See part 2 in #NC(F)-4
Additional points
The list of sources of plasticizer contamination continues to grow. Filter paper may be inexplicably contaminated with di*iso*butyl phosphate (J. Ramsey). Vulcanizing agents that may persist in rubber septa include thiobenzthiazole, which may (after, say, methylation) interfere in GC analysis of acidic drugs

such as barbiturates; its N is AFID-responsive (A. Clatworthy). Interference in GC or HPLC can arise, due largely to tributyl aconitate, if nitrogen used in drying down is conducted through PVC rather than Teflon or rubber; even a heated empty glass tube may acquire such contamination (Colin R. Jones).

If drying down is to be conducted with a nitrogen or air stream [cf. #NC(F)-7], use of a bubbler to monitor the rate calls for careful choice of liquid. A suitable innocuous liquid for a bubbler is ethylene glycol, which further serves to ensure dryness of the gas stream (E. Reid & A. D. R. Harrison).

To the list of compound types that may be affected by solvents may be added a nomifensine glucuronide which, by inadvertent hydrolysis, can give μg/ml rather than the true ng/ml values for 'free' nomifensine in plasma (J. Chamberlain), and methadone which, if oxidized during extractions, may break down to a cyclic metabolite (A. Clatworthy). Solvent extraction (2-phase) and associated problems feature especially in #0 [parts (iv) & (vi) with a 'checklist'], in #D-3 and #D-4, and in part 2 of #NC(F)-4. Dichloromethane of supposedly high purity, used to extract lipophilic analytes from plasma for RP-HPLC, has been found to contain intractable contaminants, and a further interfering substance is formed in variable amount during drying down [1]. Methyl t-butyl ether (b.p. 55°), which is not prone to form peroxides, may be advantageous as an extracting or HPLC solvent [2] but is hopelessly expensive.

Erratic results were obtained in the Editor's laboratory when, with flour as substrate, spiking-in of the analyte (a peptide) was 'realistically' done prior to, rather than after, initial extraction with ethanolic HCl.

The risk of ECD contamination if silyl derivatives are prepared for GC needs consideration (J. Chamberlain, B. Scales). An ECD kept connected inadvertently during 'Silyl-8' treatment of a GC column showed only transient saturation (J. P. Leppard). Purified halosilyl derivatives of steroids are suitable for GC-ECD (B. Thomas, in Vol. 7, this series). Benzodiazepines may be analyzed by GC-ECD as silyl derivatives (de Silva); agents such as t-butylsilylimidazole may be removed by water before GC.

Editorial work on authors' texts has led to the thought that 'silanizing', and 'siliconizing' terms for treating glass surfaces to reduce adsorptivity (p. 199) may sometimes be confused.

[1] Masoud, A. N., Dubes, G. R. and Krupski, D. M. (1980) *Clin. Chem.*, **26**, 1363–1364.

[2] Little, C. J., Dale, A. D., Shatley, J. L. & Wickings, J. A. (1979) *J. Chromatog.*, **169**, 381–385.

Cumulative Index of Compound Types (covering Vols. 5 & 7 also)

Generic names, as in the 'Merck Index' (which gives formulae), are used preferentially. Metabolites are not specifically listed. Some compounds fall under a class title, e.g. Phenylethylamine. See opposite for the CHEMICAL CLASSIFICATION: this reflects some common features of analytical relevance. The style policy for Index entries is as follows.

PRIMARY *vs.* CARRIED–FORWARD ENTRIES	ASSAY GUIDANCE OR CITATION FOR 'REAL' SAMPLES as specified parenthetically (*see below*)	BACKGROUND INFORMATION, RELEVANT TO SAMPLE HANDLING– e.g. metabolism, extractability	BEHAVIOUR RELEVANT TO MEASURE- MENT, e.g. R_T, derivatization
How page– indexed if in present volume	e.g. 15 (air); 20– (bld) where ' – ' denotes a major entry	e.g. 17	e.g. 19 (i.e. as in previous column)
How page– indexed if in Vol. 5 or Vol. 7	e.g. *v.5: 201 (urn);* *v.7: 191– (bld)* where ' – ' denotes a major entry	e.g. *v.5: 21; v.7: 192*	*Not* re-indexed; see Cumulative Index in Vol. 7

Sample type is denoted parenthetically ( ) as follows, the categories 'air', 'aq' and, in part, 'sld' being especially of environmental interest: –

air
aq = *water, effluents, beverages*
bld = *blood, usually plasma*
sld = *solids other than animal tissues, e.g. foodstuffs, faeces or soils*

slv = *saliva*
tss = *tissue (animal) samples*
urn = *urine*

CATEGORY I [*lacking* amino group/ cyclic N *(other than imino)*]

#Ia (acid or ester)
See also Acids *entry in General Index and in Vol. 7 Compound Index*

Adipic acid: 158–(tss)
Aspirin: *v.5: 35 (bld, urn)*

Barnon [formula on p. 143]: 141–(sld)

#1a, *continued*
Carbofuran: 155 (sld)
Cephalosprins: *see* #IIa (even if of #Ia type)
Cephradine: *v.5: 227– (bld, urn); v.7:* *341– (bld, urn)*
Chlorpropamide: *see* #Iz (though $pK_a < 5$)
Clofibrate: *v.7: 320 (bld)*
Cromoglycate: *v.7: 36, 333 (urn)*
Cypermethrin: 129, 132– (bld)

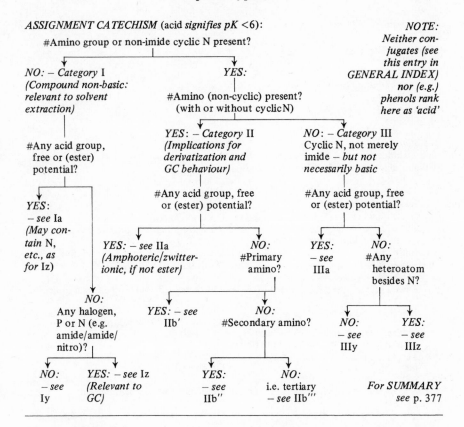

ASSIGNMENT CATECHISM (acid *signifies pK* <6):

#Amino group or non-imide cyclic N present?

NO: – Category I *(Compound non-basic: relevant to solvent extraction)*

YES:

#Amino (non-cyclic) present? (with or without cyclic N)

YES: – Category II *(Implications for derivatization and GC behaviour)*

NO: – Category III Cyclic N, not merely imide – *but not necessarily basic*

#Any acid group, free or (ester) potential?

YES: – *see* Ia *(May contain N, etc., as for* Iz)

#Any acid group, free or (ester) potential?

#Any acid group, free or (ester) potential?

YES: – see IIa *(Amphoteric/zwitterionic, if not ester)*

NO: #Primary amino?

YES: – *see* IIIa

NO: #Any heteroatom besides N?

NO: Any halogen, P or N (e.g. amide/amide/ nitro)?

YES: – see IIb′

NO: #Secondary amino?

NO: – *see* IIIy

YES: – *see* IIIz

NO: – *see* Iy

YES: – see Iz *(Relevant to GC)*

YES: – *see* IIb″

NO: i.e. tertiary – *see* IIb‴

NOTE:
Neither conjugates (see this entry in GENERAL INDEX) nor (e.g.) phenols rank here as 'acid'

For SUMMARY see p. 377

#Ia, *continued*

Dicapthon: 99 (aq)
Dicrotophos (Bidrin): 136 (sld)
Di-(2-ethylhexyl) adipate (DEHA): 158– (tss)
Di-(2-ethylhexyl) phthalate (DEHP): 167 (bld), 287 (bld)

Fenitrothion: 155 (sld)

Ibuprofen: *v.7: 301 (bld)*

Limonin: 94 (sld)

'Metabolite 32–692' [formula on p. 245]: 245– (bld)
Mevinphos: 145 (sld)

Naproxen: (346 (bld); *v.7: 321 (bld)*
N-Nitrosamino acids: 124– (sld)

'Org 4122' (a hydroxamic acid; formula on p. 215): 214– (bld)

#Ia, *continued*

Organophosphorus compounds (if not individually listed): 172; and *see* General Index entry

Penicillic acid: 114 (sld)
Phenoxyacetic acids: 77 (aq)
Phthalates/Plasticizers: *see these General Index entries, and* Di-(2– *above*
Ponceau 4R (= 3R): 113 (sld)
Procetofenic acid/ester: *v.7: 320– (bld)*
Prostaglandins: *v.7: 142–* (PGF-M, a metabolite; *urn)*
Pyrethroids: 145 (sld)

SDS (Sodium dodecyl sulphate) & related surfactants: 161– (bld), 167 (aq)

Tienlilic acid: *v.7: 332 (bld, urn)*

Vanilmandelic acid (VMA): *v.7: 224 (urn)*

Zearalenone: 114 (sld), 155 (sld)

#Iy (no halo, N or P, nor acid/ester)

Aflatoxins, e.g. B$_1$: 19 (sld), 112– (sld), 155 (sld)

Benzene: 27 (air), 41 (air)

Cannabinols: *v.5: 243 (bld, urn)*

Danazol: *v.5: 82 (bld)*
Dicumarol: 210
Diethylstilboestrol: *v.7: 226 (bld, urn)*
Digitalis glycosides: v.5: 71– *(bld), 247– (bld, sld)*
2,6-*cis*-Diphenylhexamethylcyclotetranloxane: 288 (bld)

Ethanol: 233– *(aq etc.); v.7, 36 (bld)*
Ethoxylate (non-ionic) surfactants: 161– (bld)
Ferrocenes: *v.5: 147*
Formaldehyde: 41 (air), 54 (air), 219
Furfuraldehyde: 49 (air), 235 (oils)

Glucose: 169 (bld)

Hydrocarbons, various (incl. alkylbenzenes; *see also* Benzene, Polynuclear, Styrene): 41 (air), 64 (air), 64 (air), 73 (aq), 97 (aq)

Ketones, various: 41 (air), 47 (air), 49 (air), 73 (aq)

Patulin: 114 (sld)
Phenol: 25 (urn)
Phenols, various: 99 (aq)
Polynuclear aromatic hydrocarbons (PNA, PAH): 90, 99 (aq), 115– (sld)

Quercetin: *v.5: 31 (bld, urn)*
Quinizarin: 232– (gas oil)

Steroid hormones: 89 (bld); *see also Compound Index in v.7*
Steroids, synthetic (*see also* #Iz, & Diethyl..., *above)*: 77 (aq); *v.5, 82 (bld)*
– esters, hydrolysis: 218
Styrene: 41 (air), 49 (air), 67 (air)

Warfarin: *v.5: 29*

―――――――

#Iz [with halogen/P/N (e.g. amide)] *See also General Index entries* - Chlorocarbons; Organophosphorus; Pesticides

#Iz, *continued*

Acetaminophen (Paracetamol): *v.5: 242 (bld, urn)*
Acetanilide: *v.5: 242 (bld, urn)*
Acrylonitrile: 41 (air), 48 (air), 67 (air)
Amygdalin: 172
Arctons: 286
Arochlor 1260: 91 (aq)

Barbiturates: *v.7: 13 (bld), 301*
– Phenobarbital: 181, 183 (bld, slv), 189, 190, 337 (bld)

Chloral: *v.5: 138 (urn)*
Chlorpropamide: 182 (bld, slv)

Dieldrin: 99 (aq)
p-Dioxins, Polychlorinated dibenzo–: 205; *v.7: 161– (tss), 331*
Diphenylhydantoin: *see* Phenytoin
Disulfiram: *v.5: 20 (bld)*

Halothane: 18 (bld), 49 (air), 61 (air), 62, 67, 357

Irgasan DP 300 (a chloroether): *v.5: 221– (bld, urn, bile)*

Laetril: 172
Lindane (BHC): 99

Meprobamate: 343; *v.5: 16 (bld)*

Nitrosamines (various; as contaminants), e.g. *N*–Nitrosodiethylamine: 100 (aq), 119– (sld)

Ochratoxin: 113– (sld, tss), 155 (sld), 167 (tss)

PCBs: *see* Arochlor
Pentazocine: 106 (tss)
Phenacetin: *v.5: 242– (bld, urn); v.7: 13 (bld)*
Phenylbutazone: 107 (bld); *v.7: 191* (bld)
Phenytoin: 107 (bld), 178 (slv), 182 (slv), 189, 190, 266 (bld), 337 (bld & [metabolites] urn); *v.5: 204 (bld)*
Propoxur: 155 (sld)

Steroids, synthetic halogenated (e.g. Triamcinolone): *v.5: 82 (bld), 141– (ointments)*
– hydrolysis of esters: 218

TCDD: *see* Dioxins
Tolbutamide: 181, 185 (bld, slv); *v.7: 191 (bld)*

#Iz, *continued*

VCM (Vinyl chloride monomer): 27 (air), 47 (air), 62 (air)

———

CATEGORY II (amino)

#IIa (acid/ester group *also present*)

Amino acids: 170 (sld), 174; *v.7: 191 (bld)*
Amoxicillin: 201 (urn)
Ampicillin: *v.7: 15 (bld)*

Catecholamines: *see* Phenylethylamine
Cephalosporins (if in category IIa): *v.5: 15 (bld); v.7: 342 (bld, urn)*
Cephradine: *v.5: 227- (bld, urn); v.7: 341- (bld, urn)*

Methotrexate: 89 (bld); *v.5: 31, 82 (bld)*

Niflumic acid: *v.5: 211- (bld, urn)*

Penicillamine: *v.7: 226*
Propantheline: *v.5: 91 (bld)*

———

#IIb′ (primary amino; *no* acid/ester)

Benzidine: 99 (aq)
β-Cetrotretine: *v.7: 226*

Chloranilines: 91 (aq)
Creatinine: 171 (bld), 223 (urn)

Gentamycins: 89 (bld); *v.5: 235- (bld)*

Kanamycins: *v.5: 235- (bld)*

Metoclopramide: 25 (bld); *v.7: 72, 315- (bld)*

Nomifensine: 371

'ORG 6001' (an amino–steroid): *v.7: 154- (bld)*

Phenformin: 205
Phenylethylamine derivatives (if not in IIb″): *v.7: 191 (tss), 273 (urn)*
– Amphetamine: *v.5: 35 [actually ref. 24]; v.7: 273 (urn)*
– Noradrenaline: *v.5: 64 (bld)*
– Normetadrenaline: *v.5: 159- (urn)*

#IIb′, *continued*

Polyamines (e.g. spermidine): 123, 226 (urn)
Procainamide: 181. 185 (bld, slv)
Pyrimethamine: *v.5: 110 (urn)*
Spermidine *(see also* Polyamines), Hydroxy–N–nitroso derivatives: 123 (sld)

Sulphonamides: 266; *v.7: 13 (bld)*

Tyramine: *v.5: 172 (urn)*

———

#IIb″ (sec. amino; *no* acid/ester)

Albuterol (Salbutamol): 206 (bld)
Atenolol: 296; *v.7: 71 (bld), 121 (bld)*
Atrazine, hydroxy metabolite: 94

Catecholamines: *see* Phenylethylamine
Chlordiazepoxide: *v.7: 13 (bld), 21*
Chlorhexidine: 313; *v.7: 328- (aq, bld, sld, tss)*

Desimipramine: *v.5: 128 (bld)*
Diphenylhydramine: *v.5: 164 (urn)*

Ethambutol: *v.5: 231- (bld, urn)*

Fenfluramine: *v.5: 105 (urn); v.7: 273 (urn)*

Hibitane: *see* Chlorhexidine

'ICI 72,222' (a catechol diether with an amide group): *v.7: 283- (bld, urn), 344*
Iproniazid: 19
Isocarboxazide: 19

'Kabi 1847' (a cyclopentadiene derivative): *v.7: 122- (bld)*

Labetalol: *v.7: 228- (bld, urn)*
Lidocaine: *v.7: 191 (bld)*

Methaqualone: *v.5: 179- (bld), 241 (bld); v.7: 191 (urn)*
Metaprolol: 345; *v.7: 131 (bld)*

Norzimelidine: 271-

Phenylethylamine (N-alkylated) derivatives: 346; *v.7: 273 (urn)*
– Adrenaline: *v.5: 64 (bld)*
– Ephedrine: 363; *v.5: 35 (urn)*
– Metadrenaline: *v.5: 159- (urn)*

#IIb″, *continued*
Pinodol: 244- (bld, urn)
Practolol: *v.7: 72 (bld), 324- (bld, urn)*
Propranolol: 181, 267, 269 (bld)

Streptomycin: *v.5: 235- (bld); v.7: 13*

Terbutaline: *v.7: 142- (bld)*
Terolidine: 284 (bld), 342; *v.7: 124 (bld)*
Timolol: *v.7: 327* (bld, aqueous humour)

#IIb‴ (tert. amino or quat. ammon.; *no* acid/ester)

Acetylcholine: *v.5: 97 (tss)*
Amitryptline: 107 (tss), 108 (bld); *v.7: 325- (bld)*
Aprindine (Aprinidine): 286 (bld)

Chlordimeform (Chlorphenamidine): *v.7: 328 (sld)*
Chlorpheniramine: 93 (urn)
Chlorpromazine: *see* Phenothiazines
Clomipramine (Chloroimipramine): 285 (bld); *v.5: 97 (bld)*
Cyclobenzaprine: *v.5: 247 (bld)*
Cyclophosphamides: *v.7: 192 (bld)*

Dibenzazepines: 271; *and see* Clomipramine, Imipramine, Trimipramine, and (#IIb″) Desimipramine), (#IIIy) Carbamazepine

Dothiepin: 336

Emepronium: 285 (bld), 341, 343; *v.7: 124 (bld)*

Imipramine: *v.7: 38 (bld)*

Meth(a)done: 25 (bld), 370
Minocycline: 266

Nortryptline: *v.7: 325- (bld)*

Oxyphenonium: 277- (bld)

Paraglyine: 19
Phenothiazines (*see also* #IIIz): 348
 – Chlorpromazine: 16 (bld), 20 (bld), 27, 178; *v.5; 13 (bld), 18 (bld), 185- (bld), 191- (bld, etc.); v.7: 110- (bld), 338*
 – *N*-oxygenated products: 348: *v.5: 191- (bld etc.)*
Pirimicarb: 148

#IIb‴, *continued*
Ranitidine: 217- (bld, urn)
Recipavrin: 342

Tetracyclines: *v.7: 16 (bld)*
Tricyclics: 266, *and see* individual compounds, *also* Dibenzazepines
Trimipramine: 275 (stomach contents)

Zimelidine: 271-

CATEGORY III (non-imide hetero-*N*; *no* amino)

#IIIa (acid/ester group *also present*)

Atropine: *see* #IIIz

Dicloxacillin: 266

Flavoxate: 21, 25

'ICI 74,917' (an amide): *v.7: 286- (air, bld, urn), 344*

Indomethacin: 210, 292 (bld); *v.7: 327 bld, aqueous humour)*

Kynurenic acid: 224 (urn)

Meperidine (Pethidine): 181

Nicotinic acids: *v.5: 98; v.7: 258, 301*

Penicillins: 172 (culture media), 179, 266; *v.5: 15 (bld); v.7: 13, 280*
Pethidine: *see* Meperidine
Porphyrins: *v.5: 45- (urn)*

Xanthurenic acid: 224 (urn)

#IIIy (only N-hetero, *not* merely imide; *no* acid/ester)

Acetazolamide: *v.5: 132 (bld)*
Allopurinol: *v.7: 277 (urn)*
Antipyrine: 181, 183 (bld, slv)
Artane: *v.7: 171- (bld)*
Atropine [arguably in ester class] : 13 (bld); *v.5: 35*

#IIIy, *continued*

Benzodiazepines (*see also* Chlordiazepoxide, #IIb'): 105 (tss), 107 (bld), 371; *v.7: 39 (bld), 115 (bld), 120 (bld), 336 (bld)*
- Bromazepam: *v.7: 23- (urn)*
- Carbamazepine: 183 (bld, slv); *v.5: 42- (bld)*
- Clonazepam: 202 (bld); *v.5; v.5: 203- (bld); v.7: 21- (bld)*
- Diazepam: 102 (bld), 185 (bld, slv); *v.5: 18 (bld), 203- (bld); v.7: 13 (bld), 21, 120*
- Flunitrazepam: *v.7: 21 (bld)*
- Lorazepam: *v.7: 120 (bld), 121 (bld)*
- Midazolam: 210 (urn)
- Oxazepam: *v.7: 120 (bld), 121 (bld)*

Biperidin: *v.7: 167- (bld)*

Carbamazepine: 183 (bld, slv); *v.5: 42- (bld)*
Chlorthalidone: 253- & 264 (urn), 266, 268 (bld)
Clonidine: *v.7: 109 (bld), 142- (bld)*

Dihydroergocristine: 87 (urn)
Doxanthrazole: 210

Ethosuximide: 183 (bld, slv), 190

Halquinol: *v.5: 217- (sld)*

Indapamide: 259 & 264 (urn)

Methyprylon: *v.7: 13 (bld)*
Metofoline (Metopholine): *v.7: 13 (bld)*
Metronidazole: *v.5: 18 (bld etc.)*
Mianserin: *v.7: 154- (bld)*
Molsidomine (a sydnone): *v.7: 336 (bld)*

#IIIy, *continued*

Nicotinamide: *v.7: 121 (bld)*
Nitroimidazoles: *v.7: 15 (bld, urn), 17 (bld)*
Nitrosamines, heterocyclic: 121 (sld)
N-Nitrosopyrrolidine: 123 (sld)

'Org 3509' (formula on p. 215): 214- (bld)
Ornidazole: *v.7: 17 (bld)*
Oxipurinol: *v.7: 277 (urn)*

Pentazocine: 266
Phencyclidine: 205, 281 (bld, urn)
Pilocarpine: *v.7: 327 (bld, aqueous humour)*
Primaclone (Primidone): 185 (bld, slv), 190

Quinidine: 269 (bld); *v.7: 191 (bld)*
Quinine: *v.7: 13 (bld)*

Tartrazine: 113 (sld)
Theophylline: 181, 182 (bld, slv)

#IIIz (heteroatom besides N; *no* acid/ester)

Acetazolamide: 266

Bendroflumethazide (Bendrofluazide): *v.7: 135- (bld)*
Butaperazine: 178 (slv)

'Kabi 1702' (a saccharin-like amide): *v.7: 121- (bld)*

Morphines: *v.5: 38 (bld), 68 (bld), 69 (urn); v.7: 13 (bld)*

Phenothiazines (*see also* #IIb'''): *v.7: 13 (bld), 226- (bld)*
- Fluphenazine: *v.5: 115- (bld, sld, urn); v.7: 37 (bld etc.)*

SUMMARY	I	II	III
Amino ?	no	√	no
Non-amide hetero-N ?	no	maybe	√
Acid or potential acid ?	√ = Ia	√ = IIa	√ = IIIa
− no! (see p. 373, for clarification of definition)	Halo, P or N ? − no: Iy −√ = Iz	Primary amino ? √ = IIb' If no: sec. = IIb'', only tert. = IIb'''	Hetero-atom besides N ? − − no = IIIy, √ = IIIz

See Key p. 372

General Index

The entries in this Index, concerned especially with phenomena and points of technique, have been made comparable with those in the Indexes to Vols. 5 & 7 so as to facilitate consultation. The Index lists some classes of analytes, but not individual organic compounds (these are in the foregoing type-of-compound Index), nor all applications of a technique such as GC.

Page entries such as 25- signify that ensuing pages are also relevant, i.e. the - denotes a major entry.

Accuracy (*vs.* precision): 17, 299
Acetonitrile for extraction: e.g. 20, 27, 111, 125, 134, 155, 195
Acids (organic) in water: 75
Adsorbents in sample preparation (*see also* Automatic, Bonded, Cartridges, Charcoal, Ion-exchangers, Porous, XAD): 19, 26, 137, 243-
- activated carbon: 24, 79, 86, 121
- cellulose: 24, 244
- Florisil: 19, 24, 115, 137-, 150
- for trapping volatile organics: 37, 55, 96
  - desorbability (*see also* Thermal): 48, 100
  - humidity effects: 47, 67
Adsorptive losses (*see also* Binding, Losses): 27, 61, 163, 199, 343, 348
Air analysis (*see also* Adsorbents, Bags, Co-condensation, Gases, Passive, Pumps, Threshold): 25-, 51-
- calibration (trace organics): 26, 39-, 59, 66
- field tests: 36, 40
- inorganic contaminants (*see also* Gases,) 34
- IR and other assessment modes: 18, 36
- sampling tactics: 36, 41, 51-
- storage of collected air: 26, 48, 60
  - in rigid vessels: 52, 61-

Albumin: *see* Plasma
Amino-acid removal: 226
Ammonia-electrode applications: 168-
'Amyl' alcohol: 20
Anaesthetics in blood: 18
Anionic compounds (*see also* Acids. Ion-exchangers): 73-
- in blood, affecting extraction: 270-
Anticoagulants (*see also* Heparin): 195, 295, 350, 369
Automated methods, especially for sample preparation (*see also* Centrifugation, Chromatography, Fluorimetry, TLC): 240-, 264, 304
- nomenclature: 229
- with a solid-phase extraction step: 138, 221-, 230, 264
- with a solvent-extraction step (*see also* Solvent): 229-
  - continuous-flow: 221-, 228, 253, 261
  - flow-injection: 229

Bags for air collection: 26, 52, 60, 61
Bakery products: 112
Basic compounds (amines), approaches/problems: e.g. 192, 199, 270-, 284-
Beverage samples (e.g. beer): 90, 228, 233, 235
Binding of analytes (not particularly drugs in body fluids): 19, 101-

Binding of drugs, especially to body-fluid proteins (*see also* Enzymic, Erythrocytes, Ionization, Plasticizers): 102, 179-, 192, 210-, 270
– availability to enzyme electrode: 219
– determination: 210
– during storage (*see also* Storage): 25
– liberation (*see also* Enzymic): e.g. 199
– temperature effect: 266
– to salivary mucoproteins: 183
Blood samples (*see also* Plasma, Storage): e.g. 195-
– collection (and artefacts): 267-, 348-
– constituents affecting extraction (*see also* Ion-pair): 270
Bonded-phase materials (e.g. C-18) for sample preparation: 24, 138-
Bubblers (impingers) for air–contaminant collection: 41, 54

Calibration curves (*see also* Standards): 16, 302, 304, 305, 312, 369
– for instruments: 301
– possible inconstancy: 16, 313, 345, 362, 370
– precision assessment (*see also* Statistical): 320-
Carbonic anhydrase (*see also* Erythrocytes), inhibition (drug assay): 259
Cartridges/short columns in liquid sample work-up (*see also* Bonded, Trace): 21, 138-, 220, 244, 247
– difficulties: 21, 247
Centrifugation, e.g. for phase separation: 195
– advantages of low temperature: 195
– automatic: 229-, 241-, 263
Cereal samples (*see also* Bakery, Crops): 131-, 150, 238
Charcoal (*see also* Adsorbents) 55, 244, 246
– for volatile organics (*see also* Passive): 44, 49, 56
Chelation, inadvertent: 200
Chloro compounds (*see also* Halo, Solvents): 18
Cholesterol (endogenous; *see also* Lipids) as an analytical 'marker': 110
Cholinesterase inhibition as means of assay: 19, 168, 170, 264
Chromatography (*see also* GC, HPLC, TLC): e.g. 18, 137, 197, 200
– automated, especially for sample preparation: 221-
– for sample clean-up (cf. Bonded, Cartridges, Trace): 137-, 278
Co-condensation to trap volatiles: 59, 67, 100

Co-distillation for analyte clean-up: 18, 25, 119, 121, 154
– with steam: 122, 133, 154
Collection of samples: *see* Air, Blood, Storage, Vessels
Colour materials in foods: 111-
Confidence limits: *see* Precision, Statistics
Conjugates: 197, 201, 337, 370
Contaminants (analytical; *see also* Fatty, Plasticizers, Vessels): e.g. 156, 356
Co-precipitation, inadvertent: 195, 199
Crop samples, incl. vegetables: 128-, 134-, 150
– homogenization: 133
Crown ether-catalyzed derivatization: 276-, 279

Data handling (instrumental): 241, 305-, 308, 317
Data interpretation/presentation (*see also* Records, Statistical): 17, 351
Deproteinization (*see also* Acetonitrile): 20, 24, 101, 192-, 195
Derivatization (*see also* Crown ether, GC, HPLC): 18, 120, 189, 193, 197, 200
– fatty acid interferences: 276
– fluorigenic: 240, 260
Desalting: 221, 226
Design of assays (*see also* Internal, Replication): 217, 299, 303-, 323, 328, 350-
Detectability/sensitivity: 17, 300, 302
– appraisal/criteria: 302, 305, 311-, 362, 370
– in HPLC: 214-
– in relation to measurement mode: *see* Electrochemical, Fluorimetric
Detector tubes for air contaminants: *see* Air, field tests
Detergents: *see* Surfactants
Deterioration of samples: *see* Storage
Diffusion of gases/vapours (*see also* Passive): 40, 44, 46
Distillation to isolate analyte (*see also* Co-, Vacuum): 19, 26, 121
– flash vaporization: 233
Dusts: 33

Effluent samples: 75, 89, 98-
Electrochemical methods (*see also* Enzymes): 19, 169-
– in HPLC: 18, 214, 346
Emissions (e.g. in flue gases): 27
Employee monitoring (*see also* Occupational): 25, 34
Emulsion formation/breakage: 20, 230, 360
Enzyme assay: 226

Enzymes in measurement: 253–
– electrodes: 169–
– immobilized: 168–
– inhibition approach (*see also* Carbonic, Cholinesterase, Monoamine) 256
Enzymic interferences (*see also* Plasticizers), especially loss of analyte: 20, 292, 295, 370
Enzymic release of analyte: 101–, 111–, 166, 196
Equilibrium dialysis: 193, 210–
Error, sources of (*see also* Losses): e.g. 299, 341, 349
Erythrocytes (*see also* Blood, Haemolysis) and drug analytes: 196, 265–, 369
– binding to constituents: 265
– carbonic anhydrase: 266
– equilibria/disturbances: 265–, 354
– losses onto: 195, 199
Esters of drugs, hydrolysis assessment:218
Evaporation steps/losses: 23, 69, 99, 121, 200, 360, 363–, 366
– apparatus: 363, 366
– automatic: 229, 231–, 246
– bubbler in gas stream: 371
– contamination of gas stream: 371
– Kuderna Danish: 23, 99, 122, 151
Exclusion separations, e.g. 'gel filtration': 94, 137, 167, 195
– with a lipophilic gel: 155, 195
Extraction (*see also* Automated, Ion-pair, Shaking, Solvent):
– apparatus: 150
– disturbances: *see* Fatty, Phthalates

Faecal samples: 196
Fatty acids as interferants (*see also* Lipids): 189, 274, 356
– extraction disturbance: 271–, 274–, 350, 356
– circumvention (*see also* Lipids): 356
– occurrence/concentration: 275
Fluorimetric methods: 168–,
– automated: 241, 253–, 259–
– in HPLC: 111–, 201, 214
– post-column ion-pairing: 93
Food samples (*see also* Crop, Colour, Horticultural): 19, 119–, 150–, 158–, 166
Forensic samples: 101–, 195, 166
Freeze-drying of samples (*see also* Plasma): 125, 294
– water samples: 69, 83
– *vs.* freeze-concentration: 83
– *vs.* resin adsorption: 84
Fruit samples/constituents: 94, 131–, 150

Gases (permanent), assay (*see also* Nitrogen, Sulphur): 52–
– sulphur dioxide in beverages: 235
GC (Gas chromatography) (*see also* Porous): e.g. 197, 235
– back-flushing: 97
– capillary-column: 17, 73, 79–, 267
– SCOT: 160, 278
– WCOT: 74
– colorimetry on peaks: 236
– derivatization: 124, 197, 201, 342, 371
– with iodo group: 162
– on-column methylation: 190, 276, 336–
– degradation artefacts: 339
– detectors (*see also* MS): 18, 201
– contamination: 201, 371
– Coulson type: 122
– for phosphorus: 129, 201
– thermal energy type: 122–
– headspace: 18, 19, 77–, 154
– followed by Tenax: 68
– peak-size changes: 16, 345
Glassware, cleanliness etc. (*see also* Losses, Silanization, Vessels): 27, 275, 277, 280, 366
– chromic treatment: 27
Good Laboratory Practice (GLP): 16, 298–, 370

Haemolysis: 268, 348
– influence on drug assay: 268
Halo compounds (*see also* Chloro):
– ECD responsiveness: 18
– in air: 48, 91
– in water: 67, 73, 77, 95, 97–
Headspace analysis: *see* GC
Heat-desorption: *see* Thermal
Heparin perturbation of drug levels (*see also* Anticoagulants): 269, 369
Homogenates/Homogenization: *see* Crops, Tissues
Horticultural produce: *see* Crops, Fruit
HPLC (*see also* Bonded, Detectability, Ion–): e.g. 201, 214, 218
– automated (*see also* Chromatography): 215
– column switching/by-passing: 215
– detection other than UV (*see also* Electrochemical, Fluorimetric):
– thermal energy analyzer: 119–
– visible range: 111–
– displacement mode: 89
– load volume: 17, 87
– packings, especially unorthodox: 18, 112, 115, 138
– pre-column role (*see also* Trace): 18, 89–, 95, 100, 215

HPLC, *continued*
- reactions/derivatization: 93
 - post-column: 93, 115, 201
- sample-preparation role: 115, 128, 138-

Immunoassays: 19
Impingers: *see* Bubblers
Infrared (IR) methods: 18, 36, 40, 238
Instrument care/checking: 300, 301, 304
Internal standards (*see also* Radio): 160, 299, 302, 336, 341-, 352, 361, 369
- in capillary GC: 160
- in HPLC: 346
- possible non-validity: 220, 352
Ion-exchangers: 24, 197, 224, 244
- automated use: 222-
- for colour materials: 111-
- for water samples: 74, 100
- used in an organic solvent: 112
Ionization as affecting analyte behaviour:
- protein binding: 270
- saliva *vs.* plasma: 180
- solvent extraction: 21, 108, 188-, 195, 286
 - anomalous acid-pH behaviour: 21
Ion-pair chromatography: 217
Ion-pair extraction: 23, 33, 93, 220, 245
- HPLC post-column: 93
- inadvertent: 271-
- reversal (displacement): 275, 356
- with tetraphenylborate: 23
Irradiation to confer detectability: 121
Isotopes: *see* Radio, Stable
Kidney samples: 113, 166

Lability of some analytes (*see also* Solvents): 20, 106, 200, 300, 348-
Lipids in samples (*see also* Fatty, Plasma): 139, 151
- disturbance of analyte extraction: 220, 270-, 288
- removal (incl. fatty acids): 113, 115, 137, 167, 189, 195, 277, 356, 369
 - automatic: 264
 - Folch extraction: 289
Lipophilic analytes: e.g. 20, 284-
Liver samples: e.g. 102-, 158, 166
Losses, causes of (*see also* Co-precipitation, Enzymic, Evaporation, Solvents, Storage): e.g. 163, 195, 199-, 292, 299, 341-, 357, 365

Mass spectrometry: *see* MS
Metabolites (*see also* Conjugates, Pesticides): e.g. 24, 217, 220, 271-, 337, 348-
- occurrence, in internal standard context: 337

Metabolites, *continued*
- of inhaled stubstances: 25, 35
- *N* & *S* oxides, antioxidant effect on: 348
- polarity: 18
- radioisotope usefulness: 200
- stable-isotope usefulness: 148
Metals in air, e.g. Pb: 34
Mixing (*see also* Shaking): 360
Monoamine oxidase inhibition, for assay: 19
MS (Mass spectrometry) (*see also* Stable, Statistical): e.g. 121, 167, 205-
- GC-MS: e.g. 70, 98, 122, 126, 205-, 274, 281, 302, 344
 - of volatiles: 79
 - operational options: 122, 205
Mutagens: *see* Water
Mycotoxins: 19, 111-

Nitrogenous trace constituents, e.g. nitrite in foods: 122, 125
- oxides in air: 53, 54
Nitrosamines: 100, 119-
Nomenclature: 10, 15, 16-, 21, 227, 314, 322, 336, 351
Nuclear magnetic resonance (NMR): 148

Occupational hygiene: 25-, 32-, 51
Organophosphorus compounds: 129, 145, 168, 170, 172
Outliers ('rogues'): 16, 352, 369

Passive sampling of air: 36, 43-, 57-, 67
Partition between two phases (*see also* Solvent): 21-, 188-
- determination/concentration influence: 21, 137, 220
PCBs: 91
Pesticides (*see also* Cholinesterase, Organophosphorus, Residue): 99, 128-, 150, 167, 172, 220, 264, 277
- metabolites: 19, 128-, 148
- tightly bound: 19
Phospholipids/Phosphatides (*see also* Lipids), influence on extraction: 272, 288
Phthalates (*see also* Plasticizers):
- determination in plasma: 287
- extraction disturbances from: 272
- in water: 89, 90
- interferences from: *see* Plasticizers
Plasma as type of blood sample: 195, 348, 354
- dilution for extraction: 284-, 286, 347
- drying/semi-drying for extraction: 287
- *vs.* serum: 348, 354
- *vs.* whole blood: 21, 192-, 267-, 296, 354, 369

Plasma proteins (*see also* Binding, Deproteinization):
- disturbance of extraction: 266
- lipoprotein complications: 288
- precipitation *vs.* pH: 21, 286
Plastic materials (*see also* Vessels):
- attack by solvents: 232
- vapour losses (*see also* Bags): 61-, 357
Plasticizers (*see also* Phthalates):
- determination in tissues: 20, 158-
- interferences from: 158, 200, 267, 272, 280-, 357, 371
- ion-pairing by breakdown to acid: 357
- protein-bound drug displacement: 267, 357
Porous polymers (e.g. Porapak; *see also* Tenax) for trapping: 19, 26, 45, 54, 96-, 100
- *vs.* freeze-drying, for water: 83
- thermal desorption: *see* Thermal
Precision/reproducibility: 17, 45-, 299-, 320-, 328, 352
- obsessiveness: 27, 32-, 326
Proteins: *see* Deproteinization, Plasma
Pumps for air sampling: 26, 57
Purging of water samples: 67, 69, 77-, 96-

Radioimmunoassay: *see* RIA
Radioisotopes:
- in drug studies: 16, 200, 303-, 352, 361
- in pesticide field studies: 134-
Records/reporting: 17, 27, 302, 305, 308
- as free compound *vs.* salt: 28, 300
Recovery assessment: 302
Red cells: *see* Erythrocytes
Replication (*vs.* singlicates): 302, 305, 318, 325, 351, 369
Residue analysis (*see also* Pesticides), laboratory practices: 156
RIA (Radioimmunoassay): 197, 303, 304
Rogues: *see* Outliers

Saliva samples (*see also* Transport): 59, 176-, 193, 219, 348
- *vs.* plasma, e.g. pH influence: 179-
Sample collection (fluids; *see also* Air, Blood, Water): 347-
Sample preparation (*see also* Adsorbents, Automatic, Solvent, etc.): e.g. 18-, 166, 193-, 244-
- micro-scale: 359, 361
- minimization: 217
- strategy for crops etc.: 129-
- *vs.* end-step chosen: 197
Sampling of solids: 19, 131
Screening (toxicological): 102, 195

Sensitivity (*see also* Detectability): 369
'Sephadex' gels: *see* Exclusion
Serum: *see* Plasma
Shaking, modes (*see also* Mixing): 21, 360
- time-dependence: 220, 284-
Silanization/Siliconization of glassware: 199, 344, 357, 365
- nomenclature: 372
Soil samples: 128-, 166, 173
Solid samples (*see also* Tissues): 18, 111-, 121-, 130
- usefulness of dampening: 136
- wastes: 19
Solvent extraction, one-phase (organic) systems (*see also* Acetonitrile, Deproteinization, Freeze-drying): 113, 124
- for field samples: 134-
- optimized water content: 136
- rendered two-phase: 21, 125, 201, 287, 289, 371
- Sohxlet: 133
Solvent extraction, two-phase systems (*see also* Centrifugation, Emulsion, Ion-, Lipids, Partition, Shaking): 20, 137, 188-
- as primary step: e.g. 133
- automated (*see also* Automated): 241, 263
- back-extraction: 22, 220, 356
- both phases organic: 115, 137, 150, 189
- boundary-sensing: 230
- choice of solvent: 107, 195, 217, 233, 288
  - density: 107, 192
- pH choice: *see* Ionization
- selectivity: e.g. 24, 277
- strategy: 21-, 129-, 188-, 195, 220, 272, 360
  - volume optimization: 22-, 220
- time-dependence: 220, 284-, 360
- usefulness of salt addition: 122, 193
- water samples: e.g. 69, 76-
Solvents, impurities/purification (*see also* Acetonitrile, Amyl): 24, 280, 356-, 358-, 366, 370
- detriment to analytes: 25, 358, 360, 371
- ethyl acetate instability: 359
Sparging: *see* Purging
Specificity (*see also* Accuracy): 298-
Spiking (*see also* Tissue): 350, 372
- field-trial samples: 134
Stable isotopes in MS: 148, 167, 302, 344, 361
Standard deviation/error: 322
Standard Operating Procedures (SOPs): 17, 299
Standards (*see also* Air, Internal, Nomenclature, Spiking): 300-, 361-, 370

Statistical approaches/calculations (*see also* Calibration, Outliers, Precision): 17, 294, 299, 303, 312-, 316, 320, 352
- GC-MS: 301
Steam distillation: *see* Co-distillation
Storage of raw samples (*see also* Air, Binding, Trapping): 25, 131, 291-, 307, 349, 355-
- contamination: 281
- survival of particular compounds
- in blood samples: 25, 291-, 293, 295
- in crops etc.: 65
- temperature choice: 349, 355
- thawing: 350
- verification: 131-, 292-, 300
Sublimation: *see* Vacuum
Subtilisin: *see* Enzymic release
Sulphur dioxide: 53
- in beverages: 235
Surfactants, estimation: 167, 220
- in blood: 161
- in water: 229-
- metabolites: 164
Sweep co-distillation: *see* Co-

Tenax, use of : 19, 45, 67-, 97, 98
Thermal desorption for GC: 49, 56, 70-, 78, 96, 100
- from XAD-4 resin: 70-
Threshold Limit Values (TLVs): 27, 34-, 35, 36
Tissue samples, especially extraction (*see also* Enzymic): 19
- animal: 20, 101, 158-, 166, 195
- plant (e.g. crops): 131-, 138
  - spiking: 131, 134, 350

TLC (Thin layer chromatography), applicability: e.g. 104, 112, 162, 199
- automatic applicator: 260
- with irradiation step: 120
Trace enrichment (*see also* Water): 86-, 100, 202
- displacement HPLC: 122
Transport processes (*see also* Erythrocytes), plasma/saliva: 179-

Vegetables: *see* Horticultural
Vessels, e.g. tubes (*see also* Glassware, Losses): 29, 99, 199, 349, 356
- plasticware artefacts (*see also* Plasticizers): 267, 281, 372
Volatile analytes: *see* Air, Evaporation, Halo, Purging, Trapping, Water

Water/aqueous samples (*see also* Ion-exchangers, Phthalates, Purging, Surfactants): 68-, 76-, 96-, 100
- effluents: 84, 98-
- extraction of organics: 69, 79, 86
  - with XAD resins: 24, 79, 86, 100
- mutagen detection: 81
- seawater/marine sediments, hydrocarbons: 90
- trace enrichment: 98, 100
  - by HPLC: 86-

XAD resins in sample preparation (*see also* Water): 24, 195, 244
- purification: 24, 70, 100

Corrections to Vol. 7

Entry #D-9 in Contents list:
    Authors are Wolfgang Dünges, Karin Kiesel and Karl-Erhardt Müller.
p. 7 sensitivity *not* sensivity.
p. 170 *Delete* last three lines (are duplicated on page 171).
p. 226 *should read* Comments on the above [see also p. 335].
Transpose pp. 264 and 265.
p. 293 Title *should read* Chromatographic not Chromatofraphic.
p. 347 Dopamine: 192 should be *transferred* from #IIa to next column #IIb following Cytabarine: 192.